Springer Series in Synergetics   Editor: Hermann Haken

Synergetics, an interdisciplinary field of research, is concerned with the cooperation of individual parts of a system that produces macroscopic spatial, temporal or functional structures. It deals with deterministic as well as stochastic processes.

Synergetics – From Microscopic to Macroscopic Order

Proceedings of the International Symposium
on Synergetics at Berlin, July 4 – 8, 1983

Editor: E. Frehland

With 146 Figures

Springer-Verlag
Berlin Heidelberg New York Tokyo 1984

Priv. Doz. Dr. Eckart Frehland
Fakultäten für Biologie und Physik, Universität Konstanz, Postfach 5560,
D-7750 Konstanz, Fed. Rep. of Germany

Series Editor:

Professor Dr. Dr. h. c. Hermann Haken
Institut für Theoretische Physik der Universität Stuttgart, Pfaffenwaldring 57/IV,
D-7000 Stuttgart 80, Fed. Rep. of Germany

ISBN-13:978-3-642-69542-1   e-ISBN-13:978-3-642-69540-7
DOI: 10.1007/978-3-642-69540-7

2153/3130-543210

Preface

This volume contains the papers presented at the International Symposium on "Synergetics - From Microscopic to Macroscopic Order", held at the Wissenschaftskolleg zu Berlin (Institute for Advanced Study Berlin), on July 4-8, 1983. Furthermore, it contains a contribution of T. Ohta, who unfortunately could not participate in this meeting, on the evolution of multigene families.

The papers discuss the evolution and the function of ordered structures from small microscopic scales up to large macroscopic dimensions. On the one hand, these structures derive from physical or biological systems; on the other hand, they also affect economic, sociological and philosophical questions.

I would like to thank the Wissenschaftskolleg zu Berlin for the extraordinary support and hospitality during my one year's stay there as a fellow, which made the planning, preparation and organization of this symposium possible. I would also like to acknowledge the work of B. Fritsch and D. Dörner who actively participated in this undertaking.

I am grateful to Professor Haken, the founder of synergetics, whose participation in the planning and styling of the concept of the conference was essential. I am especially thankful to Mrs. U. Monigatti for her indefatigable help with the preparation and organization of the conference.

The financial support was provided by the "Deutsche Forschungsgemeinschaft".

Berlin, November 1983   *E. Frehland*

Contents

VII

Part IV   Social Sciences

Part V   Complex Systems

Some Introductory Remarks on Synergetics

Hermann Haken

Institut für Theoretische Physik, Universität Stuttgart, Pfaffenwaldring 57/IV
D-7000 Stuttgart 80, Fed. Rep. of Germany

I am glad to make a few introductory remarks at this meeting on
"Synergetics, from microscopic to macroscopic order". The reader of
these proceedings may be puzzled by the variety of topics treated.
Therefore especially for those who are not familiar with the general
goal of synergetics a few words of explanation might be in order.
According to its definition, synergetics deals with systems composed
of many parts which can form spatial, temporal or functional structures
on macroscopic scales. Indeed most of the contributions to these proceedings deal with such systems. Over the past decade numerous systems
in physics, chemistry and specific model systems of biology have been
studied in which macroscopic structures are formed through self-
organization. Examples are provided by chemical oscillations, where
the molecules react in such a way that macroscopically a periodic
change of color or of other properties can be observed (cf. the
contribution by B. Hess). Such processes play an important role in
biology also. Another example is provided by different kinds of wave
formations in fluid dynamics and in lasers. In all these cases a
rather coherent picture could be drawn within the past decade. When
an external control parameter such as energy input or temperature
is changed, systems may run through a hierarchy of instabilities.
At each instability point a new macroscopic pattern replaces the
old one. It is by now well known that the transition from one macroscopic pattern to another one obeys general principles which can be
formulated by order parameters and the slaving principle. I shall
not dwell here on the details of this approach because it has been
presented in two books of mine [1,2], and in quite a number of
contributions to proceedings of conferences and summer schools. I
rather wish to address myself to the problem in how far those general
concepts can be applied to other disciplines, especially those of
sociology and psychology.

Modern science is becoming more and more confronted with the
question of how to cope with complex systems. Examples are provided
by biological systems including the brain, or economy. A complete
description of the behavior of such systems seems to require an
enormous amount of information which nobody is able to handle.
Therefore it seems to be a problem of utmost important to find
means to compress information. In the realm of former work of synergetics it could be shown that a compression of information becomes
possible close to instability points where the whole dynamics of the
complex system is governed by few order parameters only.

With respect to economy such a point of view may look revolutionary because, as it seems to me, a good deal of approaches in this
field are dealing with stable situations and are devoted to the
question of how to maintain stability. From the general point of
synergetics it seems to be, however, more interesting and even more
promising to study the behavior of complex systems at points of
instability. Instability in this technical context does not mean

that an economy breaks down but that it changes its behavioral patterns. An example of this kind is provided by studies of Mensch [3] on the transitions between full employment and underemployment. Another example is provided by evolutionary models of socioeconomic systems as in the paper presented by Allen at this symposium. According to our understanding, evolution may be visualized as a sequence of instabilities at which over and over again new patterns or structures emerge.

In these evolutionary steps an important interplay between deterministic processes and fluctuations, i.e., random events, takes place. In biology these fluctuations are, of course, represented by mutations. The interplay between these two kinds of processes plays a fundamental role, for instance in Rechenberg's approach to an optimization of technical devices.

There are other fields of study in which an application of the general ideas of synergetics may be less obvious. One of them is photosynthesis where it may seem to be extremely difficult if not perhaps impossible to single out specific order parameters or use the equivalent idea of macroscopic degrees of freedom. However, also in this field the situation may change soon. For instance it has been convincingly suggested by Blumenfeld in his monographs in the Springer Series in Synergetics [4] that enzymes act as molecular machines with few degrees of freedom. Indeed also in photosynthesis the main problem seems to be to explain how the molecular machinery prevents a quick dissipation of energy into many degrees of freedom, i.e., into heat, and rather manages instead to keep the energy concentrated.

The contribution by G. Vollmer in "Synergetics, Cognition, and Evolutionary Epistemology" seems to me particularly remarkable because it elucidates the fact that synergetics is a scientific discipline whose principles can be applied to itself. Namely the evolution of new scientific paradigms is nothing else but the emergence of a new pattern of thoughts. This evolution obeys again the general rules of synergetics, namely that of accidental fluctuations, competition between collective modes, and survival of the fittest mode.

I should like to make a comment on the formation of associative memory, a concept due to Kohonen [5] on which Palm reports in his contribution to these proceedings. An important concept in that approach rests on the idea of Hebb's synapse, often invoked in theories of learning processes. According to this concept certain synapses are strengthened if they are used more than others. At first sight it seems as if such a concept is totally local and that there is no relation with the global aspects of synergetics. However, it may be worthwhile pointing out that quite close connections between these two aspects can be established. For instance, it has been shown by Häussler and von der Malsburg [6] that in retinotopic processes, which in a way are related at the microscopic level to the idea of Hebb's synapse, the concept of order parameters allows one to deduce the macroscopically evolving pattern of connections between retina and tectum. Invoking results from synergetics may possibly resolve another puzzle. So far experimentally Hebb's idea could not be confirmed. In my interpretation this may be due to the fact that in large enough nets of neurons even a tiny change of synaptic connections may produce macroscopic effects. This implies that an individual synapse may change very little which may escape experimental study. I base my interpretation on the analogy with laser physics where somewhat above laser threshold a small fraction of the total activity of each individual atom suffices to produce a coherent light wave. Thus I believe that the methods of synergetics and the more microscopic approaches used in associative memory can complement each other.

In conclusion I should like to make a rather general statement on the role played by synergetics. Synergetics can be considered as a forum where scientists from quite different disciplines meet each other to exchange their ideas on how to cope with complex systems. Looking back over the last decade I think it is fair to say that enormous progress has been achieved towards the understanding of complex systems by searching for unifying principles. I am convinced that not only for philosophical but also for quite practical and pragmatic reasons it will be necessary to search further for unifying mathematical structures without whose help we can hardly deal adequately with complex systems.

References

1   H.Haken: Synergetics. An Introduction, 3rd ed.
    (Springer, Berlin, Heidelberg, New York 1983)

2   H.Haken: Advanced Synergetics. Instability Hierarchies of Self-Organizing Systems and Devices, Springer, Berlin, Heidelberg, New York 1983

3   G.Mensch, K.Kaasch, A.Kleinknecht, R.Schnopp: Discussion paper series, Internat.Institute of Management, Wissenschafts-zentrum Berlin (1950-1978)

4   L.A.Blumenfeld: Problems of Biological Physics, Springer, Berlin, Heidelberg, New York 1981

    L.A.Blumenfeld: Physics of Bioenergetic Processes, Springer, Berlin, Heidelberg, New York 1983

5   T. Kohonen: Associative Memory, Springer, Berlin, Heidelberg, New York 1977

6   A.F.Häussler and C.von der Malsburg, J.Theor.Neurobiol. in press

Part I

Order, Chaos, Indeterminancies

Time Pattern Transitions in Biochemical Processes

Benno Hess and Mario Markus

Max-Planck-Institut für Ernährungsphysiologie, Rheinlanddamm 201
D-4600 Dortmund 1, Fed. Rep. of Germany

1. Introduction

Historically, science has learned from the analysis of equilibrium
states. The biochemist extracts reaction mechanisms from the study
of isolated reactions under closed conditions. Thus, an enzyme-ca-
talyzed reaction runs into equilibrium in a first-order approach
whenever substrates or products are added. The rate laws describe the
overall process fitting to the kinetics and the given chemical potential sets the direction of flux. The closed case is only a sophisticated approach of a simple experimental design. The open case is the
case of nature: it means that all enzymes in a biological process are
constantly, stochastically or periodically activated by substrates
which are produced by the environment or by precursor enzymes and
transformed so that they are picked up by other enzymes within a reaction sequence. Each enzyme in a sequence becomes an element in a mul-
tienzymic network.

Steady state in an open enzymic reaction within a network can be
maintained as long as the substrate is constantly supplied and the
product is constantly consumed. The rate for Michaelis enzymes is
pseudo first order because in general in living cells the substrate
concentration is smaller than the Michaelis constant. Any fluctuation of the substrate supply is linearly relayed because of the
simple proportionality of the rate and the substrate concentration
yielding a high stability in any possible dynamic states for a controlled chemical potential.

The overall non-linearity of biochemical processes results from the
fact that these processes are organized in form of reaction cycles
in which biochemical intermediates are cyclicly regenerated, yielding multiple feedback interactions. In addition, the hierarchical
principle of organization presents an ordering function which is reflected by the time constants as well as the spatial sizes involved
on the various levels of organization, be it subcellular, cellular,
multicellular or supercellular structures. In the multienzyme network of cellular metabolism control of turnover is exerted by a few
enzymes, which regulate conversion rates at critical locations in
metabolism. Regulation is due to a biochemical mechanism which is
described by the allosteric interaction of the enzymes with controlling ligands or to the more complex chemical modification of enzymes
which are themselves catalyzed reversibly by specific enzymes [1].

The allosteric enzyme is the prototype of functionality in the biochemical process. In addition to the catalytic function it allows
control of turnover upon interaction with controlling ligands in
terms of positive or negative feedback or feedforward, resulting in
a sigmoidal relationship between the activity and concentration of

ligands. The activity of allosteric enzymes is another source of
non-linear behaviour of biochemical and biological processes.

Analysis of the dynamic states being maintained in an active biochemical process readily reveals that not only classical steady
states described by a stable node in a phase representation but
also complex transitions (overshot - undershot, stable focus) and
oscillatory states (limit cycles) frequently occur. Indeed, oscillations of the concentration of cellular metabolites such as in
the case of glycolysis have been described over recent decades.
Classical limit cycle behaviour of glycolysis in yeast cells is observed well in the physiological turnover range. The dynamic properties of glycolysis can also be tested in a glycolyzing yeast extract.
Of special interest is the ability of synchronization of the glycolytic oscillation with an external periodic substrate addition. It
has been found that this chemical process readily responds to entrainment with one-half and one-third of the glucose injection frequency. Also, the response of glycolytic oscillations towards stochastic perturbation could be described [1,2].

Recently theoretical and experimental analysis focussed on more
complex dynamic phenomena such as chaos, polyrhythmicity and hyperstability, which will be described in the model studies of the
following sections of this presentation.

2. Two-Enzyme Model of Glycolysis and Techniques of its Analysis

Earlier studies have shown that biochemical oscillations observed
in cell-free extracts of yeast can be simulated well on the basis
of an open allosteric model representing essential kinetic features
of the enzyme phosphofructokinase (PFK). Indeed, it could be shown
experimentally that this enzyme being coupled to the other glycolytic enzymes is responsible for the generation of oscillatory turnover of the whole reaction pathway.

In an effort to understand more complex dynamic states of glycolysis,
we recently studied an extended glycolytic model, which is composed
of two enzymes, namely PFK and its phosphorylating counterpart pyruvate kinase (PK), each enzyme obeying a two-substrate kinetic
mechanism [3-5].

Fig. 1 shows the reaction pathway with the substrate phosphoenolpy-
ruvate (PEP) and fructose-6-phosphate (F6P) supplied as input and ADP

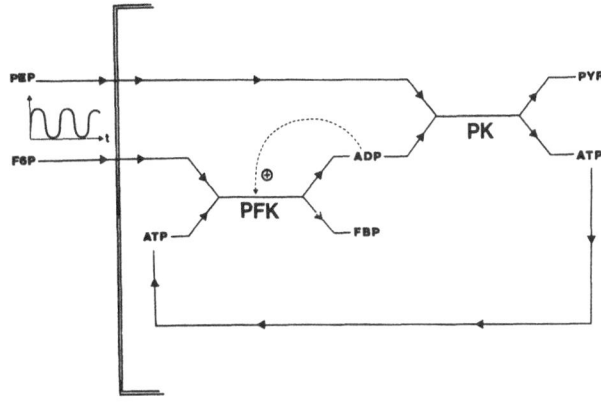

Fig. 1
Diagram showing the
essential features
of the model. In ad-
dition, we considered
the inhibition of PFK
by PEP, the inhibition
of PK by ATP and the
interaction of Mg^{2+}
and K^+ with PK and
with the ligands

and ATP as intrinsic coupling components. The intermediate part of the glycolytic pathway is replaced by the input of PEP. The products fructose-1,6-bisphosphate (FBP) and pyruvate accumulate. These products do not affect the kinetics of the enzymes over a wide range of concentrations, although FBP acts as strong activator of PK. Therefore, the initial concentration of FBP was set at saturating value with respect to its interaction with the enzyme PK: [FBP] = 1 mM. The total concentration of potassium $[K_{tot}]$ and of magnesium $[Mg_{tot}]$ remain constant in time.

The system is described by the following equations:

$$\frac{d[F6P]}{dt} = \bar{V}_{in} + A \sin\omega_e t - V_{PFK} \tag{1}$$

$$\frac{d[PEP]}{dt} = \bar{V}_{in} + A \sin\omega_e t - V_{PK} \tag{2}$$

$$\frac{d[ADP]}{dt} = V_{PFK} - V_{PK} \tag{3}$$

$$\frac{d[ATP]}{dt} = V_{PK} - V_{PFK} \tag{4}$$

where $\bar{V}_{in}$ is the constant component of the input flux (mean input flux). V_{PK} and V_{PFK} are the enzymic rates, and the sinus modulation term contains the amplitude (A) and the frequency ω_e (period T_e). Substraction of (1) from (2) and addition of (4) leads to:

$$[PEP] + [ATP] - [F6P] = C . \tag{5}$$

Addition of (3) and (4) leads to:

$$[ADP] + [ATP] = ADN . \tag{6}$$

C and ADN are constants in time. Due to the two conservation laws (5) and (6), we are left with only two independent variables. We choose [FBP] and [PEP].

The following dimensionless quantities are introduced in a similar way as in [6]:

$$\nu = \bar{V}_{in}/V_{max(PK)} \quad , \quad \alpha = A/V_{max(PK)} \quad , \qquad \varepsilon = V_{max(PFK)}/V_{max(PK)} ,$$

$$v_{PK} = V_{PK}/V_{max(PK)} \quad , \quad v_{PFK} = V_{PFK}/V_{max(PFK)} \quad ,$$

$$\tau = t\, V_{max(PK)}/K_{PEP} \quad , \quad \tilde{\omega}_e = \omega_e K_{PEP}/V_{max(PK)} \quad ,$$

$$[pep] = [PEP]/K_{PEP} \quad , \quad [f6p] = [F6P]/K_{F6P}, \quad Q = K_{PEP}/K_{F6P} \quad ,$$

$$[adp] = [ADP]/K_{PEP} \quad , \quad [atp] = [ATP]/K_{PEP},$$

$$adn = ADN/K_{PEP} \quad and \quad c = C/K_{PEP}.$$

Here, K_{PEP} is the dissociation constant of magnesium-free PEP with the R state of the PK (0.31 mM [7]) and K_{F6P} is the dissociation constant of total F6P with the R state of PFK (0.0125 mM) [8].

8

For calculations, we use the rate law for PK from E. coli derived in [7] depending on [PEP], [ADP], [ATP] and [Mg^{2+}]. The influence of the formation of complexes of potassium and magnesium ions is treated as given in [9]. For the PFK, we used a rate law derived in [8] depending on [F6P], [ATP], [ADP] and [PEP], under the additional assumption of random substrate binding with dissociation constants independent of the binding order.

In order to obtain an autonomous system, we introduce a new phase variable $z = \tau \tilde{\omega}_e$ and write:

$$Q \frac{d[f6p]}{d\tau} = \nu + \alpha \sin z - \varepsilon v_{PFK} , \qquad (7)$$

$$\frac{d[pep]}{d\tau} = \nu + \alpha \sin z - v_{PK} , \qquad (8)$$

$$\frac{dz}{d\tau} = \tilde{\omega}_e , \qquad (9)$$

$$[pep] + [atp] - Q[f6p] = c , \qquad (10)$$

$$[adp] + [atp] = adn. \qquad (11).$$

Here, the following parameters might be changed experimentally to influence the dynamics of the system: ν, α, ε, $\tilde{\omega}_e$, c, adn, [Mg_{tot}] and [K_{tot}].

Due to the conditions given by (10) and (11), the four variables [pep], [f6p], [adp] and [atp] can be displayed using only two dimensions by a suitable modification of the known geometrical methods [10]. Here, the four variables are given along the four edges of a trapezium as shown in Fig. 2. The limits between which the variables [pep] and [adp] are displayed within the trapezium are:

$$[pep_1] \leqslant [pep] \leqslant [pep_2] \qquad\qquad [adp_1] \leqslant [adp] \leqslant [adp_2] .$$

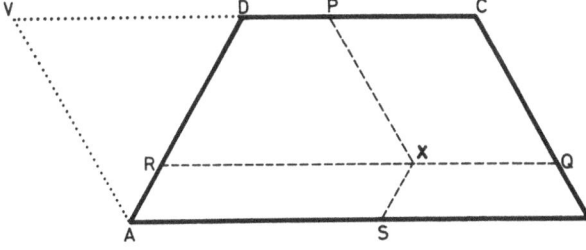

Fig. 2
Trapezium used for a two-dimensional display of the four metabolite concentrations [PEP], [ADP], [ATP] and [F6P]

The trapezium of Fig. 2 is constructed in such a way that

$$\overline{AB} = [pep_2] - [pep_1], \quad \overline{AD} = \overline{CB} = [adp_2] - [adp_1], \quad \sphericalangle DAB = \sphericalangle CBA = 60^o.$$

The auxiliary line $\overline{VA}$ is drawn so that $\sphericalangle DAV = 60^o$. For given [adp] and [pep] of the trajectory to be displayed, we draw the point X such that

$\overline{VP}$ = [pep]-[pep$_1$] and $\overline{QB}$ = [adp]-[adp$_1$].

Simple geometrical considerations lead to

$\overline{AS}$ = Q([f6p]-[f6p$_1$]) and $\overline{DR}$ = [atp]-[atp$_1$],

where [f6p$_1$] and [atp$_1$] are the minimum displayable values of [f6p] and [atp] and are given by

[f6p$_1$] = ([pep$_1$]-[adp$_1$]+ adn - c)/Q and [atp$_1$] = adn - [adp$_2$].

On this basis, labels for each of the four metabolite concentrations along the edges of the trapezium are obtained as shown in Fig. 3 and 4. The continuous lines in both figures as well as the dotted line in Fig. 3 show periodic solutions. The actual solution of (7) to (11) is a trajectory screwing its way into infinity in the z direction. The curves shown are projections of the solution on the plane of the trapezium, which is perpendicular to the z axis. Because such projections are confusing in case of quasiperiodic or chaotic solutions with an ever increasing number of crossovers of projected trajectories, a stroboscopic phase portrait [11] is of use. In this representation, the solution is plotted by points taken at equidistant values $z = \omega_e n \tau_s$ (n=1..., N) on the plane perpendicular to the z axis. τ_s is the time between plotted points. This display technique permits to discriminate between different types of solutions. A finite number p of points on the stroboscopic portrait indicates that the solution is periodic with period $p/q\tau_s$, where q is an integer to be determined. A closed curve on the stroboscopic plot indicates a quasi-periodic solution. The picture resulting from chaos is quite different: the points seem to be randomly distributed on curves with a finite width (see Fig. 4). Diagnostics of dynamic processes may be complemented by determining power spectra [12], time correlation functions [13], or Liapunov numbers [14](see below). In addition, solutions can be

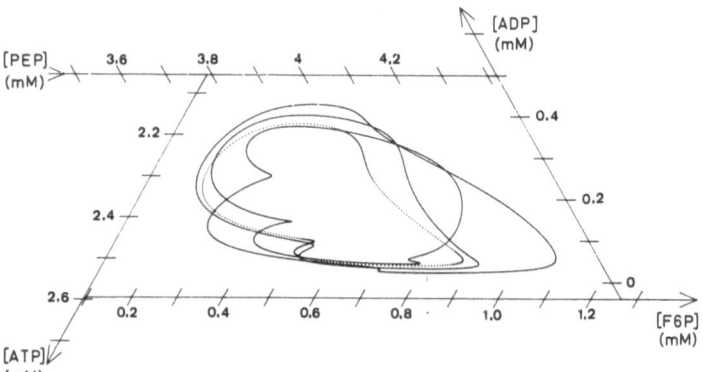

Fig. 3
Two periodic orbits coexisting in phase space at A = 1.05 mM/min and ω_e = 14.0 min^{-1} (assuming V$_{max(PFK)}$ = 3.3 mM/min). The period of the dotted line is 3 times and that of the continuous line 10 times the input flux period. The dots are plotted each hundredth of the input flux period, i.e., each 0.27 seconds

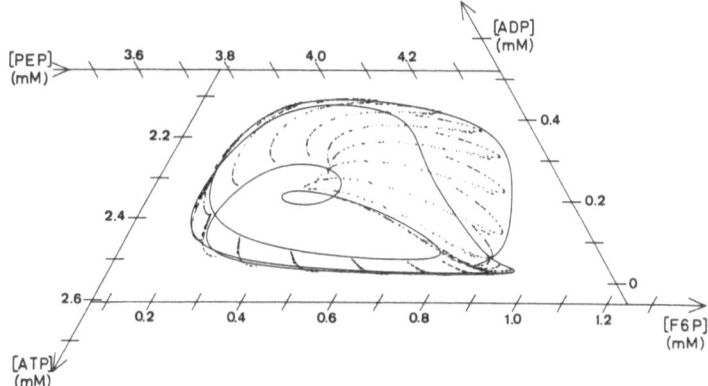

Fig. 4
Coexistence of an orbit with a period six times the input flux period (continuous line) and chaos (dots). A = 1.14 mM/min and ω_e = 8.88min^{-1} (assuming $V_{max(PFK)}$ = 3.3 mM/min). The dots are plotted each tenth of the input flux period, i.e., each 4.25 seconds

displayed in a compact form by a coordinate transformation such that two metabolite concentrations are plotted on a plane and the time is proportional to the rotation angle of this plane around some axis. This technique displays periodic solutions in form of closed curves whereas chaotic attractors appear as quasi-two-dimensional closed strips (details will be given elsewhere).

3. Results

In the absence of periodic excitation (ω_e = α = 0) the two-enzyme model displays only point attractors and limit cycles, as expected. Numerical integration of (7) to (11) shows that the system has an unstable singularity with a limit cycle at ν = 0.12, ε = 0.34 , c = 18.5, adn = 8.24, $[Mg_{tot}]$ = 5 mM and $[K_{tot}]$ = 200 mM. These are experimentally feasible conditions. With periodic excitation, the system is rather sensitive to variations of α and $\tilde{\omega}_e$. Upon variation of these parameters, keeping the other parameters constant, a plethora of bifurcations to different types of solutions is obtained. In the α-$\tilde{\omega}_e$ plane, patches of chaos, quasiperiodicity and periodicity are observed including period-doubling cascades [15]. Periodic entrainment has been found theoretically and in biochemical experiments during investigations of glycolysis [2]. Also, in a sequence of enzymic reactions with two positive feedback loops, the coexistence of two limit cycles in phase space with the same set of bifurcation parameters has been described in terms of bi-rhythmicity [16] .

In the system described here, we found up to 4 coexisting periodic solutions in phase space, e.g., at α = 0.0456, $\tilde{\omega}_e$ = 0.207, the periods being 2, 7, 9 and 11 times the input period. This phenomenon may be called "tetrarhythmicity" or more generally "polyrhythmicity". A large amount of overlaping patches on the α-$\tilde{\omega}_e$ plane was obtained, the patches corresponding to different types of solutions (periodic, quasiperiodic and chaotic). A typical coexistence of two periodic orbits is shown in Fig. 3. Fig. 4 demonstrates the coexistence of one periodic orbit and chaos.

The coexistence of two or more periodic orbits can be used as a switching device. The switch - which is phase dependent - can be simulated by fast additions or substractions of metabolites in accordance with the conservation laws. By this technique, new initial conditions for the system of differential equations are generated. A recipe for a switching process is the following: a transition from an orbit 1 to an orbit 2 is obtained by addition PEP and F6P at a time where the [ADP] of both orbits are at the same level and the [PEP] of orbit 1 is smaller than the [PEP] of orbit 2 by a difference δ. In such an experiment, it is important to add PEP and F6P so that [PEP]=[F6P]=δ. The simultaneous addition of F6P is necessary to keep the bifurcation parameter C (given by (5)) unchanged.

A switch from orbit 1 to orbit 2 may also be accomplished at a time where the [ADP] of both orbits are the same but [PEP] of orbit 1 is larger than the [PEP] of orbit 2. In such a case PEP and F6P have to be taken away from the system at equal amounts. This may well occur in nature. Switching can also be accomplished by less rigorous conditions if the system is disturbed in such a way that it falls from one orbit to the basin of attraction of the other. In such a case, however, it may take an appreciable transient time for the system to settle on the new limit orbit.

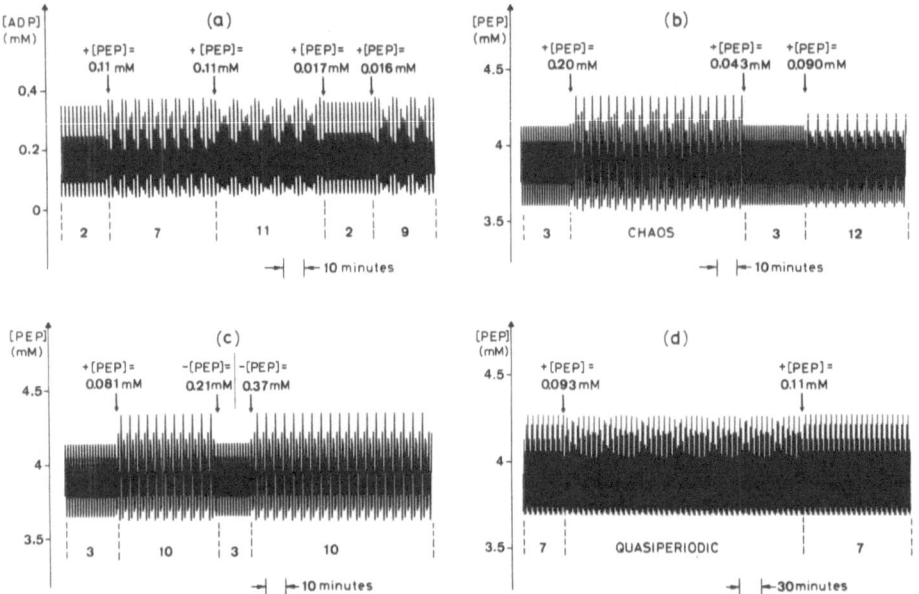

Fig. 5
Switching between different oscillatory modes by addition (or substraction) of PEP by fast pulses. Simultaneous to each addition (or withdrawal) of PEP, an equal amount of F6P has to be added (or withdrawn). The numbers below each wave pattern indicate the period of the oscillation divided by the input flux period. For the determination of the time scale we set $V_{max}(PFK)$ = 3.3 mM/min. The conditions are: (a) A = 0.44, ω_e = 6.47, (b) A = 1.14, ω_e = 14.1, (c) A = 1.05, ω_e = 14.0, (d) A = 0.47, ω_e = 6.68. (A is given in mM/min and ω_e in min^{-1}). The times (given in multiples of the input flux period and counted from the beginning of the current oscillation mode) for the additions (or substractions) are (a) 21.52, 53.02, 60.30, 25.06, (b) 57.48, 211.22, 75.31, (c) 55.45, 111.58, 34.87 and (d) 56.61, 322.92

12

Fig. 5 illustrates four examples with perturbations resulting in a change of the wave form. Because of the phase sensitivity of the process, the exact time of additions (or substractions) have to be observed and are given in the figure caption. Fig. 5a exemplifies tetrarhythmicity and demonstrates that the enzyme system can be switched back and forth between four different periodic orbits. It can be seen that a switching perturbation triggers a drastic wave form change and allows to settle on an orbit with period 2, 7, 11 and again 2, finally, in this experiment period 9 is yielded. The switching from an ordered periodic state to chaos and back into the ordered state is demonstrated in Fig. 5b. This experiment also shows the switch between periodic responses differing in amplitude and modulation (3 attractors coexist).

In Fig. 5c the switching is illustrated which occurs upon substraction of metabolites. Here, period 3 is switched over in period 10 by addition of PEP, returned to period 3 upon substraction, and finally switched back into period 10 by substraction. Fig. 5d represents reversible switching between periodic and quasiperiodic regimes. In contrast to the periodic modes, the quasiperiodic mode is characterized by frequency mixing and the chaotic mode by a continuous spectrum.

The observation of the generation of a variety of complex dynamic states maintained by a chemical system raises the question of their stability. For one-dimensional recurrences, like the logistic equation, the stability can be described by the Liapunov characteristic exponent given by

$$\lambda = \lim_{n \to \infty} \frac{1}{n} \sum_{i=1}^{n} \log_2 \left| \frac{dy}{dx} \right|_i \quad , \tag{12}$$

where i is the iteration index, n is the number of iterations and dy/dx is the slope of the next amplitude plot [14]. Chaotic behaviour leads to $\lambda > 0$, whereas periodic behaviour leads to $\lambda < 0$. Bifurcations between different modes of oscillations are characterized by $\lambda = 0$. Points where $\lambda \to -\infty$, which occur between two successive points of period doubling bifurcations, have been called "superstable" [12].

For a three-dimensional system as investigated here, (12) can be generalized [14,17,18,19] yielding three characteristic exponents, which indicate the stability along three independent directions in space. The exponent corresponding to the direction of the trajectory is 0. We found that the other two exponents are either both negative, indicating a stable orbit, or one is negative and the other one positive indicating a locally unstable orbit with the possibility of chaos.

Fig. 6 shows the frequency dependency of the maximum non-zero characteristic exponent λ_{max} for the system examined above with $\alpha = 0.118$. The chaotic region ($\lambda_{max} > 0$) at the far right side of the figure shows periodic windows ($\lambda_{max} < 0$) in a similar way as observed in other systems [12,20]. More periodic windows appear upon increasing resolution. The Liapunov dimension [21,22] of the chaotic attractors is given in our case by

$$D_L = 2 + \frac{\lambda_{max}}{|\lambda_{min}|} \quad . \tag{13}$$

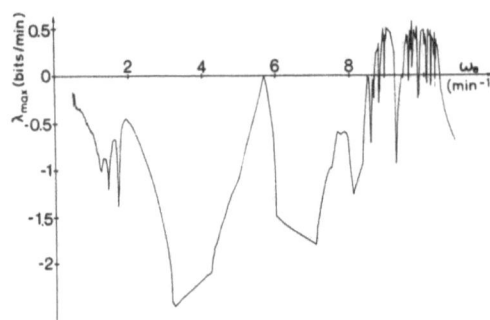

Fig. 6
Spectrum of the maximum Liapunov exponent (abscissa: input flux frequency). For dimensioning we set $V_{max(PFK)}$ = 3.3 mM/min

MORI's conjecture [23] predicts the quantity D_L to be equal to the fractal dimension of the attractor. For the chaotic solution shown in Fig. 4, we found λ_{max} = 0.20 and λ_{min} = -1.15. We thus obtain D_L = 2.17. This means that the attractor is almost a surface (dimension 2.), and the stroboscopic portrait consequently consists of almost one-dimensional curves, as seen in Fig. 4.

The stable region in Fig. 6 (λ_{max} < 0) shows deep minima in λ_{max}. Like in an investigation on the stability of integration algorithms [20], the exponents do not become $-\infty$, as in superstable points, but nevertheless guarantee a relaxation from orbit perturbations that is orders of magnitude faster than at neighbouring regions. The points at which such minima occur, which we may call "hyperstable", may be of physiological significance where highly stable biorhythms are required.

4. Comments

The studies of the dynamics of a biochemical model system reveal a surprising richness of time-independent as well as time-dependent oscillation patterns. It is remarkable that such a richness is displayed by a simple model simulating a short cut of glycolysis and consisting solely of a sequence of two enzymes. The mechanistics of the overall dynamics of glycolysis which is ubiquituous in living nature can be considered as general and thus of great significance. In addition, biological relevance relies on the fact that the physiologically typical levels of rates and concentrations are obtained in the calculations, although the level of PEP might still be considered high.

In addition to the phenomenon of entrainment yielding harmonic and subharmonic responses in biochemical experiments and simulations reported earlier [2], we here see periodic responses of highly stable performance and chaotic states. The system displays a multitude of dynamical properties such as a) oscillations with single and mixed frequencies with harmonics and subharmonics yielding time-independent oscillation patterns, b) time-independent oscillation patterns resulting from amplitude modulation and c) chaotic attractors leading to time-dependent oscillation patterns.

The general relevance of entrainment with respect to synchronization of cellular functions and/or multicellular performance or species interaction has been stressed before [2]. Here we would like to point to the possible function of chaos and of patternization of timely metabolic performance. Quasiperiodic entrained and chaotic response of the self-sustained oscillations of Nitella flexilis and the Onchidium

14

giant neuron have been observed [24,25]. In a different context, chaos has been associated with pathological conditions like different kinds of arhythmic behaviour. However, the chaotic state may be of physiological importance because here fluctuations, which appear in a time scale that can be biochemically relevant, make an organism flexible to the response towards external influences, permitting it to choose by trial and error. Such fluctuations may optimize its relationship to the environment not only with respect to phenomena like taxis and signal reception, but also in terms of evolution.

Periodic and chaotic states can be compared in relation to their stability. While in a chaotic system the fluctuations are amplified ($\lambda_{max} > 0$), a periodic state is characterized by a suppression of fluctuations ($\lambda_{max} < 0$) which is especially strong in hyperstable states. Conditions for hyperstability may be required to stabilize physiological rhythms. Here, however, it should be stressed that the chaotic states also display stability along certain directions of space.

The surprising sensitivity of the time patterns towards metabolite perturbations is of further interest. As shown, the patterns may be changed back and forth almost instantaneously by pulses of very small quantities of metabolites. This switching phenomenon has the character of a hysteresis loop. It could well be that such switching processes might be realized in the control of time patterns forming the basis of the informational processes in neural networks.

The different possible patterns that a system with given enzymic properties and metabolite concentrations might display - we may call this the "pattern-library" of the system - is genetically controlled and may change during the evolutionary process as well as by repression or depression upon interaction with the environment even in one single cell. It should be stressed that the switching is strongly phase-dependent, as demonstrated in biochemical experiments by direct titration in which switching could readily be achieved [26,27].

Acknowledgements
This article is an abbreviated form of a lecture given at the Conference on "Synergetics - from Microscopic to Macroscopic Order" and focusses on the phenomenon of chaos. We thank Dr. Th. Plesser for valuable discussions and Heiko Becher for his assistance in the numerical computations.

References
1. B. Hess: 8. Fritz Lipmann-Vorlesung, Hoppe-Seyler's Z. Physiol. Chem. 364, 1 (1983)

2. A. Boiteux, A. Goldbeter and B. Hess: Proc. Natl. Acad. Sci. USA 72 (10), 3829 (1975)

3. Th. Plesser, in: VII Int. Konf. über Nichtlineare Schwingungen ed. by G. Schmidt, Vol. 2, 273 (Akademie-Verlag, Berlin, 1977)

4. M. Markus and R. Schueller: 15th FEBS Meeting, Brussels, Abstract S-17/WE-194 (1983)

5. M. Markus, H. Becher and B. Hess: Hoppe-Seyler's Z. Physiol. Chem. 364(9), 1177 (1983)

6. J.G. Reich and E.E. Sel'kov: FEBS Letters 40, S119 (1974)

7. A. Boiteux, M. Markus, Th. Plesser, B. Hess and M. Malcovati: Biochem. J. 211, 631 (1983)

8. D. Blangy, H. Buc and J. Monod: J. Mol. Biol. 31, 13 (1968)

9. M. Markus, Th. Plesser, A. Boiteux, B. Hess and M. Malcovati: Biochem. J. 189, 421 (1980)

10. P. Schuster, K. Sigmund and R. Wolff: SIAM J. Appl. Math. C 37(1), 49 (1979)

11. N. Minorsky: Nonlinear Oscillations (R.E. Krieger Publ., Huntington, N.Y., 1974)

12. P. Collet and J.-P. Eckmann: Iterated Maps on the Interval as Dynamic Systems (Birkhäuser, Boston, 1980)

13. S. Grossmann und S. Thomae, Zeitschrift für Naturforschung 32a, 1353 (1977)

14. R. Shaw: Z. Naturforsch. 36a, 80 (1981)

15. M. Feigenbaum: J. Stat. Phys. 19, 25 (1978) and 21, 669 (1979)

16. O. Decroly and A. Goldbeter: Proc. Nat. Acad. Sci. USA 79, 6917 (1982)

17. G. Benettin, L. Galgani and J.-M. Strelcyn: Phys. Rev. A14, 2338 (1976)

18. I. Shimada and T. Nagashima: Progr. Theor. Phys. 61, 1605 (1979)

19. E. Ott: Rev. Mod. Phys. 53(4), 655 (1981)

20. M. Prüfer, Report Nr. 85, Forschungsschwerpunkt Dynamische Systeme, University of Bremen (1983)

21. J. Kaplan and J. Yorke: in Functional Differential Equations and Approximation of Fixed Points, ed. by H.O. Peitgen and H.O. Walther (Springer, Berlin, Heidelberg, New York, 1979)

22. J.D. Farmer: Physica 4D, 366 (1982)

23. H. Mori: Progr. Theor. Phys. 63, 3 (1980)

24. H. Hayashi, S. Ishizuka, M. Ohta and K. Hirakawa: Physics Letters 88A, 435 (1982)

25. H. Hayashi, M. Nakao and K. Hirakawa: Physics Letters 88A, 265 (1982)

26. B. Chance, B. Schoener and S. Elsaessar: J. Biol. Chem. 240, 3170 (1965)

27. B. Hess and A. Boiteux: in: Regulatory Functions of Biological Membranes, ed. by Johan Järnefelt (Biochim. Biophys. Acta Library, 1968) Vol. 11, 148

Chaotic Behaviours Observed in Homogeneous Chemical Systems

C. Vidal

Centre de Recherche Paul Pascal, F-33405 Talence, France

1. Introduction

Dynamical systems theory is a branch of mathematical physics, growing rapidly nowadays, which deals with ordinary differential equations and with discrete applications as well. It has introduced into science several new concepts, such as weak stability, strange attractor, mixing of trajectories, and so on. The aim of this lecture is to show how far these ideas prove to be relevant in the study of chemically reacting systems, evolving far from equilibrium. To this end, an overview of the main results obtained over the past four years is presented in what follows. As a conclusion I will mention some still unanswered questions and try to guess how the future could look in this field of experimental research.

2. Evolution equations of a CSTR

The laws of thermodynamics tell us that a chemical system left to itself evolves irreversibly toward an equilibrium state, corresponding to an extremum of a given thermodynamic potential which depends on the conditions applied. Moreover we learn from irreversible processes in thermodynamics that this equilibrium state is always reached monotonically - and even exponentially. If one is interested in non-monotonic behaviour one must :

   i) either look to the first stages of evolution, while the system is still far from equilibrium,

   ii) or keep the system out of equilibrium by maintaining an energy and/or a matter flux through it.

Transient regimes are sometimes useful to observe, but quantitative studies can be carried out only on permanent ones. This is the reason why the second point is often preferred to the first. The standard device used in chemistry to impose non-equilibrium conditions is a continuously stirred tank reactor (CSTR), assumed to provide an instantaneously uniform (in concentration and temperature) medium. Figure 1 presents an oversimplified scheme of such a reactor. Concentration changes $\frac{dx_i}{dt} = \dot{x}_i$ of species i are given by the balance equation :

$$\dot{x}_i = F_i(x_j) + \frac{J}{V}(x_i^o - x_i) \qquad i,j = 1,\ldots, N \qquad (1)$$

N : total number of reacting species.

The second term simply represents the flow through the reactor (when no volume variation comes from the reaction), while the first one depicts the contribution of the reaction itself. Most often the chemical mechanism is unknown ; nevertheless it is expected to involve a set of elementary steps, briefly stated :

17

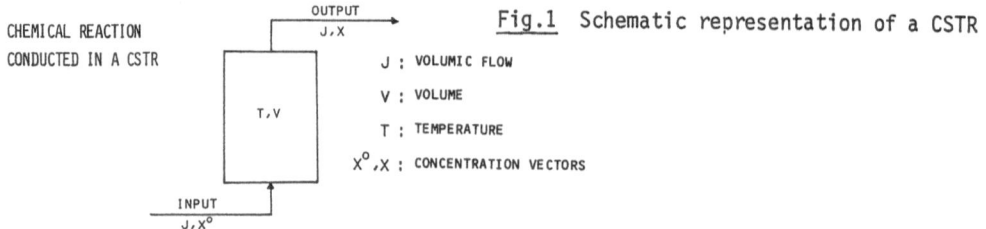

CHEMICAL REACTION
CONDUCTED IN A CSTR

OUTPUT
J,X

Fig.1 Schematic representation of a CSTR

J : VOLUMIC FLOW

V : VOLUME

T : TEMPERATURE

X^o,X : CONCENTRATION VECTORS

INPUT
J,X^o

$$\nu_{jr}\, X_j \xrightarrow{k_r} \nu'_{\ell r}\, X'_\ell$$

$j, \ell = 1, \ldots, N$; $r = 1, \ldots, R$

R : total number of elementary steps
k_r : rate constant of step r.

Therefore, according to the kinetic mass action law, $F_i(x_j)$ has the form :

$$F_i(x_j) = \sum_{r=1}^{R} (\nu'_{ir} - \nu_{ir})\, k_r \prod_j x_j^{\nu_{jr}}$$

that is to say, it contains essentially terms which are non-linear with respect to the dynamic variables x [1]. Thus the behaviour of a CSTR is accounted for by the set of N non-linear differential equations (1) and, even more, several experimentally controllable parameters are at the experimenters' disposal, namely:

i)   any of the inlet concentrations x_i^o,

ii) the temperature T, since the rate constants k_r vary according to the Arrhenius law,

iii) last, but not least, the ratio J/V = μ which is the reciprocal of the mean residence time.

Parameter μ is of special interest because it allows one to change the relative magnitude of the linear flow term at the same time and in the same way in all the N differential equations. One can thus easily understand why it has been systematically used in experiments so far [2].

We already learn something from the structure of equation (1). Indeed, if μ becomes so large ($\mu > \mu_2$) that $F_i(x_j)$ is negligible, there is an obvious stationary solution x^S, merely

$$x_i^S = x_i^o \qquad i = 1, \ldots, N$$

whose physical meaning is clear : when the residence time is too short, the chemical reaction cannot actually take place to any significant extent. On the other hand, $\mu = 0$ corresponds exactly to the equilibrium situation (after cancellation of the transients) where :

$$\dot{x}_i = 0 \quad \Rightarrow \quad x_i^S = x_i^e \qquad i = 1, \ldots, N.$$

Close enough to equilibrium, a stationary state does exist along the so-called thermodynamic branch (see figure 2). This stationary state may become unstable be-

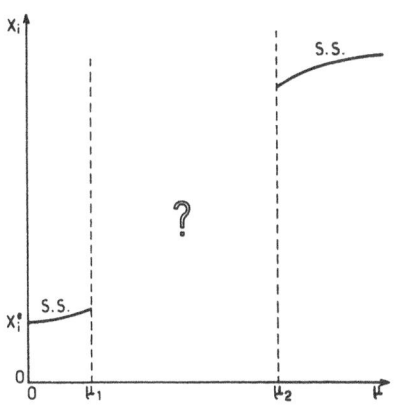

Fig.2 Dynamic features predicted by equation (1)

yond a critical threshold μ_1, thanks to the non-linearity of equation (1). In between the two limits (μ_1, μ_2), certain chemical systems may thus exhibit non-stationary features, i.e. periodic and/or aperiodic regimes. In fact reactions capable of developing this kind of behaviour seem to be rather few.

3. Non-monotonic chemical reactions

The first reaction of this type was discovered by BRAY while studying the stability of hydrogen peroxide solutions in presence of additional compounds. A small amount of iodate ions induces regular oscillations, even at room temperature [3]. Half a century later, BELOUSOV [4] found, also by chance, another oscillating reaction : the oxidation of organic substances (e.g. citric or malonic acid) by $Br\,O_3^-$ ions in presence of a catalyst such as the redox couple Ce^{3+}/Ce^{4+}. Meanwhile oscillations were also discovered in the gaseous oxidation of hydrocarbons, H_2S, and similarly during the 30's [5]. More attention was paid to these homogeneous periodic reactions when it became clear that similar oscillations occur in the metabolism of living systems : photosynthesis [6] and glycolysis [7] were the first two identified.

One important question which comes to mind is : what prerequisites must be fulfilled by a chemical reaction to give rise to periodicity ? The only thing known with certainty up to now is the need for an autocatalytic process or, more generally stated, for a feedback loop. Other conclusions, although interesting, are rather specific to restricted media [8]. Recently it was shown that bistability, not necessary nor sufficient by itself, may be helpful for designing new chemical oscillators [9]. The first successful procedure applied to the systematic search of periodic reactions is based on this idea ; it has led to more than twenty different mixtures, involving $Cl\,O_2^-$ as a common compound and named chlorite oscillators [10].

Nevertheless the more widely studied and understood chemical system of this type is the bromate oscillator, which will be referred below as the BELOUSOV - ZHABOTINSKY reaction (abbreviated BZ). A large amount of work has been done to build up its detailed mechanism : in the model labelled FKN - from FIELD, KÖRÖS and NOYES [11] - N = 14 and R = 13 ! A simplified version with N = 3 and R = 5 has been proposed [12] to allow for numerical computation and, even, analytical calculation of the main features [13].

19

In practice the reaction is most often monitored by potentiometric (redox potential, Br⁻ sensitive electrode) or spectrophotometric (optical density at 340 nm) techniques. In both cases, no unusual experimental care is required to obtain a very good signal to noise ratio.

4. Some explainable results

Although theory [14] predicts the very existence of aperiodic regimes, the first task is to observe them and to establish their genuine aperiodicity. Preliminary observations of irregular regimes in the BZ reaction were first reported by SCHMITZ et al. [15,16]. By Fourier transformation of the signal, our group [17,18]. was able to characterize these regimes and to identify aperiodic behaviour. In particular we clearly demonstrated that noise of the recorded signal could not account for the emergence of a broad band in the power spectrum. Even at a glance at Fig. 3a this is almost obvious.

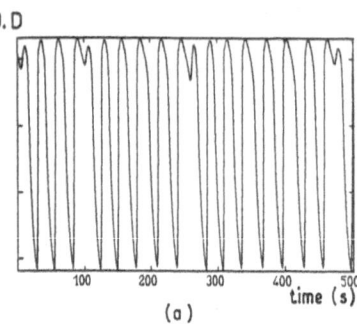 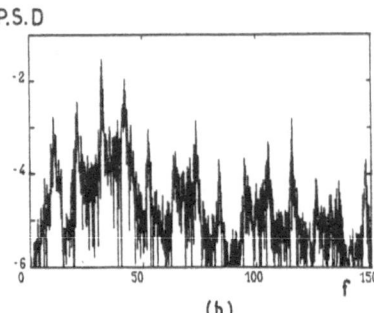

Fig.3 Time series (a) and corresponding power spectrum (b) of an aperiodic regime of the BZ reaction (from ref 28)

Since the largest part of the mathematical analysis is devoted to the properties of phase-space trajectories, it seems desirable to look at them. Unfortunately there are two problems. The first is that one measures only a single phase variable ; in addition, the number of degrees of freedom itself is unknown. According to a conjecture suggested a few years ago, the topological properties of the true attractor would be preserved by images reconstructed in spaces involving as independent coordinates :

- either $\chi(t)$, $\chi(t+\tau)$, $\chi(t+2\tau)$, ...
τ being an arbitrary time interval [19]

- or $\chi(t)$, $\dot{\chi}(t)$, $\ddot{\chi}(t)$, ... [20] .

Following this second way, the attractor image of an aperiodic regime is shown in Fig.4, where trajectories go clockwise [21]. It is seen that nearby trajectories at the bottom of the picture may either enter in the eddy, make an undefined number of rotations and then leave it, or elso go directly to the left. Therefore it is not possible to predict how the distance between two points will evolve : this attractor thus exhibits sensitivity to initial conditions.

Such a conclusion is strongly supported by the first return map deduced from a POINCARE section [22], as shown in Fig.5. The points fall along a nice curve and

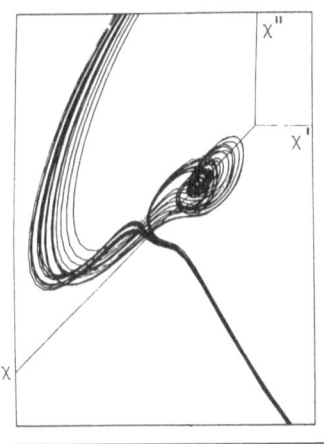

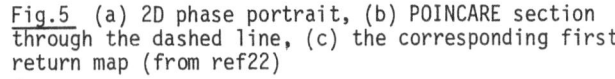

Fig.4 Image of an attractor in $(\chi, \dot{\chi}, \ddot{\chi})$ space representation displaying sensitive dependence on initial conditions (from ref21)

Fig.5 (a) 2D phase portrait, (b) POINCARE section through the dashed line, (c) the corresponding first return map (from ref22)

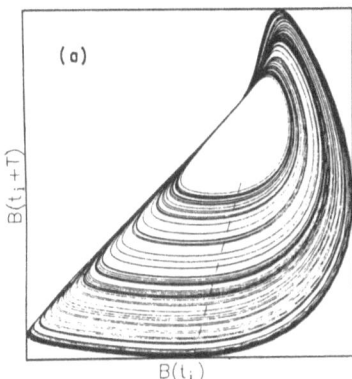

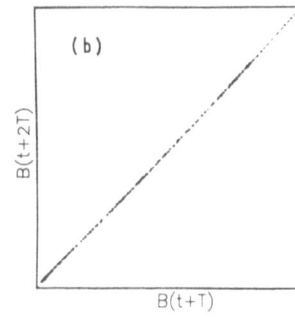

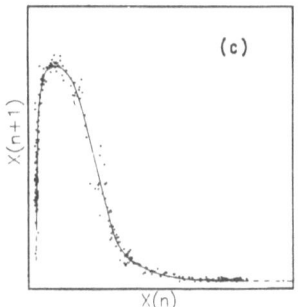

are not distributed all over the plane, indicating a deterministic rather than a stochastic behaviour. Moreover, the curve has a good shape : it actually displays a bump, associated with non-linear effects. Briefly speaking the BZ reaction provides us with an experimental illustration of a strange attractor, even though there is of course no way to prove that these attractors are really strange. One should not be surprised that no sheet structure is noticeable in the POINCARE section of Fig.5, which looks like an ordinary line. Perturbation experiments [22] point out that the system is highly dissipative : accordingly contraction in phase-space is very strong and the fractal dimension of the attractor is expected to be only slightly higher than 2. It turns out that its fine structure is unobservable, as is true for the celebrated Lorenz model.

Once chaotic behaviour is known to exist in chemical systems, the next question naturally deals with its onset. Theory predicts that three routes may lead to chaos in dissipative systems having a few degrees of freedom. These routes, involving different types of bifurcation, are usually named *scenarios*, according to ECKMANN [23]. Let us recall very briefly their essential feature :

- the RUELLE-TAKENS-NEWHOUSE scenario (1971) asserts that a strange attractor may occur after three successive HOPF bifurcations,

- the FEIGENBAUM scenario (1978) involves a cascade of period doubling pitchfork bifurcations,

- the POMEAU-MANNEVILLE scenario (1980) is associated with a saddle-node bifurcation giving rise to intermittency.

These routes towards chaos are observed in chemical systems and are more or less characterized. Better than a rather long discussion, Fig.6 and Table I illustrate to that.

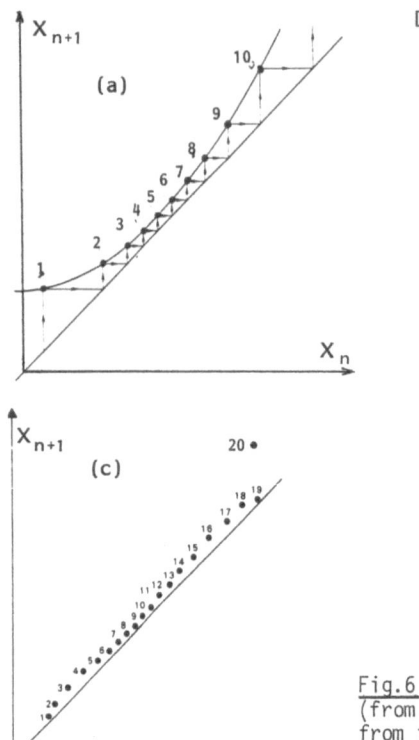

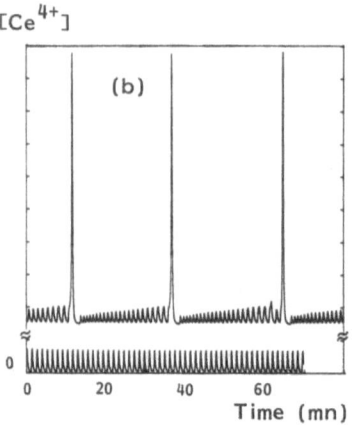

Fig.6 Intermittency, (a) theoretical prediction (from ref24), (b) experimental observation coming from the recorded time series (c) of a regime exhibited by the BZ reaction (from ref25)

Period	Sequence	Pattern
1	$\cdots$	0
2	R	0-1
2×2	RLR	2-0-3-1
$2^2 \times 2$	RLR^3LR	2-6-0-4-3-7-5-1
10	RLR^3LRLR	2-8-6-0-4-3-9-5-7-1
6	RLR^3	2-0-4-3-5-1
5	RLR^2	2-0-4-3-1
3	RL	2-0-1
2×3	RL^2RL	2-5-3-0-4-1
9	RL^2RLR^2L	2-8-5-3-0-6-4-7-1
5	RL^2R	2-3-0-4-1
4	RL^2	2-3-0-1
2×4	RL^3RL^2	2-6-3-7-4-0-5-1

Table 1 A set of dynamic regimes of the BZ reaction indicating subharmonic bifurcation cascade and several components of the U sequence (from ref26)

In Fig.6, graph (a) shows that right after a saddle-node biburcation, the system requires a considerable length of time to cross the narrow channel created between the first bissectrix and the application curve [24]. Taking the amplitude of oscillations as the relevant dynamical variable (Fig.6b) it is seen in Fig.6c

that this phenomenon actually takes place in a BZ experiment. A more thorough investigation supports this conclusion better [25]. Then a set of dynamical regimes,
also displayed by the BZ reaction, is collated in Table I [26]. The first steps of
the period-doubling cascade are quoted there ; however, because of the high conver-
gence rate linked to the universal number δ = 4.6692..., no more than three bifurcations can be observed. Interesting too is the identification of several elements
of the so-called U sequence [27]. Although the slow drift in flow rate due to pe-
ristaltic pumping does not allow one to establish definitively the order of the reported periodic regimes along the μ axis, it is worth noting that these data are
extremely consistent with the very existence of such a universal sequence. Lastly,
work is now in progress to look at the oldest proposed scenario, which involves
quasi-periodicity. According to preliminary results obtained in Bordeaux, one can
reasonably expect that experimental support will be obtained quite soon.

5. Some others

Not all the reported observations are well understood currently or, say, explained
in a satisfactory manner. To illustrate this point, I will simply give two examples.

Cascades of bifurcations have been observed many times in chemical systems.
First of all with the BZ reaction, and on both sides of the control parameter μ :
that is to say at short residence times [16,18,28] near μ_2, and also at very long
residence times [22]. Fig.7 gives an example of a sequence of 22 different regimes,
some of them periodic (above the axis) and the others aperiodic (below the axis).
Each periodic regime corresponds to a pattern of L large amplitude oscillations
followed by ℓ small amplitude oscillations, pattern quoted as L^ℓ in Fig.7. (The
difference between large and small amplitude can be seen in Fig.3a). Most often an
aperiodic regime is a mixing at random of the two nearby (the previous and the next)
periodic regimes along the μ axis. This kind of behaviour, sometimes named mixed-
mode oscillations, seems to be rather widespread in chemical systems. Indeed it
has also been reported for a chlorite oscillator [29] and even for the gaseous chemical oscillator CH_3 CHO + O_2 [30]. Preliminary attempts have been made to account
for what seems to be a very typical feature [31,32]. However a general interpretation, not dependent on the detailed chemical mechanism, is still lacking and should
perhaps be sought in the structure of equation (1). A theoretical search, taking into consideration the particular form of non-linearity involved there, would be welcomed.

The second example appears in Fig.8, where the Fourier spectrum of a dynamical
regime exhibited by the BZ reaction is represented at three different frequency
resolutions. With the poorest one (Fig.8a) only three large peaks, three harmonics,
are seen. Increasing the resolution about 6 times leads to a set of well-ordered
peaks, previously undetectable (Fig.8b). Changing again the frequency resolution,
one gets 12 harmonics of a shorter basic frequency (Fig.8c). In other words, we
encounter three sets of peaks, packed like a *matriochka* [33]. Owing to the fact

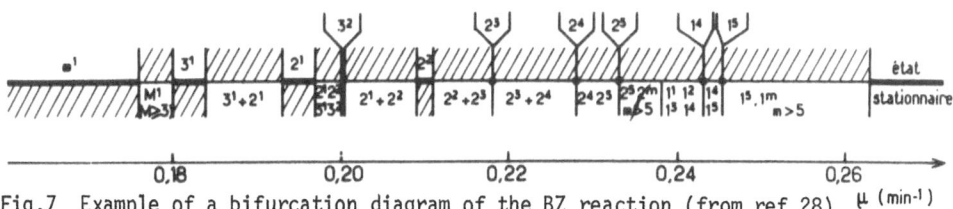

Fig.7 Example of a bifurcation diagram of the BZ reaction (from ref 28) μ (min-1)

PSD

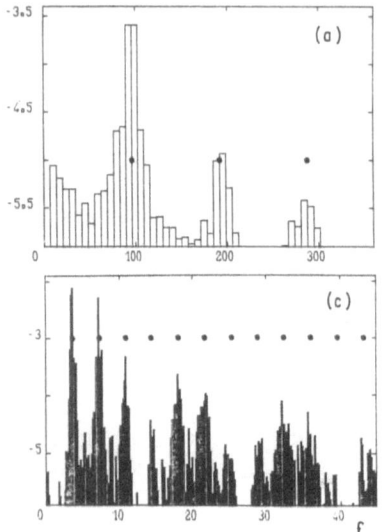

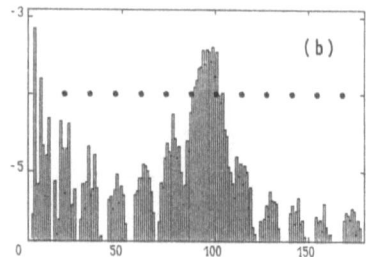

Fig.8 Power spectral density (logarithmic scale) versus frequency (mHz) of a dynamic regime having a structured Fourier spectrum (from ref33). Frequency solution:
(a) 6.75 mHz, (b) 1.07 mHz, (c) 0.15 mHz.
Abscissa (mHz) of black dots:
(a) 96 n   n = 1,..., 3
(b) 19.5 + 13.5 n   n = 0,..., 11
(c) 3.6 n   n = 1,..., 12

that a similar observation has also been made in RAYLEIGH-BENARD experiments [34], the need for a general explanation is once again obvious.

6. The power of chaos

Up to now we have discussed qualitative features of dynamic regimes and of transitions between them. In a further stage one would like to reach a more quantitative approach. In this respect an important point is certainly the measure of chaos, since the level of disorder does not necessarily have the same extent in all the aperiodic regimes. Of course theory provides us with several quantities which might be used to estimate the power of chaos. Nonetheless their determination from experimental time series is not so straightforward.

For example, if one agrees with the conjecture that a typical chaotic attractor has a fractal structure, then the fractal dimension of the attractor could be a relevant parameter. Unfortunately, in highly enough dissipative systems (like a chemical reaction) where contraction in any direction perpendicular to the flow is strong, the measure of the fractal dimension seems to remain out of reach. It has even been shown numerically by GREENSIDE et al. [35] that box-counting techniques fail in that case.

The largest LYAPUNOV characteristic exponent is another useful quantity. Not only a positive exponent λ tells us that nearby trajectories diverge in phase-space, but also, the greater is λ, the faster the system forgets its initial conditions, so that chaos is more important. However, when starting from experimental time series, computational difficulties are met here again. Different attempts have already been performed to determine λ : all of them rely on a next amplitude map drawn from a POINCARE section of the attractor, rather than on the trajectories themselves. The last reported value 0.3 ± 0.1 seems to be the best one [36]. But the uncertainty remains so large that LYAPUNOV exponents determined in such a way do not allow us really to discriminate between different aperiodic regimes. Algorithms involving a measure of the metric entropy [37] are now in progress and bring some hope for the future.

24

Other suggestions have also been made and, in particular, one can imagine the evaluation of the power of chaos of a dynamic regime through its Fourier spectrum. Among the various quantities which can be computed, we have shown that an entropy-like function, defined as :

$$H = \sum_i p_i \log p_i$$

p_i : power density at frequency $i \cdot \Delta f$
Δf : frequency resolution

seems to give a rather good picture of λ's variations [38]. Since H is quite easy to calculate, it might be used until a more tractable and refined method of measuring λ has been designed.

7. Concluding remarks

Chemical oscillators have thus proved to be useful in broadening the experimental field of research on dynamical systems, mainly restricted for a long time to hydrodynamics (if one excepts computer simulations, of course).

As expected, they have led to observations which support previously developed theoretical considerations. But they have also pointed out other phenomena which are not readily understandable at the present stage of dynamical systems theory. Thus, in so doing, chemical systems actually stimulate further progress.

Considering the various contributions coming from several fields, gathered over the past ten or fifteen years, on systems having a small number of degrees of freedom one can wonder about the future of this research domain. An important point, which has not yet been investigated enough - at least experimentally - is the influence of *noise*, in the widest sense of the word. Both internal fluctuations and external noise are thought capable of changing drastically the behaviour of a system evolving out of equilibrium [39,40]. For instance, as recently emphasized by FARMER [41], sensitive dependence to noise may exist without sensitive dependence to initial conditions, as soon as the attractor is locally unstable. This calls our attention to the care which must be taken in interpreting experimental observations, because noise is always present. To clarify these questions, experiments especially designed to determine the influence of internal fluctuations, and of external noise as well, are henceforth necessary. Chemical systems can and must be used to this end. Should I say that our group in Bordeaux is now busy in this direction ?

References

1. C. Vidal, "Chaos and order in Nature" (H. Haken, ed.) Springer-Verlag 1981, p. 69.
2. C. Vidal, "Non-linear phenomena in chemical dynamics (C. Vidal, A. Pacault, eds.) Springer-Verlag 1981, p. 49.
3. W.C. Bray, J. Am. Chem. Soc. 43, 1262 (1921).
4. B.P. Belousov, "Sb. Ref. Radiat. Med." Medgiz 1959, p. 145.
5. It must be reminded that we are dealing with homogeneous reactions. Therefore periodic processes resulting from a competition between a physical effect and a chemical reaction are disregarded : e.g. periodic dissolution of metals in acidic media, periodic decomposition of formaldehyde in concentrated sulphuric acid (Morgan reaction).

25

6. A.T. Wilson, M. Calvin, J. Am. Chem. Soc. 77, 5948 (1955).
7. L.N.M. Duysens, J. Amesz, Biochim. Biophys. Acta, 24, 19 (1957).
8. J. Chopin-Dumas, P. Richetti, "Non-linear phenomena in chemical dynamics" (C. Vidal, A. Pacault, eds.) Springer-Verlag 1981, p. 213.
9. J. Boissonade, P. de Kepper, J. Phys. Chem. 84, 501 (1980).
10. M. Orban, C. Dateo, P. de Kepper, I.R. Epstein, J. Am. Chem. Soc. 104, 1913 (1982).
11. R.J. Field, E. Körös, R.M. Noyes, J. Am. Chem. Soc. 94, 8649 (1972).
12. R.J. Field, R.M. Noyes, J. Chem. Phys. 60, 1877 (1974).
13. J. Tyson, "The Belousov-Zhabotinsky reaction" Springer-Verlag (1976).
14. D. Ruelle, Tans. N.Y. Acad. Sci. 35, 66 (1973).
15. R.A. Schmitz, K.R. Graziani, J.L. Hudson, J. Chem. Phys. 67, 3040 (1977).
16. J.L. Hudson, M. Hart, D. Marinko, J. Chem. Phys. 71, 1601 (1979).
17. C. Vidal, J.C. Roux, A. Rossi, S. Bachelart, C.R. Acad. Sci. Paris, 289C, 73 (1979).
18. C. Vidal, J.C. Roux, S. Bachelart, A. Rossi, Ann. N.Y. Acad. Sci. 357, 377 (1980).
19. D. Ruelle, private communication.
20. N.H. Packard, J.P. Crutchfield, J.D. Farmer, R.S. Shaw, Phys. Rev. Lett. 45, 712 (1980).
21. J.C. Roux, A. Rossi, S. Bachelart, C. Vidal, Phys. Lett. 77A, 391 (1980).
22. J.C. Roux, H.L. Swinney, "Non-linear phenomena in chemical dynamics" (C. Vidal, A. Pacault, eds.) Springer-Verlag 1981, p. 38.
23. J.P. Eckmann, Rev. Mod. Phys. 53, 643 (1981).
24. Y. Pomeau, P. Manneville, Physica D 1, 219 (1980).
25. Y. Pomeau, J.C. Roux, A. Rossi, S. Bachelart, C. Vidal, J. Physique Lett. 42, L271 (1981).
26. R.H. Simonyi, A. Wolf, H.L. Swinney, Phys. Rev. Lett. 49, 245 (1982).
27. N. Metropolis, M.L. Stein, P.R. Stein, J. Comb. Theory, A15, 25 (1973).
28. C. Vidal, S. Bachelart, A. Rossi, J. Physique, 43, 7 (1982).
29. M. Orban, I.R. Epstein, J. Phys. Chem. 86, 3907 (1982).
30. P. Gray, J.F. Griffiths, S.M. Hasko, P.G. Lignola, Comb. and Flame, 43, 175 (1980).
31. K. Tomita, I. Tsuda, Progr. Theor. Phys. 64, 1138 (1980).
32. C. Lobry, R. Lozi, "Non-linear phenomena in chemical dynamics" (C. Vidal, A. Pacault, eds.) Springer-Verlag 1981, p. 67.
33. C. Vidal, A. Rossi, "Non-linear phenomena in chemical dynamics" (C. Vidal, A. Pacault, eds.) Springer-Verlag 1981, p. 277.
34. M. Dubois, private communication.
35. H.S. Greenside, A. Wolf, J. Swift, T. Pignataro, Phys. Rev. A, 25, 3453 (1982).
36. J.C. Roux, R.H. Simonyi, H.L. Swinney, Physica D, in press.
37. J.D. Farmer, Physica D, 4, 366 (1982).
38. A. Lafon, A. Rossi, C. Vidal, J. Physique, 44, 505 (1983).
39. G. Nicolis, I. Prigogine, "Self-organization in non-equilibrium systems", Wiley & Sons (1977).
40. W. Horstemke, R. Lefever, "Noise induced transitions" Springer-Verlag, 1983.
41. J.D. Farmer, private communication.

Wave-Number Selection in Rotating Couette-Taylor Flow

Guenther Ahlers and David S. Cannell

Department of Physics, University of California
Santa Barbara, CA 93106, USA

In nonlinear dissipative systems subjected to an external stress, R, a transition frequently occurs to a state of reduced symmetry having a spatial structure with a characteristic wavelength when R exceeds a critical value R_c. Examples are Rayleigh-Bénard convection [1], Taylor-vortex flow [2-4], certain chemical reactions [5], flame-front propagation [6], and crystal growth [7]. The equations of motion, with boundary conditions corresponding to systems of infinite spatial extent, usually possess a continuum of linearly stable solutions [8], corresponding to a band of wave numbers having a width which varies as $\varepsilon^{1/2}$ ($\varepsilon \equiv R/R_c - 1$).

The above mathematical result suggests that it should be possible to prepare time-independent stable states experimentally for, say, Rayleigh-Bénard convection or Taylor-vortex flow when the stress parameter (Rayleigh or Reynolds number) corresponds to a small positive value of ε. However, the stability analysis predicts a *band* of stable states, and thus for $\varepsilon > 0$ it is not clear which, if any, wave vector is chosen by a real system in preference to all the others within the stable band. For a *finite* system, one of a finite number of allowed states often is expected to be selected by the *boundaries* and the previous history [9], but as the system becomes very large the number of possible states increases. Thus, the finite system is multistable with a barrier to transitions between adjacent states which may well vanish as the number of states and the size of the system diverge. Unless some mechanism selects one state out of the band of linearly stable states in preference to all the others, one would expect such a system to become more and more susceptible to external perturbations as its size grows. For the extremely large systems, it would become questionable whether a time-independent state can exist, for instance, in the presence of thermal noise [10].

These theoretical considerations can be compared with recent experiments on heat conduction by a horizontal layer of fluid heated from below (Rayleigh-Bénard convection) [11]. Those measurements indicate that stationary states do indeed exist for positive ε provided the system is of modest lateral extent and ε is not too large. However, for very large systems, nonperiodically *time*-dependent flow was observed even for ε as small as 0.1, and no stationary states were ever reached for $R > R_c$. It was suggested recently that the time-dependence in Rayleigh-Bénard systems of large lateral size might be explained by extremely long-lived transients associated with the adjustment of both the direction and the characteristic width of the convection roll pattern [12]. In numerical calculations, this essentially two-dimensional process was shown to last many times the horizontal thermal diffusion time (which is the obvious large time scale in the problem). In practice, it is virtually impossible to prolong the real experiment as long as required (according to the calculations [12]) for the transients to decay. But regardless of whether the observed phenomenon is a transient or a stationary random process, the existence and strength of any selection mechanisms which tend to choose one state out of the linearly stable band in preference to the others would be expected to have a strong influence on the observations. We summarize in this paper three experiments pertinent to wave-number selection processes in a dissipative system far from equilibrium.

For our investigation, we chose rotating Couette-Taylor flow in a fluid between concentric cylinders with the inner one rotating [2-4]. In that case, Taylor vortices form for Reynolds numbers larger than critical. This system is simpler than Rayleigh-Bénard convection because any adjustment processes leading to transients are essentially one-dimensional in character since the direction of the flow field is fixed by the cylindrical symmetry. Adjustment processes involve only changes in the widths of the Taylor vortices. This process is diffusive in character and its time scale is given by $t_v \equiv H^2/\nu$, where H is the length of the system and ν is the kinematic viscosity. Indeed, experimentally it was shown by SNYDER [13] that disturbances of the system decay with a characteristic time $t_s \approx 0.15\ t_v$. Thus, we do not expect the complications of long-lived transients in this system which have been suggested in the Rayleigh-Bénard case [12].

Our inner cylinder radius was 3.09 cm, and the radius ratio was $\eta = 0.892$. The end boundaries were rigid and nonrotating. The aspect ratio $L \equiv H/d$ (d is the difference between the cylinder radii) could be varied by moving one of the end boundaries. The inner cylinder speed and the temperature were controlled to at least $\pm$ 0.01% and $\pm$ 5 mK, respectively. The fluid was 30% glycerol in water by volume with 0.6% of a "Kalliroscope" flake suspension added for flow visualization. The characteristic diffusion time d^2/ν was about 5 sec. A signal could be derived from a moveable probe which detected the scattering of laser light by the flakes. Digital records of the signal could be obtained with the help of a dedicated mini-computer. Wavelengths could be measured visually with a cathetometer. A schematic diagram of one version of the apparatus is shown in Fig. 1.

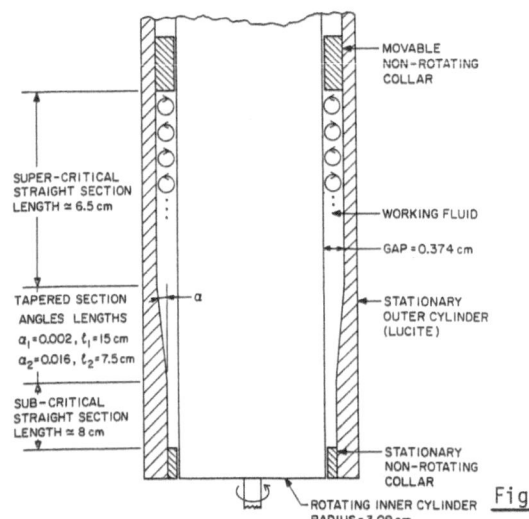

SUPER-CRITICAL
STRAIGHT SECTION
LENGTH ≈ 6.5 cm

TAPERED SECTION
ANGLES LENGTHS
$a_1 = 0.002$, $\ell_1 = 15$ cm
$a_2 = 0.016$, $\ell_2 = 7.5$ cm

SUB-CRITICAL
STRAIGHT SECTION
LENGTH ≈ 8 cm

α

MOVABLE
NON-ROTATING
COLLAR

WORKING FLUID

GAP = 0.374 cm

STATIONARY
OUTER CYLINDER
(LUCITE)

STATIONARY
NON-ROTATING
COLLAR

ROTATING INNER CYLINDER
RADIUS = 3.09 cm

Fig. 1. Schematic diagram of the apparatus

The first experiment we wish to report on was devoted to establishing the existence and width of a band of stable states as a function of Reynolds number R [14]. For that purpose, the tapered section shown in Fig. 1 was eliminated and the bottom collar terminated the straight section of uniform gap d = 0.374 cm and aspect ratio L near 54. A state of wavelength $\lambda \approx 2$(in units of d) was established by crossing R_c slowly from below. After an integer number of vortex pairs N about equal to L/λ had formed, R was increased to about 1.12 R_c^∞ ($\varepsilon \equiv R/R_c^\infty - 1 \approx 0.12$ and R_c^∞ is R_c for the infinitely long system). Thereafter, the upper boundary was moved slowly so as either to increase or decrease L. During this process, N remained constant provided the boundary movement was sufficiently slow, and thus λ changed. For the final state, the average wavelength of the vortices in the system was measured with the cathetometer after a steady state had been established. For this purpose, typically two pairs at each end were

28

excluded because the end vortices are anomalous in width. Thereafter, ε was decreased in small steps, typically of size δε = 0.005. After each step, a time
period at least equal to t_s and not shorter than 300 d^2/ν (1/2 h) was permitted
to elapse. If during this time the number of vortices did not change, the state
was considered stable. Eventually, a sufficiently small value of ε was reached
for a transition to occur. The transition increased N by one when λ > 2 and decreased N by one when λ < 2. Whenever visual observations of the transition were
made, it was found that a pair was lost or gained somewhere in the interior of
the system and never near the ends. We thus presume the observed instability
to be a bulk effect and not boundary induced. During the transition, a pair was
added or lost without any noticeable departure from cylindrical symmetry (i.e.,
never by defect formation). These qualitative observations are consistent with
theoretical predictions for the Eckhaus instability [15].

In Fig. 2, we show as solid circles the transition points obtained [14] by
the above procedure as a function of wavelength. The predicted Eckhaus instability is shown as a solid line (the dashed line represents the marginal stability curve where the linear growth rate of Taylor vortices vanishes). For large
λ, the experiment agrees quite well with the predicted Eckhaus boundary, but for
λ < 2 there is a considerable discrepancy. In the experiment, it never was possible to quite reach the predicted stability limit for small λ.

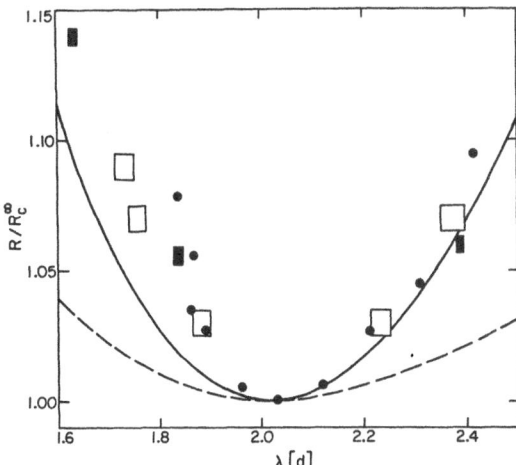

Fig. 2. Range of stable wavelengths
for Taylor vortex flow as a function
of R/R_C^∞. Solid line: predicted Eckhaus instability. Dashed line: locus
of vanishing linear growth rate of
Taylor vortices. The various individual symbols correspond to experimental determinations of the stability
limit and are discussed in the text.
For large λ, they are consistent
with the Eckhaus instability, but
for small λ the predicted limit of
the stable region could never be
reached

During the above work, it became apparent that the vortex pair widths in our
system became nonuniform over a time interval of many hours. This effect was
traced eventually to an accumulation of the Kalliroscope tracer particles near
the center of the Taylor vortices and a corresponding depletion at the inflow
and outflow boundaries. The resulting periodic concentration variation in a
vertical column couples to the gravitational field. As a result, the vortices
near the bottom became compressed and those near the top were expanded. For small
L, this effect is hardly noticeable, but for our L ≈ 54 we found that the average
λ in the top half of the system was about 2% larger than the average λ in the
bottom half after a day had elapsed.[1] We were concerned that the asymmetric
behavior displayed by the solid circles in Fig. 2 might be associated with this
inhomogeneity. Therefore, we converted our apparatus to operation with the cyl-

---

[1] It was easy to re-establish a uniform Kalliroscope concentration by briefly
entering the wavy Taylor vortex flow regime. Thereafter, the wavelength throughout the column became uniform on the time scale t_s.

inder axis horizontal. This eliminated the gravity-induced inhomogeneity since the gravitational field now was normal to the cylinder axis. In this configuration, we obtained the rectangles shown in Fig. 2. For these data, the size of the symbol corresponds to the resolution of ϵ and λ with which the measurement was made. The open rectangles are for L = 30, and the solid ones correspond to L = 22. These additional data also show that the predicted Eckhaus line cannot be reached experimentally on the small λ side. For $\lambda > 2$, all data are consistent with the prediction. In order to obtain more information about the nature of the asymmetry, we are now conducting a study of the dynamics of the transitions on the two sides.

For our present purpose, we need not be concerned with the difference in the experimentally accessible and theoretically predicted stable-state bandwidth discussed above. Within the experimentally accessible band, we can conduct a variety of wave-number selection experiments, and we report on two that have been carried out so far.

First, we consider an experiment on wave-number selection under *static* conditions [16] which was suggested by the theoretical work of KRAMER et al. [17] and of EAGLES [18]. For this purpose, we used apparatus in which the gap between the cylinders was uniform over a certain axial distance from the moveable collar and then tapered gently and linearly to smaller values as shown schematically in Fig. 1. In this geometry, it was possible to have $\epsilon > 0$ in the uniform wide section and $\epsilon < 0$ near the bottom of the taper. The mean steady-state axial wavelength of 6 - 7 vortex pairs in the wide straight section was measured with a cathetometer (excluding three vortices nearest the collar). By moving the collar in small steps and waiting for a steady state after each step, the wavelength was measured as a function of the length L of the straight section. For a wedge angle $\alpha = 0.002$ (see Fig. 1) and $\epsilon \gtrsim 0.01$, we found that the wavelength relaxed back to a unique value after each collar movement, regardless of whether this movement increased or decreased L. This adjustment occurred by a smooth, continuous addition or expulsion of vortices through the wedge. The unique, selected wavelength is shown in Fig. 3 as crosses. The experiment clearly demonstrates that a particular wavelength within the stable band is preferred by the system and will be chosen if a smooth adjustment process is allowed to occur, consistent with the prediction of KRAMER et al. [17].

Finally, we consider an experiment on wave-number selection under *dynamic* conditions [19] which was conducted near the collar in the straight section of the apparatus. Adjacent to this collar, one or a few vortices form even for $\epsilon = \epsilon_0 < 0$. If ϵ is increased suddenly from ϵ_0 to $\epsilon_1 > 0$, the now unstable Couette flow (no vortices) existing throughout most of the apparatus gives way to stable Taylor vortex flow which propagates inward from the previously sub-critical vortices adjacent to the end. The regions containing the two different flow states are sep-

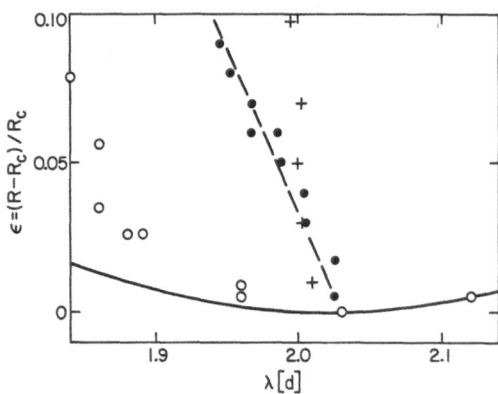

Fig. 3. Selected wavelengths and range of stable wavelengths for Taylor vortex flow as a function of ϵ. Solid line: Eckhaus instability. Open circles: some of the data from Fig. 2. Crosses: unique λ selected under static conditions in the presence of a gentle spatial ramp. Solid circles: unique λ selected under dynamic conditions in a propagating vortex front. Dashed line: locus of maximum growth rate (displaced to the right by 0.020 which is less than allowed systematic errors for λ)

arated by a propagating vortex front. The front position changes linearly in time [19], i.e.,it has a time-independent velocity.[2] We measured the width of two vortex pairs as soon as they were of sufficient strength to be visually detectable in the vortex front. The results are shown as a function of ε_1 in Fig. 3 as solid circles. With increasing ε_1, the selected wavelength decreases linearly and differs from the unique λ that is selected by the gentle spatial ramp (crosses in Fig. 3). Within experimental error, it coincides with the wavelength of maximum growth rate calculated [20] for our radius ratio from the linearized Navier-Stokes equation (dashed line in Fig. 3). The ε dependence of this dynamically selected wavelength is similar to that observed in "sudden start" experiments of BURKHALTER and KOSCHMIEDER [21], although those experiments apparently resulted in inhomogeneous growth throughout the system rather than in front propagation. It is interesting to note that the wavelength well behind the front (i.e.,at later times) tended to relax towards larger values rather closer to the statically selected wavelengths in the presence of the spatial ramp. However, it was not possible to follow this adjustment process to completion because inhomogeneous nucleation throughout the column intervened and caused irregular and irreproducible behavior at later times.

We have summarized in this review three experiments relevant to the wave-number selection problem in rotating Couette-Taylor flow. The first established that any one of a continuum of states over a range of wave numbers k can be created at a given ε by appropriate manipulation of the boundaries. At small k (large λ), the limit of the experimentally accessible range is consistent with predictions for the Eckhaus instability. However, for large k (small λ) the experiment was never able to quite reach the predicted Eckhaus limit. We speculate that the difference between experiment and theory is attributable to the fact that the theory predicts stability with respect to *infinitesimal* fluctuations, whereas the experiment naturally is subject to finite (albeit small) fluctuations. This explanation and the data would imply that the Taylor vortices are quite immune to finite-amplitude fluctuation near the small k limit of the band and more susceptible to them near the large k limit. We hope that a study of the dynamics of the transitions will shed further light on this conjecture.

The remaining two experiments were devoted to studying selection processes within the experimentally stable band. We found that a very gentle spatial variation of the stress parameter R from above R_c to below R_c permitted the continuous acquisition by or expulsion of vortices from the supercritical region and thus allowed the selection of a unique wave number within the band under steady-state conditions. This wave number differed from the one selected under dynamic conditions by a propagating vortex front. The latter coincided within experimental resolution with the wave number of maximum growth rate.

This work was supported by National Science Foundation Grant No. MEA81-17241.

References

1. A. Schlüter, D. Lortz, F. Busse: J. Fluid Mech. 23, 129 (1965)
2. G. I. Taylor: Philos. Trans. Roy. Soc. London, Ser. A223, 289 (1923)
3. D. Coles: J. Fluid Mech. 21, 385 (1965)
4. R. C. DiPrima, H. L. Swinney: in *Hydrodynamic Instabilities* and *Transition to Turbulence*, ed. by H. L. Swinney and J. P. Gollub (Springer, Berlin 1981)
5. For a recent review, see C. Vidal, A. Pacault: in *Evolution of Order and Chaos*, ed. by H. Haken (Springer, Berlin 1982) p. 74
6. G. I. Sivashinsky: Annu. Rev. Fluid Mech. (to be published)
7. J. S. Langer, H. Müller-Krumbhaar: Acta Metall. 26, 1681, 1689, 1967 (1978)
8. See, for instance, Ref. 1

---

[2] An important unresolved problem is that the observed front propagation velocity disagrees with recent predictions by G. Dee, J. S. Langer: Phys. Rev. Lett. 50, 383 (1983) and by E. Ben-Jacob, H. R. Brand, L. Kramer: (unpublished).

9. M. C. Cross, P. G. Daniels, P. C. Hohenberg, E. D. Siggia: Phys. Rev. Lett. 45, 898 (1980) and J. Fluid Mech. 127, 155 (1983); M. C. Cross, P. C. Hohenberg, S. Safran: Physica (Utrecht) 5D, 75 (1982)

10. For a discussion of the existence of long-range order and broken symmetry in dissipative systems, see P. W. Anderson: in *Order and Fluctuations in Equilibrium and Nonequilibrium Statistical Mechanics*, XVIIth International Solvay Conference on Physics, ed. by G. Nicolis, G. Dewel and J. W. Turner (Wiley, NY 1981),p. 289

11. G. Ahlers, R. P. Behringer: Phys. Rev. Lett. 40, 712 (1978) and Proc. Theor. Phys. Suppl. 64, 186 (1978); G. Ahlers, R. W. Walden: Phys. Rev. Lett. 44, 445 (1980); H. Greenside, G. Ahlers, P. C. Hohenberg, R. W. Walden: Physica (Utrecht) 5D, 322 (1982); R. P. Behringer, G. Ahlers: J. Fluid Mech. 128, 219 (1982)

12. H. S. Greenside, W. M. Coughran, Jr., N. L. Schryer: Phys. Rev. Lett. 49, 726 (1982); H. S. Greenside, W. M. Coughran, Jr., in print

13. H. A. Snyder: J. Fluid Mech. 35, 273 (1969)

14. G. Ahlers, D. S. Cannell, M. A. Dominguez-Lerma: Phys. Rev. A 27, 1225 (1983)

15. S. Kogelman, R. C. DiPrima: Phys. Fluids 13, 1 (1970); W. Eckhaus: *Studies in Nonlinear Stability Theory* (Springer, NY 1965)

16. D. S. Cannell, M. A. Dominguez-Lerma, G. Ahlers: Phys. Rev. Lett. 50, 1365 (1983)

17. L. Kramer, E. Ben-Jacob, H. Brand, M. C. Cross: Phys. Rev. Lett. 49, 1891 (1982).

18. P. M. Eagles: Proc. Roy. Soc. London, Ser. A 371, 359 (1980)

19. G. Ahlers, D. S. Cannell: Phys. Rev. Lett. 50, 1583 (1983)

20. M. A. Dominguez-Lerma, G. Ahlers, D. S. Cannell: to be published

21. J. E. Burkhalter, E. L. Koschmieder: Phys. Fluids 17, 1929 (1974)

Periodic and Chaotic Patterns in Selected Quantum Optical Systems

Lorenzo M. Narducci, Donna K. Bandy, Jin Yue Gao[+]

Physics Department, Drexel University, Philadelphia, PA 19104, USA

Luigi A. Lugiato

Istituto di Fisica, Universita' di Milano, I-20133 Milano, Italy

1. Introduction

The existence of cooperative behavior in nonequilibrium systems has been the focus of growing interest over the last 15 years, an interest which is shared by many different disciplines under the unifying heading of synergetics [1].

Quantum optics has been an active contributor to the study of cooperative effects, beginning with the pioneering advances in laser theory by Haken [2], and Lamb [3] and their respective collaborators, and continuing with the more recent discoveries involving bistable, self-pulsing, and chaotic behavior in various nonlinear optical systems [4].

Just as the laser came to be known as the trailblazer of synergetics in the late 1960's and early 1970's, the field of optical bistability has been recognized as its worthy successor, both on account of the rich variety of phenomena that it has generated and of its promising technological spin-offs. In this presentation, we wish to summarize some of the most interesting recent developments related to the emergence of instabilities and to the creation of periodic and chaotic structures in bistable optical systems and in their active counterpart, the laser with injected signal.

In its simplest form, a bistable optical system is comprised of an optical medium, such as for example, a gas of atoms, placed within an interferometric cavity and driven by a steady coherent beam of light. One of the atomic transitions with frequency ω_A is selected to be resonant or nearly resonant with one of the empty cavity modes (frequency ω_C), while the external field with carrier frequency ω_0 is also adjusted to be either at or near resonance with the atoms. This configuration is not unique, and bistable action has been observed with other quite different arrangements (see, for example, the so-called hybrid system discussed in Section 3); it is, however, a common starting point for many theoretical models, and it will serve us well for much of the following discussion.

The behavior of an empty cavity offers no surprises: for a fixed setting of the input field and cavity frequencies, the output intensity varies linearly with the intensity of the input signal. Things become more interesting when the cavity is filled with the optical medium. In this case, the behavior of the system is largely controlled by the bistability parameter $C=\alpha L/2T$, where αL is the line-center absorption coefficient of the atomic sample, and T is the mirrors' transmittivity. For increasing values of C, the system is capable of displaying differential gain and, eventually, bistable action (Fig. 1). The high and low states of transmission in a bistable system are equivalent to the logical states 1 and 0. Thus, bistability is looked upon as a promising tool for the execution of switching and logic operations in all optical systems.

[+]Permanent Address: Physics Department, Jilin University, Changchun People's Republic of China

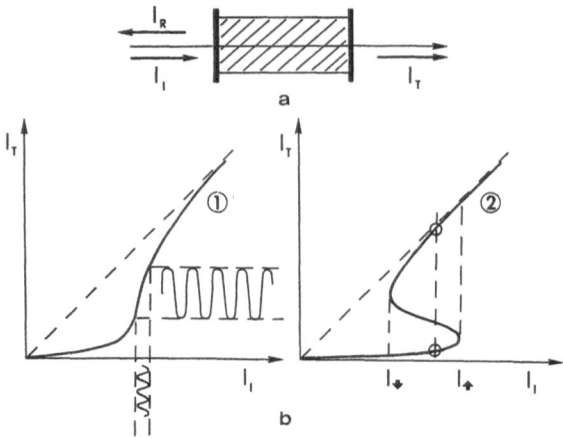

Fig. 1 (a) Schematic representation of a Fabry-Perot cavity containing an optical medium; I_I, I_T and I_R denote the incident, transmitted and reflected intensities, respectively. (b) The transmitted intensity is plotted (qualitatively) as a function of the incident intensity for two values of the bistability parameter C. Curve 1 represents a configuration for which a small intensity modulation of the input signal is amplified at the output of the system (transistor action). Curve 2 is characteristic of bistable action with $I_\uparrow$ and $I_\downarrow$ denoting the switch-up and switch-down thresholds, respectively

The existence of this effect was predicted by Szöke and collaborators [6] in 1969, but interest in the subject underwent a truly explosive growth only after the first observation by Gibbs, McCall and Venkatesan using sodium vapor in a Fabry-Perot interferometer [7]. Shortly thereafter, Bonifacio and Lugiato [8] developed a first-principle analytic theory of bistability which, eventually, led to the discovery of unstable states along the hysteresis loop of the transmitted intensity [9].

Interest in the unstable behavior of quantum optical systems, actually, dates further back to the earlier studies of single- and multi-mode homogeneously broadened lasers [10]. Here, we begin our discussion by focusing on the behavior of a passive optical system, one in which no population inversion is imposed between the two atomic levels of interest.

2. Self-Pulsing and Chaotic Oscillations in a Ring-Cavity Bistable System

The simplest setting for a discussion of this problem is one in which the optical resonator has the shape of a ring that can be excited only along one direction of propagation (Fig. 2) [9]. The resonator contains a collection of homogeneously broadened two-level atoms whose interaction with the field is governed by the Maxwell-Bloch equations in the plane-wave approximation. Under suitable conditions, a system placed in the high transmission branch can become unstable and produce output oscillations. In the following, we consider two distinct types of instabilities with rather different underlying physical mechanisms and phenomenologies.

a) Off-Resonant Mode Instability

This is a multimode phenomenon in which the cavity mode that is closest to the center of the atomic line remains stable, while several originally unexcited sidebands can become unstable. A detailed discussion of this effect has already appeared in an earlier volume of this series [11]. Here, for completeness, we review only a few relevant points.

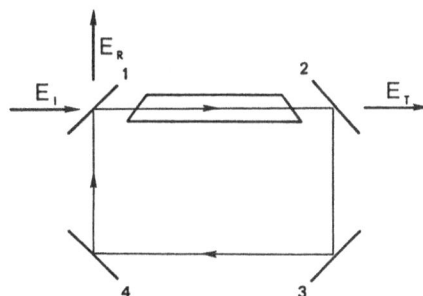

Fig. 2 Schematic representation of a ring cavity resonator. Mirrors 1 and 2 are partially reflecting, while mirrors 3 and 4 are assumed to have reflectivities R=1; E denotes the amplitude of the electric field

Consider, for simplicity, a situation where one of the cavity modes is exactly resonant with the center of the atomic line and with the carrier frequency of the input field. The existence of self-pulsing, under these conditions, requires that additional off-resonant modes be made to fall within the power broadened atomic profile, i.e., that the intermode spacing $c/\mathcal{L}$, where $\mathcal{L}$ is the length of the resonator, be smaller than the power broadened linewidth $\gamma=\gamma_\perp(1+I_T/I_S)$ ($\gamma_\perp$ is the homogeneous linewidth, and I_T and I_S are the transmitted and saturation intensities, respectively).

In this case, the resonant cavity mode is stable along both the low and the high transmission branches; the off-resonant modes, instead, just as the probe field in saturation spectroscopy, can experience gain, in spite of the fact that the atoms in the cavity are not prepared in a state of inversion. The role of the atoms, in essence, is to provide a nonlinear coupling between the strongly excited resonant mode and the originally unexcited sidebands. If the gain experienced at the off-resonant frequencies is large enough to overcome the cavity losses, the sidebands become excited and begin to beat with the stable central mode. The beat pattern manifests itself as an intensity modulation imposed on the output signal.

The results, which are summarized below, were obtained in the context of the so-called mean field limit, which in the resonant case corresponds to the requirements $\alpha L \ll 1$, $T \ll 1$, with $C=\alpha L/2T=$constant. In this case, a treatment called dressed-mode theory [12], patterned after Haken's theory of the Ginzburg-Landau equations for systems far from thermal equilibrium [13], has provided a convenient way of analyzing both the stability properties of the system and its dynamical evolution.

The stability diagram of Fig. 3 displays the domain in the control parameter space of the intermode spacing and of the output field amplitude, where spontaneous self-pulsing is predicted to develop. As already discussed in some detail in Ref. [11], the dotted region of the domain corresponds to an initial unstable state for which self-pulsing represents only a transient phenomenon, because the system eventually precipitates from the high to the (stable) low transmission branch. The dashed area of the instability domain, instead, corresponds to control parameter values for which self-pulsing persists indefinitely.

A surprising feature of this problem is the existence of a hard excitation domain (bounded in the figure by the dashed lines marked HE). In correspondence to values of the control parameters within this domain, the appropriate steady state of the high transmission branch is stable against small perturbations, but unstable for sufficiently large ones. The presence of the hard-mode domain adds another new feature to the problem, namely a new type of hysteresis cycle where self-pulsing and stationary states coexist. For additional details, we refer the reader to the original publications cited in Ref. [11].

In some sense the other side of the coin is what is commonly known as the Ikeda scenario [14]. This also involves multimode configurations, but, unlike the mean field limit, it assumes large values of αL and sufficiently large values of the detuning. An additional distinctive requirement of the setup is that the

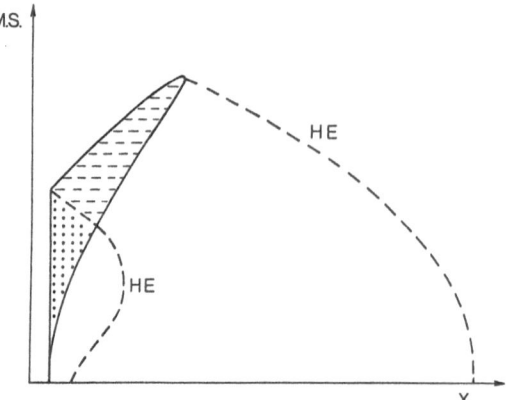

Fig. 3 Self-pulsing domain for the off-resonant mode instability. The vertical
axis is labelled by the cavity intermode spacing (IMS) measured in units of the
atomic homogeneous linewidth. The horizontal axis is labelled by the steady-state
output field amplitude corresponding to a given input field. A point in control
parameter space within the dotted area corresponds to operating conditions in the
upper branch for which self-pulsing oscillations develop spontaneously, but persist
only for a finite time, as the system eventually precipitates into the stable low
transmission branch. Operating points within the shaded area correspond to the
stable self-pulsing regime. Points within the area bounded by the dashed lines HE
(hard excitation domain) are stable against small perturbations but can be excited
into persistent self-pulsing by a sufficiently large fluctuation

round trip time of the radiation inside the cavity be much longer than the charac-
teristic atomic relaxation times $\gamma_{\parallel}^{-1}$ and $\gamma_{\perp}^{-1}$. The Ikeda scenario has played an
important role in the development of self-pulsing in bistability because it has
provided for the first time a recipe for the observation of chaotic pulsations.
In fact, not only did Ikeda show that the Maxwell-Bloch equations yield "turbulent"
solutions in this limit, but with a clever reanalysis of the situation, he also
suggested a design for a hybrid device whose operation simulates the dispersive,
long delay limit of the optical ring cavity [15].

Since the hybrid system with a delay in the feedback loop is an interesting
subject in its own right, we shall devote Section 3 to a discussion of its main
dynamical features.

b) Resonant Mode Instability

Self-pulsing in a bistable system can also be observed under single mode operation, provided that the frequency mismatch between the atomic transition, the
cavity mode and the incident field is sufficiently large [16]. (In resonance, the
single-mode model is completely stable against self-pulsing in both the low and
the high transmission branches.) In the presence of adequate frequency mismatch,
the output signal can display periodic and even chaotic oscillations which originate
spontaneously within the system.

The basic equations describing single-mode dispersive bistability in the mean
field limit are [8,18]

$$\kappa^{-1}\dot{x} = -i\theta x - (x-y) - 2Cp \tag{2.1}$$

$$\gamma_{\perp}^{-1}\dot{p} = xd - (1+i\Delta)p \tag{2.2}$$

$$\gamma_{\parallel}^{-1}\dot{d} = -\frac{1}{2}(xp^* + x^*p) - d+1 \tag{2.3}$$

where x and p denote the complex output field amplitude and the atomic polarization, respectively; y is the real incident field amplitude and d is the atomic population difference between the selected levels; θ and Δ measure the cavity mistuning ($\theta = (\omega_C - \omega_0)/\kappa$) and the atomic detuning ($\Delta = (\omega_A - \omega_0)/\gamma_\perp$), while κ is the cavity damping rate.

A typical steady-state curve linking the modulus of the output field amplitude with the incident field is shown in Fig. 4. The lower transmission branch is stable, while a segment of the high transmission curve is unstable against self-pulsing. Under adiabatic elimination conditions for the polarization (this is not a necessary requirement; it is assumed only as a convenient way to limit the size of the control parameter space), and for a fixed value of the ratio $\kappa/\gamma_\parallel$, Figs. 5a-h provide a sample of the self-pulsing oscillations of the output field. From these results, we see that upon decreasing the input field strength along the high transmission branch, the steady-state oscillations undergo successive period doubling bifurcations into a chaotic state. The emergence out of chaos develops through a sequence of inverted period doubling bifurcations that terminates when the system abruptly precipitates from the high to the low transmission branch. Each temporal solution in Fig. 5 is accompanied by a projection of the phase-space trajectory on the plane (Rex,Imx). Clearly, the periodic trajectories produce closed phase-space loops, while the chaotic solution generates complicated bands of uneven density.

As a test of the sensitivity of the solution to the initial conditions, we have constructed the Cartesian distance between trajectories that evolve from neighboring points of the phase-space under chaotic conditions. As anticipated, the distance grows exponentially in time as evidenced by the log-linear plot shown in Fig. 6.

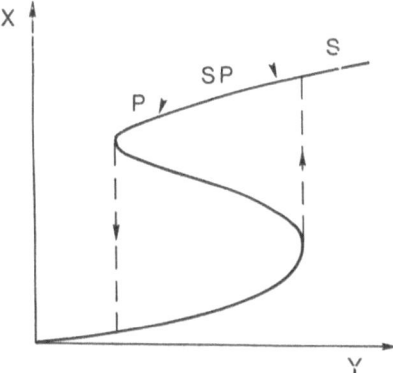

Fig. 4 Typical steady-state curve linking the modulus of the transmitted field x and the incident field amplitude y. The low transmission branch and the portion of the high transmission branch labelled by an S are stable. The arrows and the label SP indicate the unstable domain. P marks the unstable region of the upper branch which leads to precipitation to the lower stable state

Fig. 5 Self-pulsing oscillations and the corresponding phase-space projections on the Rex-Imx plane; (a and b) y=2000, 1P oscillation; (c and d) y=1350, 2P oscillation; (e and f) y=1225, 4P oscillation; (g and h) y=950, chaotic oscillations. These solutions have been obtained for C=70000 and $\kappa/\gamma_\parallel$=0.25

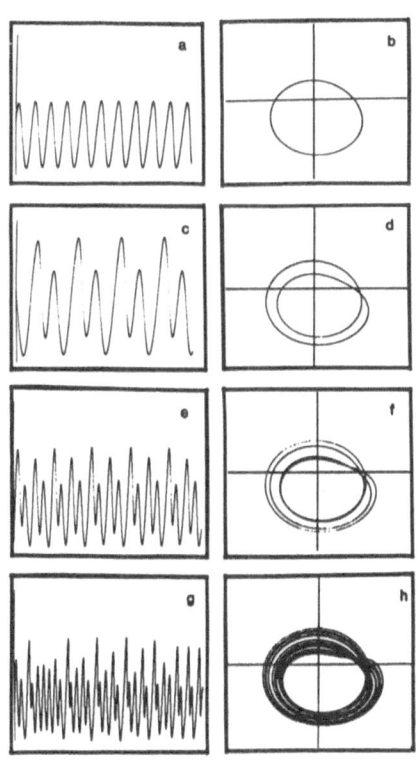

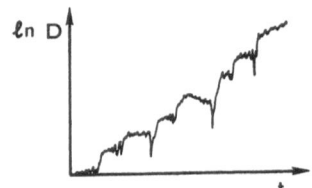

Fig. 6 The logarithm of the Cartesian distance between two initially "close" trajectories grows, on
the average, linearly with time. This plot shows
evidence of the "exponential divergence" which is
expected when a trajectory evolves on the domain
of a strange attractor

ℓn D

t

As a test of how well the sequence of observed period doubling bifurcations con-
forms with the Feigenbaum scenario, we have determined several bifurcation thresholds y; and constructed the parameters $\delta_i = (y_i - y_{i+1})/(y_{i+1} - y_{i+2})$. Our calculated
sequence of δ values is indeed consistent with Feigenbaum's universal constant
$\delta = 4.669$.

3. Self-Pulsing and Chaos in a Hybrid Bistable System with Delay

In the extreme dispersive limit (Kerr limit), Ikeda[14b] found a way to get around
the need for designing impractically long interferometric structures. His suggestion was to replace the atoms and the long ring cavity with the hybrid system shown
in Fig. 7. Although not at all obvious by inspection, the equations that control
the behavior of this system are indeed equivalent to the Maxwell-Bloch equations
in the Ikeda limit. His proposal was quickly implemented by the Arizona group [15]
which was able to show for the first time the existence of optical chaos in the output intensity, when the delay time in the feedback loop was much longer than the
system's response time. Experiments were also carried out for shorter delay times
[19] with results which are seemingly very different from those of Gibbs and collaborators. Of course, when the delay time is made comparable to the system's response
time, one cannot expect any special relation between the hybrid device and the
original ring cavity. Still, a detailed analysis of the hybrid system has revealed
quite interesting features that are worth reviewing. Thus, in this section, we
analyze the main aspects of the dynamical behavior of this delayed system whose
equations of motion take the seemingly simple form [20]

$$\frac{dV(t)}{dt} + V(t) = x(t-T) \tag{3.1}$$

$$x(t) = \frac{1}{2} y \left(1 - k \cos(\theta + V(t))\right). \tag{3.2}$$

In Eq. (3.1), $V(t)$ represents the instantaneous voltage signal applied to the modulator plates in addition to the fixed bias voltage θ; the time-dependent voltage
$V(t)$ is due to the output intensity $x(t-T)$ detected at the earlier time $(t-T)$; y
represents the constant input intensity and k is the modulation depth of the electro-
optic system. The time t and the delay T are scaled to the response time of the
modulator-amplifier-feedback loop combination. In steady state $(dV/dt=0, x(\infty)=V(\infty))$
the transmission function of the device

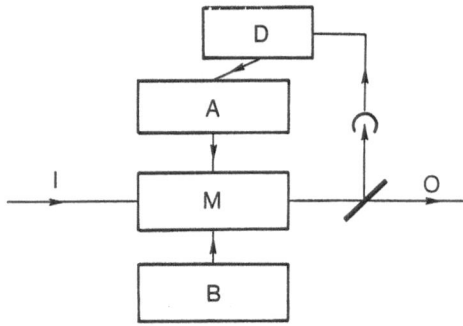

Fig. 7 Schematic representation of the
hybrid bistable system proposed in Ref.
[14b] and realized in Ref. [15]. A cw input beam (I) passes through an electro-
optic modulator (M) kept at some fixed
bias voltage (B). A fraction of the output (O) is detected, the current passed
through a delay (D) and the amplified (A)
voltage signal applied to the plates of
the modulator

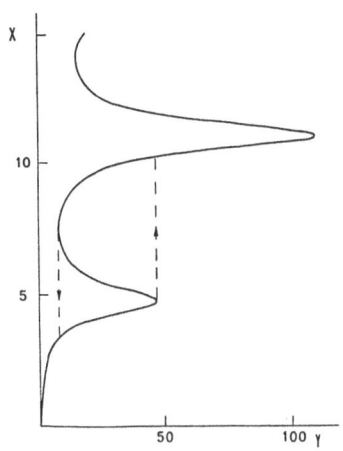

Fig. 8 Output vs. input intensity for the hybrid bistable system in steady state. In this simulation and all the subsequent figures we have selected k=0.8 and θ=90°

$$x = \frac{1}{2} y \ (1-k \cos(\theta+x))$$ (3.3)

is multistable as shown above in Fig. 8.

The stability of the steady state can be analyzed by standard linear stability methods, and the complex rate constants of the linearized modes of the system can be easily shown to be solutions of the secular equation [19,20]

$$s+1 - \frac{1}{2} y \ \sin(\theta+x) \ e^{-Ts} = 0 \ .$$ (3.4)

It is clear that an infinite number of solutions exist for the linearized problem, which is thus recognized to be infinite-dimensional by virtue of the delayed nature of Eq. (3.1). The connection between the range of instability of the system and the magnitude of the delay has been discussed in some detail in Ref. [20]. Here we only mention that for a selected range of the output field intensity, it is quite easy to find values of the delay such that the system is unstable against self-pulsing.

A convenient graphical way of displaying the roots of the secular equation (3.4) is shown in Fig. 9 where we have selected a short and a long delay time in order to emphasize one main point: for small values of T, only one mode of the system can be made unstable in the linear regime, while for large values of T, even a small change of the input intensity can cause many of the linearized modes to become unstable. This is not unexpected because, in the long delay limit, this system is equivalent to the multimode Maxwell-Bloch problem [14d].

We now summarize the results of two typical scans in control parameter space. The first corresponds to a fixed value of the input field and a variable delay, with T of the order of unity. Figure 10 (a and b) shows the self-pulsing intensity and corresponding power spectrum for T=0.5, which is slightly above the instability threshold for the chosen input field strength. The spectrum shows peaks at the fundamental frequency ω_1 and its harmonics, as expected, but also at four other well-resolved frequencies labelled $\omega_2,...,\omega_5$. The values of these frequency components are surprisingly close to the imaginary parts of the first four stable eigenvalues, a feature that persists even well above the self-pulsing threshold. Combination frequencies (e.g., $\omega_2-2\omega_1$) are also easily recognizable.

At T=0.9 a period doubling bifurcation occurs with a clearly visible frequency component at $\omega/2$ (Fig. 10c,d). This is followed by a second bifurcation with a much less pronounced peak at $\omega/4$; no additional bifurcations, however, have been observed for larger values of T. Frequency locking is, instead, the next obvious feature (Fig. 10e,f). Note the much more regular time record. Frequency locking

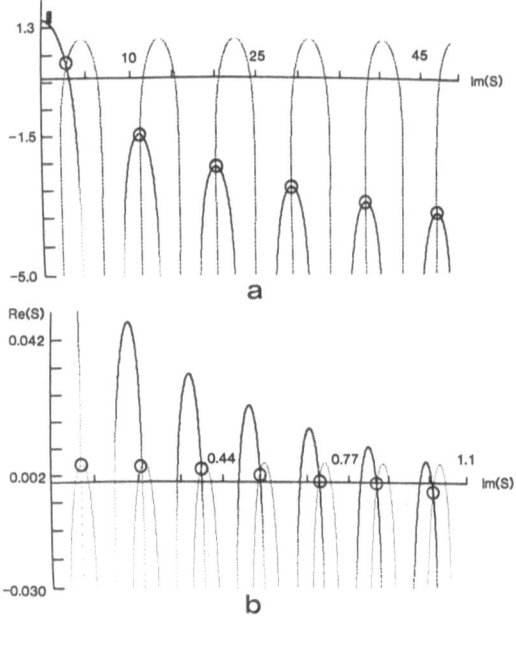

Fig. 9 The intersects (open circles) of the two families of curves shown in these figures are the roots of the secular equation (3.4). The vertical axis is labelled by Re(s) and the horizontal axis by Im(s); (a) x=3.6; T=0.5, only one unstable root (Re s>0) exists; (b) x=2.6, T=40, this condition corresponds to several unstable roots

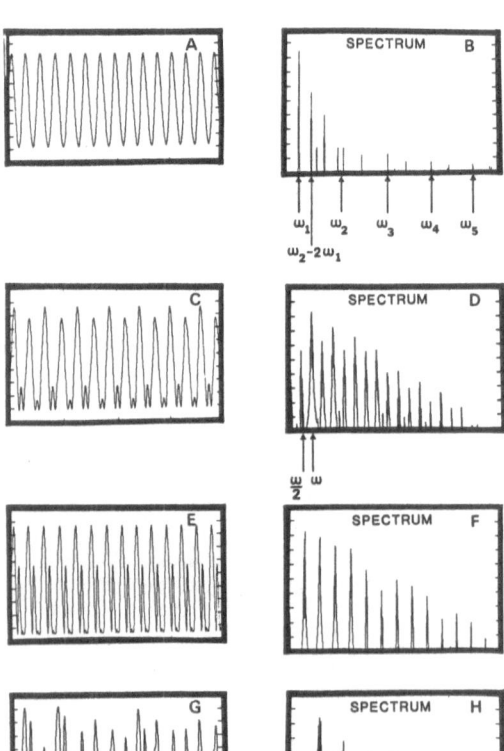

Fig. 10 Self-pulsing intensity and the corresponding power spectra for x=3.6 and (a and b) T=0.5, (c and d) T=0.9; (e and f) T=1.09; (g and h) T=1.3

is apparently triggered by the stable eigenfrequencies becoming very close to multiples of the fundamental component. This is perhaps the mechanism that is responsible for the sudden truncation of the period doubling sequence. The locking persists until the system suddenly develops chaotic oscillations and a broadband spectrum (Fig. 10g,h). The behavior of the calculated frequency components corresponding to the fundamental and the first two stable eigenvalues is shown in Fig. 11. The sharp collapse of the calculated frequencies (2 and 3) on the solid lines corresponding to the imaginary part of the eigenvalues marks the onset of frequency locking.

The second scan corresponds to a fixed large value of the delay (T=40) and variable incident intensity. The pattern is very similar to the one observed in the short delay case, except that the components previously labelled by $\omega_2,...,\omega_5$ in Fig. 10a are no longer observable. The system appears to be frequency locked right at the beginning of self-pulsing (Fig. 12a,b). Two subharmonic bifurcations are also observed (see for example Fig. 12c,d); these are followed by the sudden onset of irregular pulses and broadband power spectra.

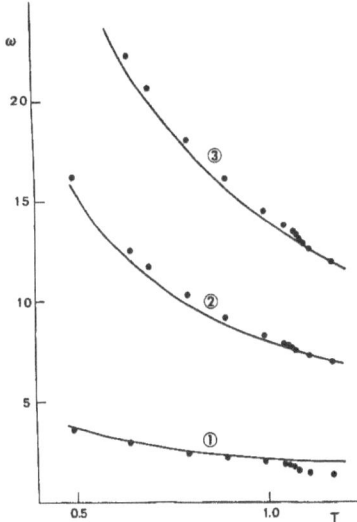

Fig. 11 The calculated frequencies of the fundamental spectral component (curve 1) and the two peaks corresponding to the first two stable eigenvalues (curves 2 and 3) are plotted as functions of the delay T (dots). The solid lines show the imaginary parts of the corresponding linearized eigenvalues

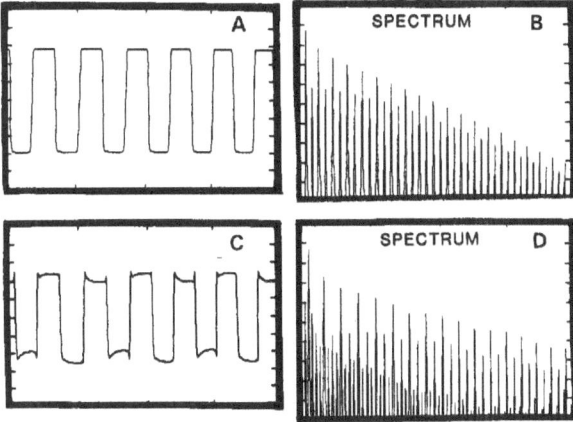

Fig. 12 Self-pulsing intensity and the corresponding power spectra for T=40 and (a and b) x=2.6; (c and d) x=2.8

The abrupt appearance of chaotic oscillations and the disappearance of the period
doubling sequence is suggestive that loss of synchronism among the strongly excited
components is at the root of the observed behavior. It is also interesting to note
that periodic behavior in the long delay limit persists even after the emergence of
three or more independent unstable modes. It is difficult to tell if this is only
a consequence of the observed frequency locking, or if it is a significant new feature of this system. Additional investigations seem to be in order, especially
since this system gives one of the few simple examples of an infinitely dimensional
problem for which a theoretical analysis can be carried out without any mode
truncation.

4. The Laser with Injected Signal

Just as the name suggests, this system consists of an ordinary laser operating
above threshold and of an external cw signal injected into the cavity at some fixed
carrier frequency. This arrangement is of interest because at an appropriate intensity level of the external beam, the laser locks in a stable way to the driving
field. Thus, several nominally identical but independent lasers can be forced to
operate in synchronism under the action of a single driving element. It is interesting to note that the laser with injected signal is the active counterpart of the
bistable device. Here we analyze a special setting in which the frequency of the
homogeneously broadened atoms coincides with one of the cavity modes (resonance).
The frequency of the external signal, of course, is different from both.

The existence of undamped oscillations in the output of this system was first
predicted by Spencer and Lamb [21]. Chaotic behavior was reported by Graham and
collaborators [22] in a single-mode ring laser after adiabatic elimination of the
atomic variables, but with sinusoidal oscillations of the pump parameter and of
the external field amplitude.

Here we consider a single-mode model without any external modulation of the control parameters [23]; we also limit our considerations to an interesting, although
unrealistically high value of the gain parameter C for the sake of displaying some
rather remarkable features of the system. The equations of motion are the usual
Maxwell-Bloch equations (2.1)-(2.3) with the population relaxation term $-\gamma_{\parallel}(d-1)$
replaced by $-\gamma_{\parallel}(d+1)$ to simulate the usual effect of the pump. In steady state,
the modulus of the output field varies as a function of the injected signal as
shown in Fig. 13. For the parameters chosen in this discussion, the entire upper
branch is stable (although this is not generally true for arbitrary values of the
parameters), while the lower branch and the negative slope part of the state equation are unstable. Thus, qualitatively, as the injected signal is turned on, one
expects self-pulsing to develop and to persist until $y > y_{thr}$, where the lower unstable branch coexists with the stable upper branch. At this point, the self-
pulsing terminates as the laser is locked in synchronism with the injected signal.

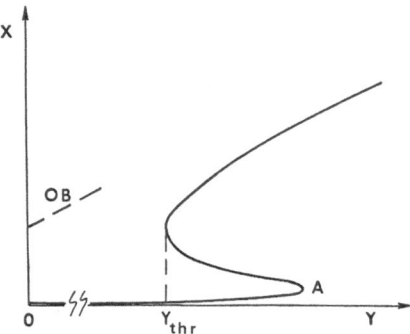

Fig. 13 State equation for the laser with
injected signal. For the chosen parameters,
the steady states in the lower branch from
0 to A are all unstable. The upper branch is
stable so that $y=y_{thr}$ corresponds to the
locking threshold. The presence of an oscillatory branch is indicated symbolically by
the dashed segment, OB

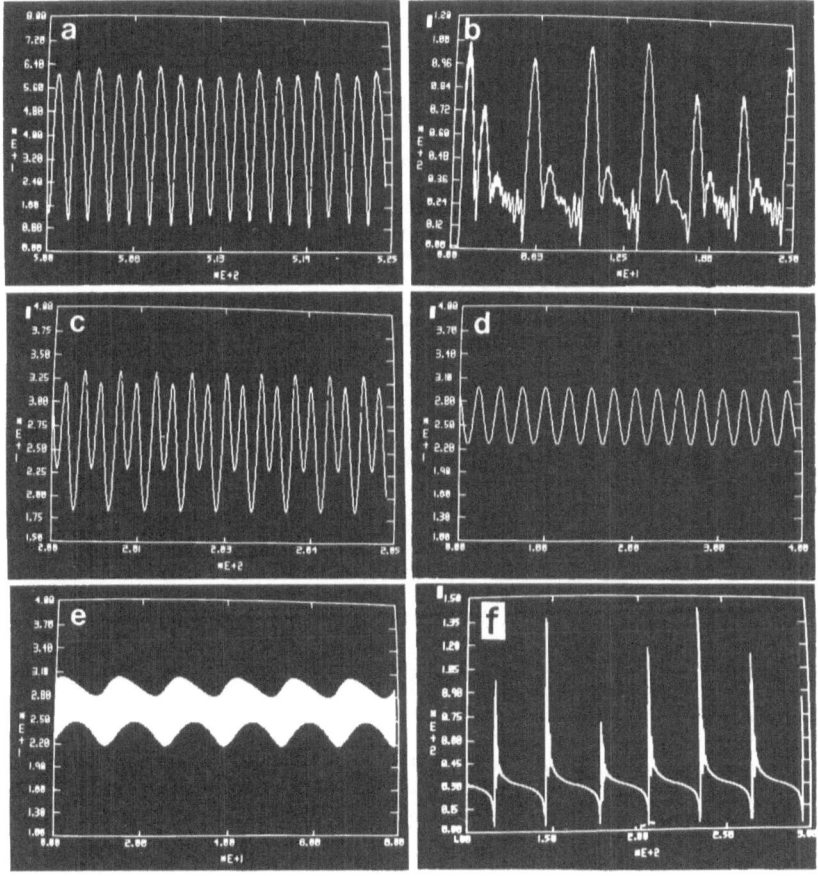

Fig. 14 Time evolution of the modulus of the output field for C=500 $\Delta=\theta=5$, $\kappa/\gamma_\perp$ =1 and $\gamma_\parallel/\gamma_\perp$ =1. The horizontal time axis is measured in units of $\gamma_\perp^{-1}$. (a) erratic behavior, y=117; (b) bursting, y=250; (c) 4P-type solution, y=279; (d) 1P-type solution, y=300; (e) envelope breathing, y=310.3; (f) spiking action, y=311

This expectation is verified by the numerical solutions of the equations of motion, which, in fact, show regular-looking oscillations of the output field for small values of y. A careful analysis reveals slight irregularities in the self-pulsing pattern as the injected field grows. These become more pronounced (see Fig. 14a) until the output breaks up into clearly nonperiodic oscillations. The chaotic pattern in its well-developed state displays large, irregular bursts followed by rapid noisy ringing patterns (Fig. 14b). The exit route from chaos consists of an inverted sequence of period doubling bifurcations; sample solutions displaying 4P and 1P periodic pattern are shown in Figs. 14c,d. At this point, the system enters a new regime where a long-lived self-pulsing envelope modulation (breathing) can be observed. This effect becomes more pronounced as the breathing pattern stabilizes (Fig. 14e). A further increase in y gradually transforms this modulation into a train of very sharp spikes with a temporal separation that, apparently, diverges as one approaches the injection locking threshold (Fig. 14f). For y >y_{thr}, finally, the system quickly converges to the stationary locked state.

The path into chaos is apparently very complex. In part, this may be so because variations of the injected field amplitude, with all the other control parameters held constant, may correspond to a complicated path in the phase-space of the system.

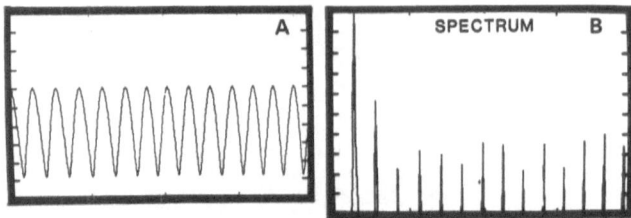

Fig. 15 Time evolution of the self-pulsing intensity (a) and the corresponding
power spectrum (b); only the fundamental frequency and its harmonics are present.
The power distribution of the harmonics is complicated; y=112.5

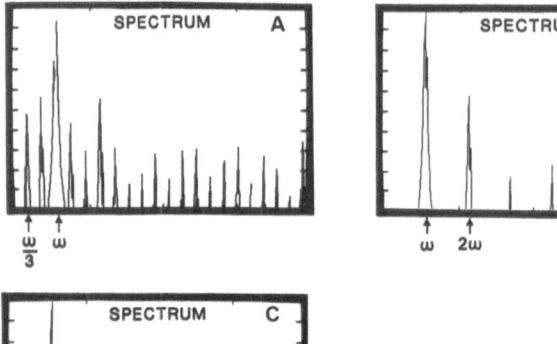

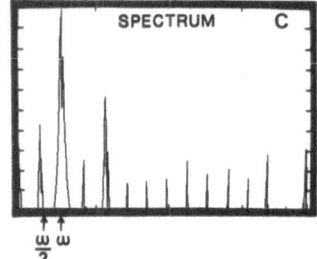

Fig. 16 Power spectrum of the self-pulsing output
in the neighborhood of the chaotic domain;
(a) y=113.4; (b) y=115; (c) y=116

The issue is not resolved at this time, so that we shall limit ourselves to a qualitative description of the behavior of the power spectrum of the output intensity
upon increasing the injected field strength in the neighborhood of the chaotic domain.
At first, the power spectrum displays only the fundamental frequency and its harmonics
(Fig. 15). It is strange that the peak heights of the higher harmonics should display
no regular pattern (one would expect a gradual decrease of the higher harmonic power),
but we have no explanation for this behavior. When y=113.4 (Fig. 16a), we note the
sudden appearance of a frequency component at 1/3rd of the fundamental frequency.
Further on (y=115, Fig. 16b), the spectrum returns to its original structure, displaying only the fundamental and its harmonic components. A subsequent increase in
the driving field to y=116 shows clear evidence (Fig. 16c) of a subharmonic frequency
at $\omega/2$ and of the associated overtones; when, however, y is increased by as little
as 0.05 units, the output power spectrum develops a chaotic broadband structure with
exponential divergence of the temporal solution.

A rather different picture is obtained for C=20 as reported in Ref. [24]. Additional studies are in progress to improve our understanding of the transition to
chaos.

Acknowledgements

This work was partially suported by a contract with the U.S. Army Research Office,
a grant from the Martin-Marietta Research Laboratories and a grant from the
Italian National Research Council (CNR).

References

1. H. Haken, *Synergetics-An Introduction,* Springer-Verlag, Berlin, 1977.

2. H. Haken, *Laser Theory in Light and Matter,* edited by L. Genzel, Handbuch der Physik, vol. XXV/2c, Springer-Verlag, Berlin, 1970.

3. W.E. Lamb, Jr., Proceedings of the International Summer School of Physics E. Fermi, Course XXI, edited by P.A. Miles, Academic Press, NY, 1965.

4. A large literature on this subject has developed in recent years. An extensive review has been prepared by L.A. Lugiato for *Progress in Optics,* edited by E. Wolf (to be published). See also C.M. Bowden, M. Ciftan, and H.R. Robl, eds., *Optical Bistability,* Proceedings of the International Conference on Optical Bistability, Asheville, NC, June 1980 (Plenum Press 1981).

5. A qualitative but informative discussion on the possible technical applications of bistable systems has appeared in E. Abraham, C.T. Seaton, S.D. Smith, Sci. Am. $\underline{248}$, February 1983, p. 85.

6. A. Szöke, V. Daneu, J. Goldhar and N.A. Kurnit, Appl. Phys. Lett. $\underline{15}$, 376 (1969); see also H. Seidel, U.S. Patent #3,610,731 (1971).

7. H.M. Gibbs, S.L. McCall and T.N.C. Venkatesan, Phys. Rev. Lett. $\underline{36}$, 113 (1976).

8. R. Bonifacio and L.A. Lugiato, Opt. Comm. $\underline{19}$, 172 (1976);
    R. Bonifacio and L.A. Lugiato, Phys. Rev. $\underline{A18}$, 1129 (1978).

9. R. Bonifacio and L.A. Lugiato, Lett. al Nuovo Cimento $\underline{21}$, 510 (1978).

10. H. Haken, Z. Phys. $\underline{190}$, 327 (1966); H. Risken, C. Schmid, and W. Weidlich, Z. Phys. $\underline{194}$, 337 (1966); H. Risken and K. Nummedal, J. Appl. Phys. $\underline{39}$, 4662 (1968); R. Graham and H. Haken, Z. Phys. $\underline{213}$, 420 (1968).

11. L.A. Lugiato, V. Benza, L.M. Narducci, in *Evolution of Order and Chaos in Physics, Chemistry and Biology,* Proceedings of the International Symposium on Synergetics at Schloss Elmau, edited by H. Haken, Springer-Verlag, NY, 1982; see also L.A. Lugiato, V. Benza, L.M. Narducci, J.D. Farina, Z. Phys. $\underline{B49}$, 351 (1983).

12. V. Benza, L.A. Lugiato, Z. Phys. $\underline{B35}$, 383 (1979); ibid. $\underline{B47}$, 79 (1982).

13. H. Haken, Z. Phys. $\underline{B21}$, 105 (1975); ibid. $\underline{B22}$, 69 (1975); H. Haken, H. Ohno, Opt. Comm. $\underline{16}$, 205 (1976); Phys. Lett. $\underline{59A}$, 261 (1982).

14. (a) K. Ikeda, Opt. Comm. $\underline{30}$, 257 (1979); (b) K. Ikeda, H. Daido, and O. Akimoto, Phys. Rev. Lett. $\underline{45}$, 709 (1980); (c) K. Ikeda and O. Akimoto, Phys. Rev. Lett. $\underline{48}$, 617 (1982); see also (d) L.A. Lugiato, M.L. Asquini and L.M. Narducci, Opt. Comm. $\underline{41}$, 450 (1982).

15. H. Gibbs, F.A. Hopf, D.L. Kaplan and R.L. Shoemaker, Phys. Rev. Lett. $\underline{46}$, 474 (1981).

16. L.A. Lugiato, Opt. Comm. $\underline{33}$, 108 (1980).

17. L.A. Lugiato, L.M. Narducci, D.K. Bandy and C.A. Pennise, Opt. Comm. $\underline{43}$, 281 (1982); L.M. Narducci, D.K. Bandy, C.A. Pennise and L.A. Lugiato, Opt. Comm. $\underline{44}$, 207 (1983).

18. R. Bonifacio and L.A. Lugiato, Lett. al Nuovo Cimento $\underline{21}$, 517 (1978); S.S. Hassan, P.D. Drummond and D.F. Walls, Opt. Comm. $\underline{27}$, 480 (1978).

19. M. Okada and K. Takizawa, J. Quant. Electr. $\underline{QE-17}$, 2135 (1981).

20. J.Y. Gao, J.M. Yuan, L.M. Narducci, Opt. Comm. $\underline{44}$, 201 (1983).

21. M.B. Spencer and W.E. Lamb, Jr., Phys. Rev. $\underline{A5}$, 884 (1972).

22. T. Yamada and R. Graham, Phys. Lett. $\underline{53A}$, 77 (1975); M.J. Scholz, T. Yamada, H. Brand and R. Graham, Phys. Lett. $\underline{82A}$, 321 (1981).

23. L.A. Lugiato, L.M. Narducci, D.K. Bandy, C.A. Pennise, Opt. Comm. $\underline{46}$, 64 (1983).

24. D.K. Bandy, L.M. Narducci, C.A. Pennise and L.A. Lugiato, Proceedings of the 5th Coherence and Quantum Optics Conference, L. Mandel and E. Wolf, Eds., Plenum Press (to be published).

Chaos in Lasers

Robert Graham

Universität Essen, Fachbereich Physik, D-4300 Essen, Fed. Rep. of Germany

1. Introduction

There are many systems in physics where macroscopic order appears by some sort of self-organization out of microscopic disorder. Hydrodynamic flows may be among the oldest and most studied examples - I only need to recall the ordered flows appearing in thermal convection or in Couette flows. Optical systems - and there first of all lasers - display this phenomenon of self-organization in an even more spectacular way, creating a highly directed and coherent emission from many independent atoms. The laser has therefore emerged as a paradigmatic model in the field of 'synergetics' [1].

Over the last decade it has become apparent that practically all systems which go 'from microscopic to macroscopic order' when driven sufficiently far from equilibrium, will undergo further transitions taking them 'from macroscopic order to macroscopic disorder' if they are driven even further away from equilibrium. 'Chaos' or 'turbulence' are other names for macroscopic disorder. Again, hydrodynamic flows are the oldest and most studied examples; however, the very subtle routes by which turbulence is reached from ordered flows could only recently be revealed by highly accurate modern experimental techniques [2].

For lasers it has also long been predicted that the highly correlated ordered emission would give way to an irregular, macroscopically disordered emission, if the pumping strength of the laser is sufficiently increased [3-7]. Very recently, several experimental groups have reported the observation of such chaotic emission in lasers [7-10]. Therefore, it seems worth while to take another look at the theory of chaos in lasers. In addition, it is of interest to examine whether lasers are just another class of nonlinear systems displaying the phenomenon of chaos, which is, of course, interesting in technical applications, or whether there is also something fundamentally new to be learned from chaos in lasers.

As a starting point of this discussion, in section 2 I first review in an intuitive way how macroscopic order is established among the radiating atoms in a laser [11]. We also introduce there the distinction between a 'good-cavity laser' and a 'bad-cavity laser'. In section 3 we discuss the important class of 'bad-cavity instabilities' in high gain single-mode lasers, which have recently been observed in some inhomogeneously broadened gas lasers.

In section 4 we consider mechanisms by which chaos can appear in multimode lasers [12]. Chaos in 3-mode lasers has also recently been observed.

In section 5 we finally comment on a fundamentally new aspect of chaos in lasers: the interplay of quantum theory and chaos in dissipative systems [13].

2. From microscopic disorder to macroscopic order in lasers

It is our purpose, in this section, to provide a physical picture of how order is established in laser-active media. For a general account of laser theory cf. [14]. The discussion we give here follows [11]. Let us consider N two-level atoms in interaction with a single mode of the electromagnetic field. We focus our attention on the population of the 2 levels and the radiating atomic dipole moments of a few of these atoms (cf. figs. 1 - 3).

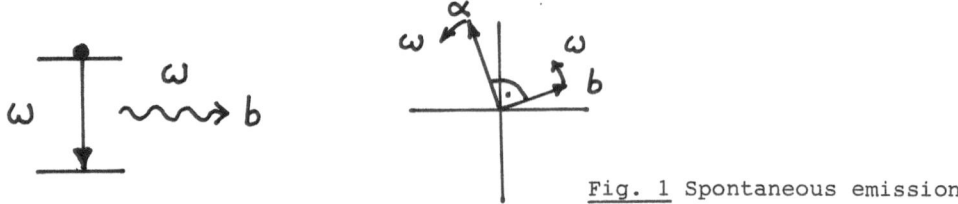

Fig. 1 Spontaneous emission

In a process of spontaneous emission (fig. 1) an atom jumps from its upper state to its lower state. It sets up spontaneously an atomic dipole moment α at frequency ω with an amplitude and a phase, which we therefore depict as an arrow in the complex plane. The dipole moment drives the laser mode b at the same frequency (we assume exact resonance). As in a driven harmonic oscillator at resonance the complex amplitude b lags behind in phase by $\pi/2$ compared to the driving dipole moment α. The phase of α is random in a process of spontaneous emission. Next, consider the process of induced absorption by an atom in its ground state (fig. 2).

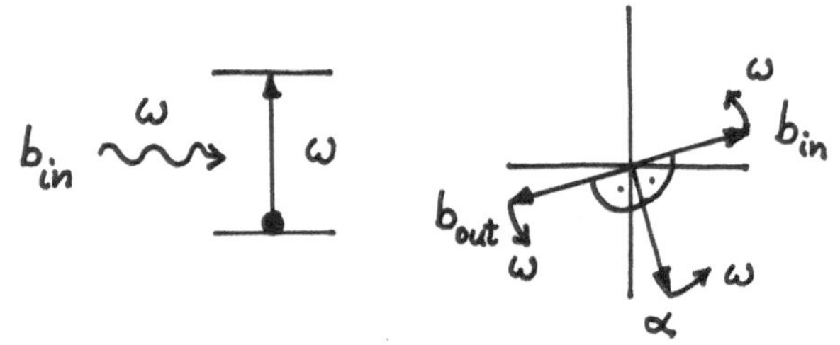

Fig. 2 Absorption

The incoming field b_{in} lifts the atom to its upper state, thereby creating an atomic dipole moment which lags behind in phase by $-\pi/2$. The dipole moment is therefore opposite to that of the atom which originally created b_{in}. The newly created dipole moment radiates a field b_{out} which lags behind in phase by $-\pi/2$ compared to α and by $-\pi$ compared to b_{in}. It therefore annihilates b_{in}, i.e., the incoming field is absorbed. Finally, consider the process of induced emission (fig. 3).

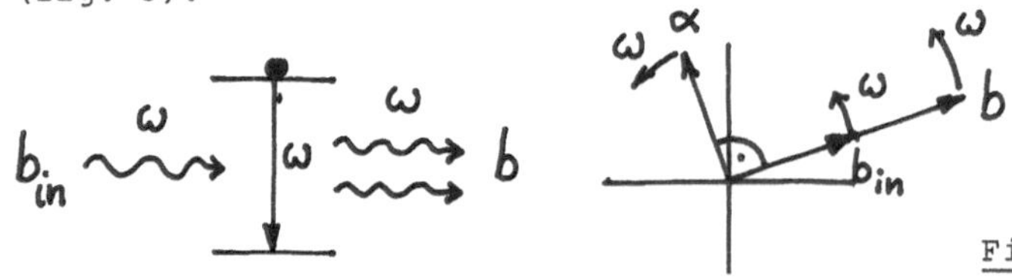

Fig. 3 Induced emission

48

Here, the incoming field b_{in} creates an atomic dipole moment α which is advanced in phase by $\pi/2$. The reason for the advance of α in phase is that the atom does not act as a driven passive oscillator, but as a driven active amplifier. Since α is advanced by $\pi/2$ compared to b_{in}, it has the same direction as the dipole moment of the atom which emitted b_{in}. Also the field b_{out} which is created by α with a phase lag of $-\pi/2$ is precisely in phase with b_{in} and therefore amplifies it.

We see immediately that we can create a uniform direction, i.e., macroscopic order, among the atomic dipole moments if we manage that the processes of induced emission outweigh the processes of induced absorption, since the latter processes create atomic dipole moments in the wrong direction. There are two possibilities for achieving that. First, we may put all atoms into their upper state initially and then watch how the processes of spontaneous and induced emission build up macroscopic order among the atomic dipole moments. If we take care to remove the electromagnetic field from the atoms before it can be reabsorbed, the net alignment of the atomic dipole moments can become very large. This is the process of superfluorescence (in the older literature also called superradiance). For a detailed theory of this process cf. [15]. Here, order is established only for a rather short time during a strong electromagnetic pulse emitted by the atoms, since it disappears as soon as most atoms have made transitions to their ground states. We shall come back to this sort of emission in a different context below.

The second possibility is to consider lasers in which we can pump the atoms continuously to their upper states, and mirrors provide feedback of the electromagnetic field. In such a laser a steady state is established in which the atomic dipole moments are partially ordered. The degree of their ordering depends on the degree to which the processes of induced emission outweigh the processes of induced absorption in the steady state.

If the laser had ideal mirrors, i.e., no losses of the electromagnetic field at all, the processes of induced emission would exactly balance the processes of induced absorption in the steady state (neglecting spontaneous emission for the moment). Therefore, in such a laser, there would be no macroscopic order at all among the atomic dipole moments. In fact, usual lasers with nearly perfect mirrors, which we call 'good-cavity lasers' in the following, operate quite close to this ideal state. Therefore they have actually very little macroscopic order among their atomic dipole moments, i.e., there is only a tiny common alignment between them. This tiny collective atomic moment is sufficient to drive a strong, coherent electromagnetic field since the losses of the field are very small by virtue of the good mirrors of a 'good-cavity laser'. Thus order in such a laser resides mainly in its electromagnetic field and is nearly absent in its atomic dipole moments.

In a 'bad-cavity laser' the situation is just the opposite. Since the losses of the electromagnetic field are high in such lasers the induced emission processes strongly outweigh the absorption processes in the steady state, and there is a strong net alignment between the atomic dipole moments in the steady state. This strong collective atomic dipole moment is needed to maintain a coherent electromagnetic field against the big losses.

The condition which distinguishes between the two cases is

$$\varkappa \ll \gamma_\perp \quad \text{good cavity}$$
$$\varkappa \gg \gamma_\perp \quad \text{bad cavity}$$

where κ is the inverse lifetime of the electromagnetic field in the cavity and $\gamma_\perp$ is the inverse lifetime of the atomic dipole moment in a single atom. At first sight it may seem anti-intuitive that order in a laser with a high-loss cavity should be higher than order in a good-cavity laser. However, it must be kept in mind that the high cavity losses must be compensated by stronger pumping, i.e., higher gain in a bad-cavity laser, in order to achieve the same output of the electromagnetic field. It is then not surprising that the order in the atomic dipole moments should be greater in a bad-cavity laser under these conditions.

Bad-cavity lasers in single-mode operation have long been predicted to become unstable if pumped sufficiently hard [3-5]. Very recently, close relatives of these instabilities occurring in inhomogeneously broadened lasers [7, 16, 17] have been observed experimentally. These are the topics to which I turn next.

3. Chaos in bad-cavity single-mode lasers

The instability of 'bad-cavity lasers' for sufficiently strong pumping follows from a linear stability analysis of their equations of motion. Their behavior in the unstable regime follows from the analogy of their equations of motion with the Lorenz model [5]. Instead of reviewing these mathematically based arguments, I turn here to a more qualitative intuitive description.

The basic reason for the instability of bad-cavity single-mode lasers can be seen in the fact that, at least for sufficiently strong pumping, they possess a second way of operation besides the usual time-independent CW operation, which we described above. This second way of operation is somewhat related to the process of superfluorescence, which we also mentioned above. It is possible only in bad-cavity lasers and not in good-cavity lasers, since superfluorescence requires that we remove sufficiently rapidly the radiated electromagnetic field from the atoms. Otherwise, as I explained above, reabsorption processes will destroy the alignment of the atomic dipole moments. The high losses of bad-cavity lasers precisely fulfill this condition and therefore can become unstable against this second form of emission. What happens then, if we pump a bad-cavity single-mode laser sufficiently hard? The continuous time-independent emission will suddenly break up into a time-dependent pulsing emission. Since in the time domain the single mode has now more than one frequency, this process has also been called 'mode splitting'. A single spatial mode oscillates at more than one frequency [18]. The time sequence of single-mode pulses is created by a strong, time-dependent collective atomic dipole moment. Successive pulses are, of course, not completely independent, since each pulse is able to deplete the upper states of atoms only partially, and the initial condition for each pulse depends on how successful the preceding pulse was in this respect. On the other hand, the correlation between two subsequent pulses is also not complete, i.e., the time sequence of pulses is not periodic, and it is exponentially unstable against small perturbations such as spontaneous emission processes of the atoms. Pulses in the sequence which are sufficiently far apart from each other are not correlated at all. These are the characteristic features of a state of macroscopic disorder or chaos.

Unfortunately, it seems that this extremely interesting state of a single-mode laser has not yet been realized experimentally so far. The reason is that the conflicting requirements of a bad cavity and of pumping very high above threshold are very hard to satisfy. More

precisely, these conditions are

$$\varkappa > \gamma_\perp + \gamma_\parallel$$

and

$$r/r_{th} > \frac{\varkappa(\varkappa + 3\gamma_\perp + \gamma_\parallel)}{\varkappa - \gamma_\perp - \gamma_\parallel} .$$

$\varkappa$, $\gamma_\perp$ have been explained above, and $\gamma_\parallel$ is the inverse relaxation time of the population difference of the atomic levels, r/r_{th} is the ratio of the required pumping rate r to the pumping r_{th} of the laser at threshold. For a bad-cavity laser, already r_{th} is very high, in comparison to a good-cavity laser.

In recent years it has been discovered [7], however, that a bad-cavity instability exists in inhomogeneously broadened lasers, which can quite easily be realized experimentally. In an inhomogeneously broadened laser, such as a gas laser, the transition frequencies of the atoms are not equal, but are distributed according to some proba-bility distribution with a certain width, the inhomogeneous linewidth of the laser. In a gas laser this distribution is a Gaussian, according to the Maxwell distribution of the molecular velocities in the gas. It turns out that the inhomogeneous depletion of the atoms by the emitted electromagnetic field greatly reduces the threshold of the bad-cavity instability. Recently bad-cavity instabilities in an inhomogeneously broadened 3.51 μ xenon laser have been observed by Gioggia and Abraham [10]. They observed the appearance of periodic states (limit cycles) and various transitions to chaos, e.g.,via the period doubling route or via the appearance of two incommensurate frequencies (2-torus). For a simple extension of the Lorenz model including effects of inhomogeneous broadening in a qualitative way cf. [19].

4. Chaos in multimode lasers

Chaos in multimode lasers should be common place. This follows already from the result of Ruelle, Takens and Newhouse [20] that a nonlinear dynamical system supporting a spectrum of sharp lines with three or more incommensurate frequencies is not 'generic'. Indeed, chaos in a 3-mode laser has been observed recently [8, 9].

A very simple model already allows to investigate the basic mecha-nism which generates chaos in a three-mode laser [12]. As an example we consider a situation where a strong central mode at frequency ω_2 oscillates simultaneously with two weaker satellite modes at fre-quencies ω_1, ω_3 (cf. fig. 4).

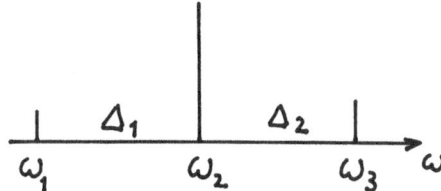

Fig. 4 Model to investigate chaos in a 3-mode laser

We focus our attention on the weak mode 1 with frequency ω_1. If its interaction with the two other modes ω_2, ω_3 could be neglected, it could be treated just as a single mode in a single-mode laser. However, there are, of course, interactions, and they modify the dynamics of mode 1 at least in two important ways:

(1) Due to nonlinear interactions the modes 2 and 3 create a field at the combination frequency

$$2\omega_2 - \omega_3 = \omega_2 - \Delta_2 \approx \omega_1 \quad .$$

This field acts as a driving force on mode 1 with a frequency close to ω_1.

(2) The gain seen by mode 1 is not constant in time, but is, instead, modulated periodically due to the presence of the other modes.

These two effects are already sufficient to generate chaos in mode 1 [12]. The real situation is, of course, even more complicated, since neither the amplitude of mode 2 nor that of mode 3 will remain constant in time, as was assumed implicitly in this simplified discussion. But this can only mean that chaos is even more likely to occur in the real 3-mode system than in the modulated single-mode model which we analyzed. Preliminary numerical results which we obtained for a 3-mode model confirm this conclusion.

In summary, it seems now well established experimentally and theoretically that lasers show transitions from macroscopic order to macroscopic disorder just like any other generic nonlinear system. The question offers itself, whether there is also something fundamentally new to be learned from chaos in lasers. My hope is that the answer is yes: optical dynamical systems like lasers could offer the opportunity to study for the first time the interplay of quantization and chaos in a driven dissipative system. We turn to this topic in the following section.

5. Quantum theory and chaos in dissipative systems

In the optical frequency range quantum effects are no longer negligible. Here lies a fundamental difference between dynamical systems in hydrodynamics, say, and in optics. Chaos in lasers, therefore, really is quantum chaos. Furthermore, unlike practically all other examples of quantum chaos studied so far (e.g., highly excited molecules or other conservative Hamiltonian systems, or externally driven Hamiltonian systems like the 'kicked rotator' [21]), lasers are dissipative non-Hamiltonian quantum systems. Virtually nothing is known about such systems under conditions where their classical counterparts display classical chaos. In the present section we therefore want to discuss some recent results about the interplay between chaos and quantum theory.

Let us begin this discussion by recalling that quantum effects influence the dynamics of a laser in essentially two ways, which are both equally important in the optical domain.

(i) First there is an influence of quantum effects on the observability of the laser. The electromagnetic field which is emitted by a laser is a quantum field, i.e., it is measured by a detector, a photodiode, in quantized lumps of energy $\hbar\omega$. For a laser emitting the intensity I with probability $W(I)dI$, the probability $P(n)$ to observe n quanta of energy $\hbar\omega$ in a certain fixed time interval T with a photodiode of a certain fixed counting efficiency η is given in [22]

$$P(n) = \int_0^\infty dI \; W(I) \; \frac{(\alpha I)^n}{n!} \; e^{-\alpha I}$$

where α is proportional to T and η and relates the average photon number n to the average intensity $<I>$.

$$\langle n \rangle = \alpha \langle I \rangle \quad .$$

Due to uncertainty relations between the real and imaginary parts of the mode amplitude it is not possible to observe the phase of the field in experiments in which photons are counted. Vice versa, it is not possible to count photons in an interference experiment in which the phase of the electromagnetic field is effectively measured (cum grano salis, since the phase is not an observable in quantum theory in the strict sense [23]). Therefore, all features of classical chaos which involve the amplitude and the phase of the electromagnetic field in an important way will be deeply modified by quantum effects. More generally, the same is true for all features of classical chaos involving noncommuting observables.

(ii) The aforementioned limitations of observability apply to all optical systems, no matter whether they are active like lasers or passive like optically bistable systems. However, the second influence of quantum theory which we wish to discuss is typical for active, pumped systems like lasers: it is the effect of spontaneous emission, which already limits the order which is attainable in lasers in the case of time-independent cw emission.

Fortunately, the quantum theory of the laser is a rather well-developed art, thanks to the pioneering work of Haken and his co-workers in the sixties [4] and also thanks to the work of many others. Therefore, a complete theoretical description of lasers is available which incorporates the quantum effects described above. This description is based on a linear 'master equation'

$$\dot{\rho} = L\rho$$

for the density operator ρ of the laser. The letter L stands for a Liouville operator acting on ρ which incorporates the conservative and dissipative parts of the dynamics. It is completely known for the laser. The only thing which is still lacking is an explicit solution of this 'master equation' in a region of parameter space where the laser is chaotic, at least in the classical limit. Such a solution would be extremely interesting. It would teach us something about the meaning of chaos in a real dissipative quantum system.

In the absence of such a solution we can still make some useful observations. First of all we can go to the classical limit in the master equation. Again, this is well known. In the classical limit the master equation is known to reduce asymptotically to a diffusion process or Fokker-Planck equation for the Wigner function [25]. The Wigner function itself is a convenient quantum generalization of a classical joint probability distribution of all observables of the system, including noncommuting variables. For the homogeneously broadened single-mode laser it is found that the resulting diffusion process for the Wigner function is equivalent to the Lorenz model with some external classical noise. The Lorenz model with external classical noise has recently been analyzed in [24]. The physical origin of the external noise is the spontaneous emission in the laser. The quantization of the electromagnetic field in this limit is completely taken into account by the expression for P(n) given above if W(I) on the right-hand side is expressed by the Wigner function. As a result, we can state that asymptotically in the classical limit the quantum noise acts just like classical noise. This feature is typical of <u>dissipative</u> quantum systems in general, since the quantum treatment automatically incorporates the fluctuation dissipation theorem, and the classical version of this theorem requires that dissipation must always be associated with classical noise in a classical system.

Unfortunately, this whole consideration is restricted to the asymptotic behavior in the classical limit, and it does not tell us anything about the modification of classical chaos by strong quantum effects. Strongly pumped lasers like a chaotic bad-cavity single-mode laser are of course near to the classical limit. Therefore, the preceding asymptotic consideration should really be applicable. Nevertheless it is very interesting to consider also the deep quantum region. Experimentally, strong quantum effects could perhaps be observed in a weakly excited chaotic laser mode occurring in a multimode laser. The example of the 3-mode laser has been discussed in section 4.

In order to understand the influence of arbitrarily strong quantum effects on a dissipative chaotic system, I have recently analyzed an exactly solvable simple model which is dissipative and displays chaos in the classical limit [13]. It is constructed from a 2-dimensional classical map (i.e., time becomes a discrete parameter n) which expands in each step all distances q by a factor of 2 (thereby modelling the sensitive dependence on perturbations in a chaotic system) and which contracts the canonically conjugate momentum by a factor of 1/2. A map of qualitatively the same form arises in the Lorenz model as the return map of an appropriately chosen cross section transverse to the strange attractor. Explicitly, the map is given by [26]

$$q_{n+1} = 2q_n \pmod 1$$

$$p_{n+1} = \frac{1}{2} p_n + g_0 \sin 4\pi q_n \quad .$$

The modulo 1 restriction on $2q_n$ describes the folding process in phase space, which, besides the local stretching of q, is the second essential ingredient for chaotic dynamics. This map gives rise to a classical strange attractor in (p,q) space, which can be calculated. Indeed, it can be shown [27, 28] that for $n \to \infty$ an initially given phase-space probability density W_0 (q,p) reduces to the form

$$W_\infty(q,p) = \lim_{n \to \infty} \frac{1}{2^n} \sum_{m=0}^{2^n-1} \delta\left[p - F_m^{(n)}(q)\right]$$

with

$$F_m^{(n)}(q) = -\sum_{\ell=0}^{n} 2^{-\ell} \sin\left(4\pi \frac{q+m}{2^\ell + 1}\right) \quad .$$

Via the δ functions, the probability density W_∞ (q,p) is concentrated on the strange attractor. In [13], I have constructed a master equation, which reduces to the above classical map in the classical limit. It takes the form

$$<q|\rho_{n+1}|q'> = \frac{1}{2} \exp\left\{-\frac{2i}{\hbar}\left[g(q/2) - g(q'/2)\right]\right\}$$

$$\times \left(\left\langle\frac{q}{2}\middle|\rho_n\middle|\frac{q'}{2}\right\rangle + \left\langle\frac{q+1}{2}\middle|\rho_n\middle|\frac{q'+1}{2}\right\rangle\right)$$

where $<q|\rho_n|q'>$ is the density matrix in coordinate representation at time n, and

$$g(q) = (g_0/4\pi) \cos(4\pi q) \quad .$$

The Wigner function of the variables q,p can be calculated from the master equation. For $n \to \infty$ it takes the form

$$W_\infty(q,p) = \lim_{n \to \infty} \frac{1}{2^n} \sum_{m=0}^{2^n-1} F_m^{(n)}(q,p,\hbar)$$

with

$$F_m^{(n)}(q,p,\hbar) = \int_{-\infty}^{+\infty} \frac{dx}{2\pi} \exp\left\{- ixp - \frac{2i}{\hbar} \sum_{\ell=1}^{n} \left[g\left(\frac{(q+\hbar x/2)(\bmod 1)}{2^\ell}\right)\right.\right.$$

$$\left.\left. - (x \to -x)\right]\right\} \quad .$$

Thus, the sharp δ functions of the classical phase-space density are replaced by the delocalized functions $F_m^{(n)}(q,p,\hbar)$ in the exact quantum results. In the classical limit we have

$$\lim_{\hbar \to 0} F_m^{(n)}(q,p,\hbar) = \delta[p - F_m^{(n)}(q)] \quad .$$

Asymptotically for small $\hbar$ we find that $F_m^{(n)}(q,p,\hbar)$ becomes proportional to an Airy function of $p-F_m^{(n)}(q)$. In short, we see that the strange attractor of the classical system is not present in the quantum system, in accordance with the uncertainty relation between p and q. This is an explicit example how quantum effects deeply modify the features of classical chaos involving noncommuting variables, here p and q. Indeed q is a periodic phase-like variable in the present model. Therefore, only quantized values of p should be observable, similar to the quantized values of the field intensity in the preceding discussion. Indeed, integrating the Wigner distribution over q we find the observable probability distribution of p as

$$W_\infty(p) = \sum_{\ell=-\infty}^{+\infty} \delta(p - 2\pi\hbar\ell)W_\ell$$

with

$$W_\ell = \lim_{n \to \infty} 2^{-n} \sum_{m=0}^{2^n-1} \left| \int_0^1 dq \exp\left[-2\pi iq\ell - \frac{2i}{\hbar} \sum_{k=1}^{n} g\left(\frac{q+m}{2^k}\right)\right]\right|^2 \quad .$$

Therefore, the observable values of p are indeed correctly quantized in units of $2\pi\hbar$, and W_ℓ gives the probability to observe ℓ quanta of this unit.

6. Concluding remarks

We have reviewed in this contribution how order on a macroscopic scale is established in lasers. We have seen why the order in bad-cavity lasers is fundamentally different from the order in good-cavity lasers, and that the bad-cavity condition opens up the possibility of single-mode pulsing. This is a macroscopically disordered or chaotic form of emission. The transition to single-mode pulsing in bad-cavity lasers is greatly facilitated by inhomogeneous broadening and has recently been observed. Other transitions to chaos

were predicted and have been observed in multimode lasers. Finally,
we have discussed the new problem of quantum chaos in dissipative
systems. Strongly pumped lasers are close to the classical limit,
and we have argued that, in addition to the quantization of photon
numbers in photo-count experiments, the quantum noise acts like
classical noise, in a description employing the Wigner function, due
to the classical limit of the fluctuation dissipation theorem. By
considering an exactly solvable model we have also given an explicit
example how the uncertainty relation modifies classical chaos involving noncommuting variables. With these considerations we have,
of course, only scratched the surface of the problem of dissipative
quantum chaos. I believe that the further investigation of this new
field under conditions where quantum effects are strong will make
the more detailed study of chaos in lasers a worthwhile subject in
the future.

References

1 H. Haken, Synergetics, Springer Series in Synergetics Vol. 1,
   Springer, New York 1976
2 Hydrodynamic Instabilities and the Transition to Turbulence,
   H.L. Swinney, J.P. Gollub ed., Topics in Applied Physics Vol. 45,
   Springer, New York 1981
3 A.Z. Grasyuk, A.N. Oravskii, Radiotekh. Elektron. 9, 524 (1964)
4 H. Haken, Z. Physik 190, 327 (1966)
5 H. Haken, Phys. Lett. 53A, 77 (1975)
6 R. Graham, Chaos in Simple Laser Systems, Workshop on Coupled
   Nonlinear Oscillators Los Alamos 1981, to appear
7 L.E. Casperson, IEEE J. Quant. Electronics 14 756 (1978); Phys.
   Rev. A21, 911 (1980); A23, 248 (1981)
8 N.B. Abraham, T. Chyba, M. Coleman, R.S. Gioggia, N.J. Halas,
   L.M. Hoffer, S.N. Lin, M. Maeda, and J.C. Wesson in 'Third New
   Zealand Symposium on Laser Physics', Lecture Notes in Physics,
   eds. D.F. Walls and J. Harvey, Springer, New York 1983
9 C.O. Weiss, A. Godone, A. Olafsson, Phys. Rev. A, to appear;
   C.O. Weiss, private communications
10 R.S. Gioggia, N.B. Abraham, Self-Pulsing Instabilities and Chaos
   in a Single-Mode Inhomogeneously Broadened Fabry-Perot Laser,
   preprint 1983
11 R. Graham, in Progress in Optics Vol. XII, ed. E. Wolf, North
   Holland 1974, p. 234
12 H.J. Scholz, T. Yamada, H. Brand, R. Graham, Phys. Lett. 82A,
   321 (1981)
13 R. Graham, Quantum Noise and Strange Attractors, preprint 1983
14 H. Haken, Laser Theory, Encyclopedia of Physics 25/2c (1970)
15 R. Bonifacio, P. Schwendimann, F. Haake, Phys. Rev. A4, 302
   (1971); A4, 854 (1971); F. Haake, this volume
16 L.M. Narducci, D.K. Bandy, L.A. Lugiato, N.B. Abraham, Stability
   Analysis of a Single-Mode Inhomogeneously Broadened Laser,
   preprint 1983
17 P. Mandel, Casperson's instability: analytic results, preprint
   1983
18 S.T. Hendow, M. Sargent III, Opt. Comm. 40, 385 (1982); 43, 59
   (1982)
19 R. Graham, Y. Cho, Opt. Comm., to appear
20 S. Newhouse, D. Ruelle, F. Takens, Comm. Math. Phys. 64, 35 (1978)
21 G. Casati, B.V. Chirikov, F.M. Izraelev, J. Ford, in Lecture
   Notes in Physics 93, Springer New York 1979, p. 334
22 L. Mandel, Proc. Phys. Soc. (London) 71, 1037 (1958); R.J. Glauber
   in Quantum Optics and Electronics, ed. De Witt, Blandin, Cohen
   Tannoudji, Gordon and Breach, New York 1965

23 P. Carruthers, M.M. Nieto, Phys. Rev. Lett. $\underline{14}$, 387 (1965)

24 M. Dörfle, R. Graham, Phys. Rev. $\underline{A27}$, 1096 (1983)

25 E.P. Wigner, Phys. Rev. $\underline{40}$, 749 (1932)

26 J.L. Kaplan, J.A. Yorke, Lecture Notes in Mathematics $\underline{730}$,228 (1979)

27 R. Graham, Phys. Rev. A, to appear

28 D. Mayer, G. Roepstorff, J. Stat. Phys. $\underline{31}$, 309 (1983)

The Falling Pencil and Superfluorescence:
Macroscopic Indeterminacies After the Decay of Unstable Equilibria

Fritz Haake

Universität-Gesamthochschule Essen, Fachbereich Physik
D-4300 Essen, Fed. Rep. of Germany

Think of a sharpened pencil placed vertically head-down on a horizontal plane. Will it fall? In practice, it always will. A student of classical mechanics should argue, though, that the upright position corresponds to an unstable equilibrium which is, once realized, infinitely long-lived. The said student could explain the fall by extraneous perturbations such as random drafts of air, vibrations of the ground,etc., or by the practical difficulties of realizing the precise upright position. A mathematician might point out that within the set of positions initially available to the pencil the strictly vertical one is a subset of measure zero. Since the tiniest initial deviation from verticality suffices to destroy the equilibrium it would thus seem hopeless that one would ever be able to realize unstable equilibrium.

For a physicist with an education in quantum mechanics the problem of the falling pencil takes on a rather different flavor. Even if it were possible by some ingenious trick to erect the pencil to the upright position with arbitrary precision we would always - as a matter of principle - see it fall eventually. According to Heisenberg's uncertainty principle it is impossible to control both the position and the velocity of any body with arbitrary accuracy. The better we control the initial position of the pencil the more we lose control over the initial velocity of, say, its center of mass. The pencil must fall in any case since both strict verticality and strict immobility would be required to secure an infinite life time for the standing pencil. In the language of classical mechanics we may attribute the fall of the initially vertical pencil to the presence of tiny random forces; in contrast to extraneous random forces due to drafts of air,etc.,the random forces connected with the uncertainty principle cannot be eliminated by ever so sophisticated experimental precautious.

Why should we bother about falling pencils? Whether or not a pencil will fall is certainly not a cause of scientific dispute. It is a quite interesting question, though, whether or not one can eliminate extraneous perturbations and secure an initially motionless upright position to a degree of accuracy sufficient to guarantee the destruction of the unstable equilibrium by quantum effects. Actually, that question has, to my knowledge, never been answered either for the pencil or for any other macroscopic mechanical or hydrodynamical system.

Quantum uncertainties and quantum fluctuations are important phenomena in the realm of individual atoms. Their influence on macroscopic bodies, on the other hand, is usually beyond the limits of observability. It would therefore be fascinating, indeed, if an experiment on, e. g., the onset of convection in a Benard cell could be so carefully designed that the initially quiescent fluid would be in an unstable equilibrium until quantum fluctuations would start the convection. As the convection pattern became visible one would then literally see quantum fluctuations, amplified from their originally tiny strength to macroscopic beauty.

Quantum optics, the field in physics which was so dramatically boosted by the invention of the laser, has recently yielded the possibility of detecting macroscopic manifestations of quantum fluctuations.

The excited state of an atom corresponding to any energy level except the lowest one is an unstable equilibrium and would thus be infinitely long-lived would not quantum fluctuations in the surrounding electromagnetic field initiate its decay. When many, say $N = 10^8$ or more, identical atoms are all brought to an excited state simultaneously, one has a macroscopic system in an unstable equilibrium. Under most circumstances such states decay both due to extraneous perturbations like collisions among atoms which produce heat and due to quantum fluctuations, i. e.,through spontaneous emission of photons. In spite of considerable practical difficulties it has proven possible to reduce all extraneous perturbations (collisions, inhomogeneous broadening) below the level of the intrinsic quantum fluctuations.

In recent experiments performed at the école Normale Supérieure in Paris [1] the initially excited atoms return to their ground states in a collective manner such that the falling pencil is not only a qualitative analogue. Rather, both problems allow for identical mathematical descriptions [2].

As the atomic excitation relaxes collectively a coherent light pulse (superfluorescence) is produced as a visible moving picture of the decay of the unstable equilibrium. On the other hand, such a pulse represents the quantum fluctuations initiating the decay, amplified from their original tiny magnitudes to a macroscopic level.

The random nature of the quantum fluctuations is not manifest, by the way, in an individual superfluorescent pulse. Each such pulse represents, while its intensity is large, coherent light as regular in its temporal structure as a monochromatic radio wave smoothly modulated to a bell shape. When the experiment is repeated a large number of times with identical preparation, however, quantum fluctuations do become manifest. No two pulses turn out alike! There are great shot-to-shot variations in the pulse shape. For instance, the delay time between the preparation of the atoms and the appearance of the maximum intensity of the radiation pulse has been shown, both theoretically [3] and experimentally [4], to have a statistical variance comparable in magnitude to the mean.

It is instructive to pursue the analogy of the collective radiation process just described with the decay of an unstable equilibrium in a mechanical system. To make that analogy as close as possible I must dispense with the falling pencil and replace it by a pendulum capable of free rotation around a fixed pivot. Imagine the pendulum immersed in a viscous fluid such that the initially upright pendulum "falls" aperiodically until it hangs vertically downward.

The orientation of the pendulum can be characterized by two angles. A "polar" angle θ measures the inclination of the pendulum with respect to the upward direction; an "azimutal" angle ϕ describes the deviation of the projection of the pendulum on a horizontal plane through the pivot (the "equatorial" plane) from an arbitrarily chosen reference direction in that plane.

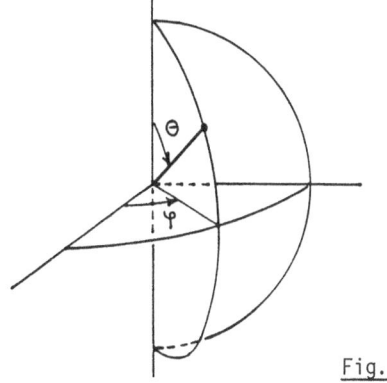

Fig. 1

The motion of the pendulum may start at some initial orientation θ_0, ϕ_0. During the fall the azimuthal angle stays fixed, $\phi(t) = \phi_0$, while the polar angle increases monotonically from θ_0 to $180°$, which latter value corresponds to the stably hanging pendulum. For a quantitative description of the time dependence of θ we must consider the equation of motion for the so-called overdamped pendulum

$$\frac{d}{dt}\theta = \sin\theta .$$ (1)

This simple differential equation expresses a balance of the friction force and the weight of the pendulum, both forces being projected on the direction of changing θ. Note that this equation correctly describes the stationary one of the two equilibrium orientations, $\theta = 0$ and $0 = 180°$.

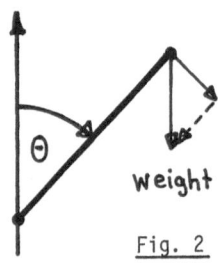

weight

Fig. 2

The solution $\theta(t)$ of the equation of motion (1) which starts out at an angle θ_0 initially is easily found to obey

$$\tan\frac{\theta(t)}{2} = e^t \tan\frac{\theta_0}{2} .$$ (2)

This again implies that the pendulum keeps standing upright forever if it is upright initially. However, any nonzero θ_0 will entail the fall. The horizontal orientation $\theta = 90°$ will be reached after a time t_f which follows from (2) as

$$t_f = -\ln\tan\frac{\theta_0}{2} .$$ (3)

The fall time t_f obviously is larger the closer the initial orientation is to the vertical one, $\theta_0 = 0$. If θ_0 is made arbitrarily small the fall takes arbitrarily long.

The collective radiation process to be discussed can be characterized by a certain abstract vector which behaves in time just like a pendulum. The "vertical" component of that so-called Bloch vector is a measure of the energy stored in the atoms. The "horizontal" components are related to the electric polarization of the atoms, the quantity which enables the atoms to generate an electromagnetic field, i. e., to radiate light.

When all the atoms are in their ground states the Bloch vector points downwards. There is no electric polarization then and the atoms cannot radiate. This state corresponds to the stably hanging pendulum. The Bloch vector points upwards, on the other hand, when all atoms are brought to an excited state associated with a particular energy level above the lowest one. For that orientation, too, the Bloch vector has no horizontal component and can therefore not radiate. We here have the analogue to the unstable equilibrium of the standing pendulum.

In the Paris experiment the length of the Bloch vector remains constant while the superfluorescent pulse is being radiated. This conservation law is, in fact, the basis for the analogy to the falling pendulum. As the Bloch vector changes its orientation its end points all remain on the surface of a sphere. The diameter of that sphere gives, in appropriate units, the total energy available for radiation, if the atoms are fully excited initially. It is proportional to the number of atoms. The radius of the sphere also gives a measure of the maximum electric polarization and thus of the maximum electric field in the radiation pulse. The conservation law in question therefore implies that the electric field rises to a maximum value which is proportional to the number of initially excited atoms. The radiation intensity, a quantity quadratic in the electric field, thus takes on a peak value I_{max} quadratic in the number of atoms, $I_{max} \sim N^2$. The quadratic dependence of the radiation intensity on the number of atoms is one of the distinctive properties of superradiant pulses.

Normal fluorescence from N identical excited atoms yields a maximum intensity linear in N. For large values of N the difference in brightness between normal fluorescence and superfluorescence is thus a considerable one. The mere proportionality of I_{max} to N in normal fluorescence implies that the atoms radiate independently from one another. Their contributions to the electric field are N in number but bear no regularity in their temporal sequence. Even though the squared electric field formally contains N^2 terms there is destructive interference between contributions from different atoms. Only the squared contribution of each atom to the electric field then remains uncancelled in the total intensity and that is why $I_{max} \sim N$ normally.

In contrast, superfluorescing atoms behave cooperatively. Their contributions to the radiation field come in strict temporal regularity such that they superimpose constructively to yield a maximum field proportional to N and thus a maximum intensity proportional to N^2.

It is interesting to realize that a superfluorescent pulse is not only much brighter but also appreciably shorter than a normal pulse radiated by N like atoms. The energy available for radiation is N times the excitation energy of a single atom. The radiation intensity is, roughly, the energy radiated per unit time. In a crude approximation which disregards the precise temporal shape of the pulse we may equate the total energy radiated with the product of the pulse duration and the peak intensity. Therefore, if a superfluorescent pulse is N times more intense than a normal one it must be shorter in time by a factor $1/N$.

As already noted above, the cooperativity of superfluorescence is intimately related to the conservation law for the length of the Bloch vector. No such conservation law holds for normal fluorescence. In that case an initially nearly upright Bloch vector of a length proportional to N not only falls, i. e., increases its polar angle θ, but also shrinks in effective length. By the time θ has grown to 90°, the value corresponding to the peak of the electric field in the cooperative case, the atoms have ceased to stay in regular phase with one another. What remains of the Bloch vector then bears little resemblence to a straight arrow of horizontal orientation. The single-atom contributions to the Bloch vector, rather than being lined up colinearly, then may be pictured as forming a chain of quite irregular shape. The end-to-end connection of this chain is oriented at θ = 90° but rather short compared to the chain length, i. e., the original length of the Bloch vector.

The folding of the Bloch vector to a random chain in normal fluorescence is usually produced by several mechanisms. One is collisions between atoms which may lead to nonradiative deexcitation and irregular reorientation of the electric polarization of the collision partners. Another mechanism lies in the thermal motion of the atoms. Due to the Doppler effect atoms moving with different velocities radiate at slightly differing frequencies even if their internal excitation states are identical; their contributions to the Bloch vector can therefore not stay aligned for long. Finally, atoms in different places see different electric fields and are thus driven out of phase with one another.

If superfluorescence as an analogue of the falling pendulum is to be observed all mechanisms breaking the conservation of the length of the Bloch vector must be rendered ineffective. In the Paris experiment collisions and Doppler broadening are suppressed by working with a so-called beam of atoms in which all atoms move with one and the same velocity in a single direction. (The technicalities of generating such beams, as well as the method used to prepare all atoms in an appropriate state of excitation, need not concern us here.) Furthermore, to make sure that all atoms see the same electric field two decisive measures are taken. First, a particular atomic spectral line, corresponding to a wavelength of the order of one millimeter, is chosen while the "active" volume containing the radiating atoms is designed to be much smaller in linear dimension. Since the electric field pertaining to a spec-

tral line varies in space on a scale given by the wavelength it remains practically constant across the active volume. The second measure is to give to the
active volume the shape of a long thin cylinder and to surround it by two parallel mirrors such that the cylinder is oriented perpendicularly to the mirrors.

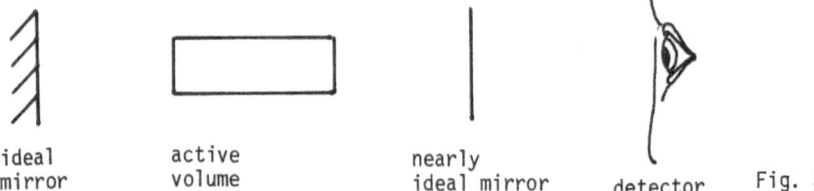

ideal   active   nearly
mirror   volume   ideal mirror   detector   Fig. 3

In between the mirrors the radiation field can exist only in certain discrete
modes. (Just as a violin string is capable of free oscillations with certain eigenfrequencies only.) By symmetry, waves travelling along the axis of the cylinder
are of interest; their wavelengths must fit an integral number of times in twice
the distance ℓ from one mirror to the next. One of the corresponding eigenfrequencies can be tuned to the transition frequency of the radiating atoms by varying the
distance ℓ. The radiation field will then oscillate in that particular mode.

It is clear, at this point, that the experiment under discussion should allow
for a simple theoretical description. The two angles θ and ϕ suffice to specify the
state of the radiating atoms and a single amplitude (a complex number, actually,
for technical reasons we don't have to go into) serves to account for the degree
of excitation of the single radiation mode. It is not obvious, of course, that the
dynamics of the radiation process is that of a pendulum. In fact, the derivation
of the pendulum equation is a technical matter of no interest here.

In the way of a qualitative explanation of the pendulum-like radiation dynamics
I should first eliminate a possible misunderstanding. A mechanical pendulum falls
due to its weight, i. e.,due to the gravitational force exerted on it by the earth.
The "fall" of the Bloch vector, on the other hand, has nothing to do with gravity.
Rather, it is caused by the electromagnetic interaction between the atoms and the
radiation field. Correspondingly, the "vertical" and "horizontal" components of the
Bloch vector do not refer to the surface of the earth. The Bloch vector happens to
live in an abstract three-dimensional space in which a Cartesian coordinate system
can be erected such that atomic excitation energy is measured along one axis and
atomic polarization along the other two axes.

Were the mirrors ideally reflecting ones the electromagnetic interaction would
cause the initially excited atoms to deliver their excitation energy to the radiation mode and, subsequent to that emission, to reabsorb that energy. One such cycle
would be followed by an identical one and an oscillation at least qualitatively
similar to that of an undamped mechanical pendulum would result. The periodic transformation of energy from atomic excitation to mode excitation and back would correspond to the cycling of potential energy of the pendulum in the earth's gravitational field to kinetic energy and back.

Actually, one of the mirrors is deliberately made semitransparent so as to
allow some leakage of radiation to the outer world (where it can be detected). Such
a leakage has the effect of damping the oscillation of the Bloch vector. After a
sufficiently long time all of the energy initially stored in the atoms will have
leaked away. The Bloch vector will then be settled at rest at $\theta = 180°$ and the
radiation mode will be devoid of energy as well. In fact, the nonideal mirror is
designed to be so bad that the motion of the Bloch vector is overdamped, i. e.,
such that energy emitted into the radiation mode leaks away faster than it could
be reabsorbed by the atoms. The Bloch vector then falls aperiodically, the angle
θ increasing monotonically from its initial value θ_0 to its stable equilibrium at
$\theta = 180°$. This overdamped motion is the physical content of the equation of motion
(1).

Recalling now that the "horizontal" component of the Bloch vector, proportional to $\sin \theta$, is the electric polarization and thus the source of the electric field radiated we may infer that the radiation intensity I will be proportional to the square of $\sin \theta$. A little trigonometry exercised on the solution (2) then yields for the radiation intensity

$$I \sim \sin^2\theta = \frac{1}{\cosh^2(\frac{t-t_f}{2})} \, . \tag{4}$$

The intensity peaks at the time t_f given by (3), previously identified as the time at which the Bloch vector passes through $\theta = 90°$.

I have up to now talked about the Bloch vector dynamics as if it were a classical process. In particular, the pendulum equation (1) contains none of the probabilistic elements characteristic of the quantum behavior of microscopic objects like single atoms. It is legitimate to use such a classical description since we are dealing with a macroscopic phenomenon involving, in a collective manner, a large number of atoms. Quantum corrections to the pendulum equation and to the intensity of radiation (4) can be shown to be of relative order $1/N$.

Quantum features enter our problem through the initial condition for the angles θ and ϕ. The energy and the polarization corresponding to a level pair of an atom are subject, like the velocity and the position of a particle, to Heisenberg's uncertainty principle, i. e., they cannot be simultaneously specified with arbitrary precision. In the experiments in question all atoms are prepared in their upper level. Such a preparation amounts to a sharp specification of the energy. The polarization of the atom is then, of necessity, random. Consequently, the total polarization of all atoms cannot be said to be strictly zero initially. Equivalently, the initial angles θ and ϕ are random numbers free to take on different and unpredictable values in every repetition of the experiment.

Only probabilistic statements about the initial polarization and the initial angle θ_0 and ϕ_0 can be made. Once we know the probability distribution of the initial angles we can calculate the average of, say, the squared polarization, which is proportional to the square of the radiated intensity, as

$$\int_0^\pi d\theta_0 \, \sin \theta_0 \int_0^{2\pi} d\phi_0 \, W(\theta_0,\phi_0) \, \sin^2\theta_0 \, . \tag{5}$$

In this expression I have written the probability density for the angles θ_0 and ϕ_0 as the product $\sin \theta_0 \, W(\theta_0,\phi_0)$.

The calculation of the distribution function $W(\theta_0,\phi_0)$ corresponding to the atomic initial state is a simple quantum-mechanical exercise. We can find it, though, without any calculation by invoking the famous central limit theorem of probability theory. (That theorem is best known in physics as the link between classical thermodynamics and the statistical mechanics of many-particle systems). The theorem yields the statistical properties of any random quantity which is made up additively by a large number N of independent contributions of equal kind: irrespective of the probability distributions of the individual contributions the sum has a Gaussian distribution with a width inversely proportional to the number N. The initial polarization is a random quantity of the kind required by the theorem since each of the N atoms contributes independently. In appropriate units the probability distribution for the initial polarization thus reads

$$W(\theta_0,\phi_0) \sim e^{-N \sin^2\theta_0} \, . \tag{6}$$

The term $\sin^2\theta_0$ in the exponent constitutes the Gaussian dependence of W on the polarization while the factor N expresses the inverse proportionality of the width of the distribution to the number of atoms. Note that W does not depend on the azimuthal angle ϕ_0 since the squared polarization does not (the length of the projection of the Bloch vector on the "equatorial" plane is independent of ϕ_0).

For a large number of atoms the width of the distribution W is tiny. Consequently, only tiny values of the polar angle are likely to occur initially. Usually, in describing macroscopic properties of many-particle systems, especially of those in thermal equilibrium, it would be perfectly adequate to neglect fluctuations as tiny as those of the initial polar angle, i. e., to replace the Gaussian distribution (6) by its limit as $N \to \infty$. Here, however, since $\theta_0 = 0$ is an unstable equilibrium it is of decisive importance that nonzero values of θ_0, however small, do occur.

Each time a collection of N atoms is excited a different and unpredictable value θ_0 will be effective. In a large number of repetitions the values of θ_0 will be distributed according to the law (6). Because of the rigid relation (3) between the fall time (the time of maximum radiation intensity) t_f and θ_0 the fall time t_f will vary randomly from pulse to pulse. By using the distribution (6) and the relation (3) we easily find the probability distribution for the fall time. In the limit of large N the simple distribution

$$W(t_f) = \frac{1}{2} N e^{-2t_f} \exp\left(-N e^{-2t_f}\right) \tag{7}$$

results. The following sketch of $W(t_f)$, drawn for $N = 10^4$, displays the remarkable fluctuations of the fall time already commented on at the beginning of this article.

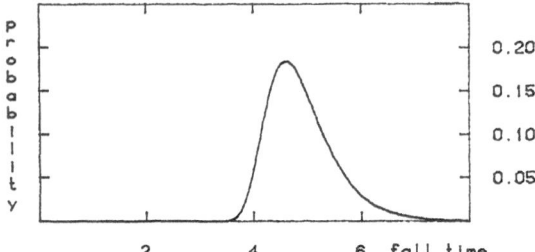

Fig. 4

The width of the fall time distribution is quite comparable in magnitude to the most probable fall time. This huge indeterminacy of the fall time is in striking contrast to the tiny fluctuations of the initial polar angle. Indeed, if the distribution of θ_0 according to (6) were drawn on the same scale as the one used in drawing $W(t_f)$ it would not be distinguishable from a straight line erected vertically on the θ_0 axis at $\theta_0 = 0$.

As another illustration of the large fluctuations in an ensemble of superfluorescent pulses the graph below shows the intensities, $\sin^2\theta(t)$, according to (4), for initial polar angles one-half, once, and twice the most probable value of θ_0, $\overline{\theta_0} = 1/\sqrt{2N}$. The delay of the pulse maximum decreases with increasing value of θ_0.

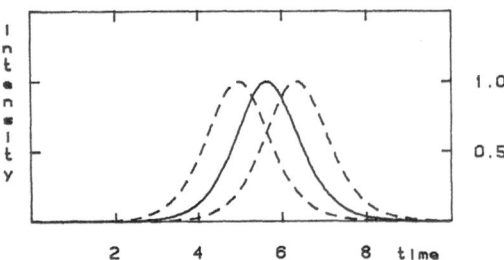

Fig. 5

Large fluctuations in macroscopically observable quantities are quite typical for the decay of unstable equilibria. There are beautiful experimental data available for a variety of such processes. The switch-on of a laser [5], the onset of Stokes radiation in stimulated Raman scattering [6], and superfluorescence from active volumes larger in spatial dimension than a wavelength [4] are the best studied cases. None of these other processes allow for a quantitative discussion as simple as the one I have discussed here, superfluorescence from an effectively point-like sample.

Unfortunately, no experiments on similar processes in mechanical or hydrodynamical systems have as yet been reported, even though such experiments should be feasible. Likewise, we are still waiting for corresponding data in first-order phase transitions ("spinodal decomposition") or instabilities in plasmas. In most of these cases the intrinsic microscopic fluctuations initiating the decay of the unstable equilibrium would be of thermal rather than quantum nature. An interplay of thermal and quantum effects, which would also be quite interesting, might most easily be studied in optical systems such as the one investigated in the Paris experiment discussed above.

It is, of course, quite tempting to extend the analogies mentioned from the many-body systems of statistical physics to the many-people systems under study in sociology and economics. Qualitatively, that is easily done: whatever may correspond to an unstable equilibrium in a many-people system (certain economic crises; a balance between hostile groups in a country?) would certainly decay in a random way; even though each such a decay process might, once well under way, be deterministic in nature, the intrinsic randomness of the initiation would make it impossible to predict, in any other than a probabilistic sense, the state of the system at some later time. The real challenge of corresponding research on many-people systems lies, of course, in identifying relevant measurable quantities and finding the laws governing their dynamics. Most interesting steps in that direction have been taken [7], some of which will be presented in other contributions to this meeting.

[1] S. Haroche, 5th Rochester Conference on Coherence and Quantum Optics, June 1983.
[2] R. Bonifacio, P. Schwendimann, and F. Haake, Phys. Rev. A4, 302 (1971); A4, 854 (1971).
[3] F. Haake, J. Haus, H. King, G. Schröder, and R.J. Glauber, Phys. Rev. A23, 1322 (1979).
[4] Q. H. F. Vrehen, in: Laser Spectroscopy IV, eds. H. Walther, and K. W. Rothe (Springer, Berlin, 1979).
[5] F. T. Arecchi and V. Degiorgio, Phys. Rev. A3, 1108 (1971).
[6] I. A. Walmsley and M. G. Raymer, Phys. Rev. Lett. 50, 962 (1983).
[7] see, e. g., W. Weidlich and G. Haag, Quantitative Sociology, Springer, Berlin (1983).

Part II

Ordered Structures
and Processes in Biomembranes

On the Formation of Transient Order in Biological Membranes

E. Sackmann

Physik Department (E 22; Biophysik), Technische Universität München
D-8046 Garching, Fed. Rep. of Germany

I DEFINITION OF BIOLOGICAL MEMBRANES

Fig.1 shows a diagram of a typical cell - such as a liver cell. As is well known, the viability of the cell depends critically on the cooperation of the subcellular organelles some of which are shown in Fig.1. Each of the subsystems as well as the whole cell is enclosed by an envelope, the membrane which fulfills a twofold task. First it is a barrier separating the organella from the cytoplasma or the cell interior from the outside world.Secondly it is the location of higher biological organisation where a manifold of elementary life processes take place.

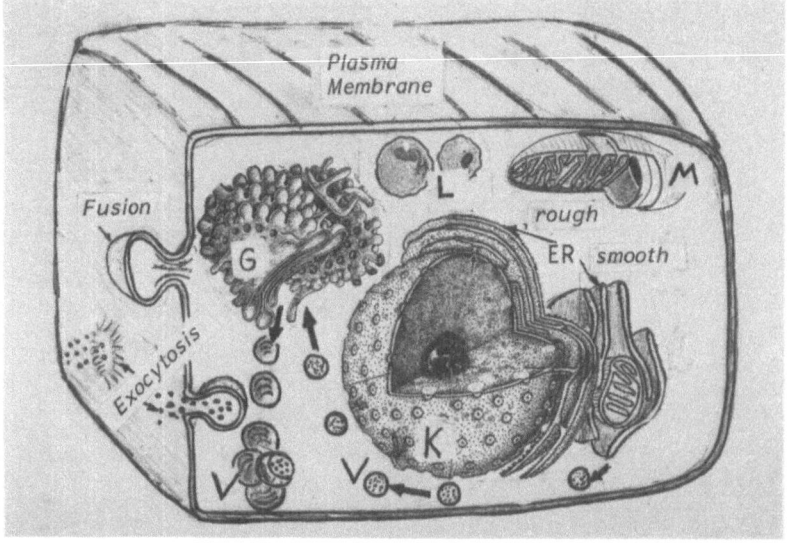

Fig. 1:

Schematic view of cell composed of subsystems such as Mitochondria (M): site of ATP production); Endoplasmatic Reticulum (ER): site of protein biosynthesis; Golgi Apparatus (G): storage, modification and transport of newly synthesized products. Lyosomes (L): vesicles filled with enzymes for protein degradation. Vesicles (V): carriers for transport (1) of molecules into and out of cell (2) of lipids and proteins from ER or G into plasma membrane.

Fig.2 shows a diagram characterizing the molecular architecture of the membranes. The basic building unit is the bilayer of lipid molecules. The biological functions, however, are performed by the proteins which are either incorporated into the bilayer or adsorbed to its surface. The first group is called the integral membrane proteins and the second one the extrinsic membrane proteins.

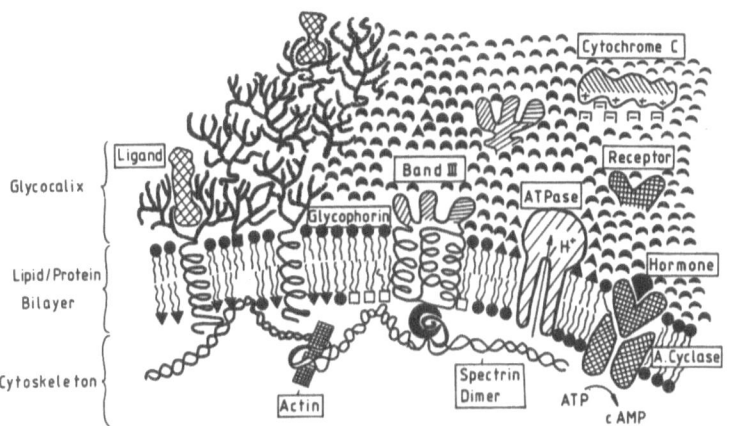

Glycocalix

Lipid/Protein
Bilayer

Cytoskeleton

Fig. 2:

Scheme of biological (plasma) membrane consisting of three coupled layers: the lipid/protein bilayer, the glycocalix and the cytoskeleton. Basic building unit of the first is the lipid bilayer. Incorporated into this are proteins as functional units comprising simple systems such as glycoproteins (acting as receptors) hormone receptors, band-III protein (an anion carrier) and multienzyme complexes such as ATP-ase (acting as proton pump) or the adenylate cyclase/receptor/hormone-complex described in the text.

The glycocalix is formed by polypeptide/polysaccharide head groups of glycoproteins and glycolipids. The cytoskeleton of erythrocytes is shown as example. The spectrin dimers are coupled to bilayer (1) by adsorption and (2) by tight binding via the protein (ankyrin) to band III, the integrated protein.

Besides simpler functional systems such as ion channels and receptors for antigens or blood group determinants (e.g. glycophorin), the membrane contains more complex functional systems composed of multienzyme complexes. Examples are (1) the ATP-ases functioning as ion pumps or (2) the hormone signal transducing system. The latter consists of the hormone receptor and the adenylate-cyclase (an enzyme complex). As illustrated in Fig.2, the coupling of a hormone to the receptor causes the binding of this complex to the cyclase system which triggers the production of cyclic AMP as 2nd messenger. This is a typical example of a transient structure formation.

II FUNDAMENTAL PROBLEMS OF ORDER FORMATION IN MEMBRANES

1) Membranes are multicomponent lipid/protein alloys: This first problem is illustrated in Fig.3. Lipids are characterized by the structure of the head group and of the hydrocarbon chains. (Fig.3a). The head group may be zwitterionic or charged. In the latter case the charging state can be modified by variations in the pH and/or by the binding of ions or charged proteins. The chains are distinguished by the number (1) of carbon atoms (the chain length) and (2) of C-C double bonds (degree of saturation). Fig.4B shows the lipid composition and distribution between the inner and the outer monolayer of the red blood cell. The phosphatidyl serine is the major charged component. Each of the four classes of lipids are composed of two different chains. We thus see that the membrane is built up of about 100 different lipids. In addition it contains some 50 different classes of proteins. Although the membrane is such an extremely complicated multicomponent system, even small changes (by one to two percent) in the natural lipid composition (such as the cholesterol-to-phospholipid ratio) may be fatal for the viability of the cell.

2) Lipids and proteins move rapidly in the plane of the membrane: With respect to lateral motion of the constituents natural membranes generally behave as

two-dimensional liquids$_7$. Lipid molecules have typical lateral diffusion coefficients of $D \approx 10^{-7}$ cm^2/sec. A consequence of the two-dimensionality is that the D value of integral proteins is proportional to the logarithm of the radius of the hydrophobic part [1]. Thus a protein with a mol wt of 100 000 diffuses only by a factor of two more slowly than the lipids. The average time it takes for a protein (e.g. a receptor/hormone complex) to move from the end of a cell to the other is of the order of a second. In plasma membrane the components may, however, be immobilized by fixation to the cytoplasmatic network.

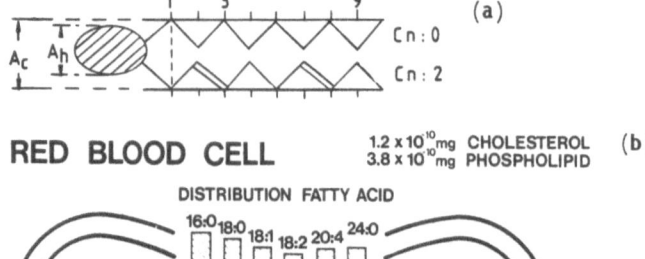

(a)

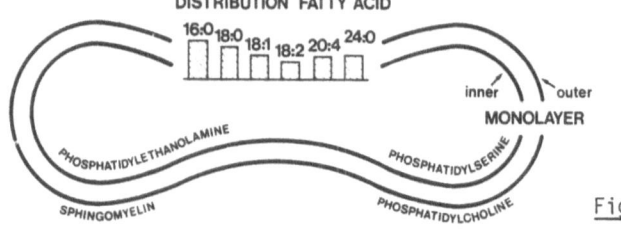

a) Phospholipids: amphiphatic molecules characterized (1) by the hydrophilic head group which may be zwitterionic, charged or simply polar and (2) by the hydrophobic chain of n carbon atoms and m C-C double bonds. Moreover, the lipids are distinguished by the relative values of the cross sections of the head (A_h) and the chain region (A_c).
b) Distribution of the four major classes of phospholipids of the plasma membrane of the red blood cell between inner and outer monolayers. 50 mole per cent of the total lipid content is cholesterol.

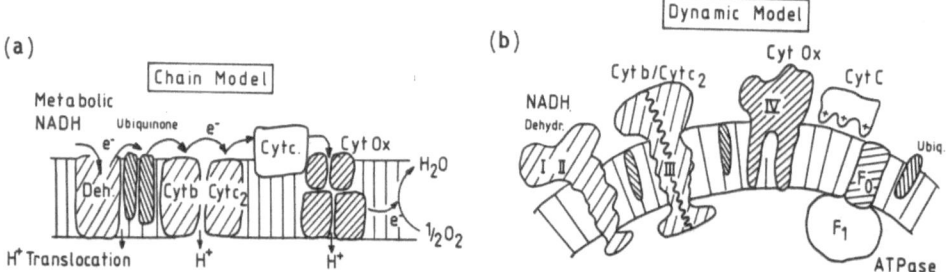

Fig. 4:
To the problem of order formation in the mitochondrial electron transfer (ET) chain. Two models are conceivable:
a) Model of ordered ET system: the enzymes involved in the transport of electrons from the end product of a metabolic pathway (stored on NADH) to O_2 form a (chainlike) complex. The electrons can move as in a conductor. Note the translocation of protons (H$^+$) from inside to outside of compartments.
b) Diffusive model: the enzymes are randomly distributed and diffuse fast within the plane of the membrane. The bulk of the mass of the integral proteins is sticking out from the bilayer into the cytoplasma. Cytochrome C is an external protein with excess positive charges interacting via Coulomb forces (for instance with the negatively charged phospholipid cardolipin). The electron transfer is a diffusion-controlled process.

3) **Membranes are two-dimensional systems**: The plasma membrane consists of three coupled layers: (1) the lipid/protein bilayer, (2) the filamentous network extending just beneath the monolayer of the cytoplasmic side of the cell-called cytoskeleton in the following (cf. Fig.2) - and (3) the gel-like network of polysaccharide and polypeptide sticking out into the external medium. The membranes of the inner organella consist only of the lipid/protein bilayer. In both cases the two-dimensionality allows for a fast exchange of molecules between the membrane and its environment (and between opposing monolayers). Small molecules exhibit exchange times of the order of 0.1 sec. The exchange of small molecules and phospholipids between the two opposing monolayers may vary between 1 and 100 000 sec.

Clearly, membranes are completely open, highly dynamic multicomponent systems. How can we then expect that ordered functional entities of some five enzymes may form within such a seemingly chaotic system? Consider the example of the electron transfer system of the inner membrane of mitochondria which consists of about seven enzymes (Fig.5). With the exception of cytochrome C these are integral proteins.

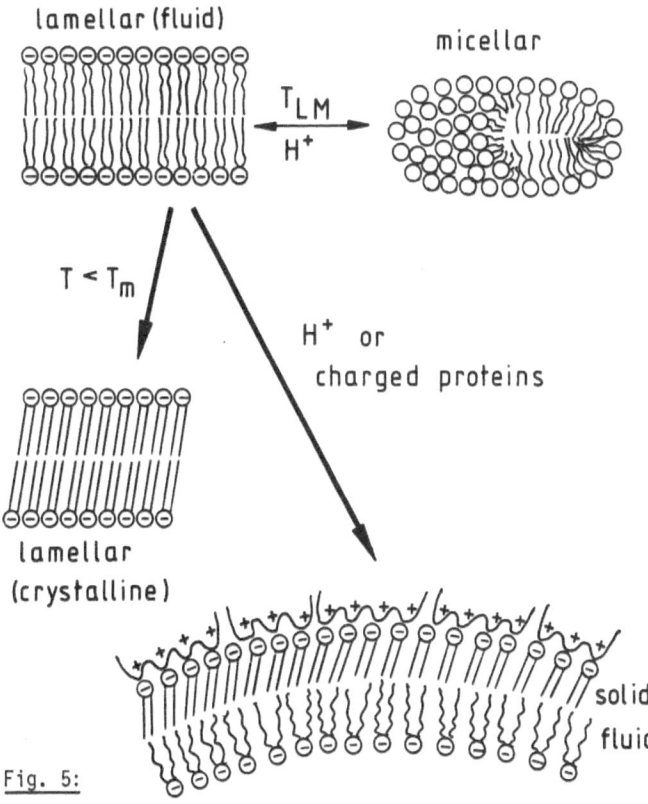

Fig. 5:

Molecular organization and structural transitions of phospholipid in water. At low lipid concentrations (0.001 M) the (isolated) lipid bilayer and elongated micelles prevail. For charged lipids, transitions between these two conformations may be induced thermally (at T_{LM}) or electrostatically (by pH or ionic strength variations). The bilayer can exist in fluid (analog: the smectic A or C) liquid crystalline states and in solid (smectic B analog) phases.

The task of this machinery is the transformation of metabolic energy into ATP, the more useful form of biological energy. This process, called oxidative phosphorylation, may be traced back to the transfer of electrons from the end product of a metabolic pathway to oxygen. The electron transfer is coupled to the translocation of protons across the membrane (to the cytoplasmic side). ATP is produced subsequently when the protons flow back in a process catalysed by another membrane-bound enzyme complex, the ATP-ase (Fig.2).

The electrons are rapidly transferred from one enzyme to the next in a defined sequence. For that reason the favorite model of the ET system is that shown in Fig.4a which assumes that the enzymes are clustered together to form a chainlike complex. One argument is that about 50% of the mass of the membrane is protein which must therefore be tightly clustered. A diffusive model is based on the finding that most of the mass of the proteins is sticking out into the cytoplasma [3] which are therefore rapidly diffusing (Fig.4b). The electron transfer is thus diffusion controlled. Nevertheless, it appears that specific complexes must be formed in order to allow for an effective electron translocation between matched enzymes. Thus the question arises how such a temporary structure formation is achieved in the highly complex membranes. Some possibilities are discussed now in view of our present knowledge about the physics of model membranes.

III   LIPIDS IN WATER FORM A VARIETY OF LIQUID CRYSTALS

Phospholipids aggregate in water at very low concentrations ($\approx 10^{-10}$M). A manifold of liquid crystalline phases are formed. Two of these, the (isolated) bilayer and the elongated micelle (Fig.5) which are formed at low lipid concentrations, are of primary interest from the point of view of membranology.

Isolated bilayers exhibit structural phase transitions from fluid (= liquid crysatalline) phases to quasi-two-dimensional crystalline phases which are associated with chain melting. A most interesting aspect is that for charged lipids the transition may be triggered by changes in the pH or by adsorption of ions or charged proteins. This enables conformational changes at constant temperature.

The bilayer micelle transition temperature depends on the ratio of the cross sections of the head and the chains. The chain-melting temperature, T_m, is a complicated function of the chain length and the lateral packing density. The latter depends on the degree of saturation and the charging state of the head group. The latter point allows for the charge-induced chain-melting transitions.

IV MEMBRANES EXHIBIT LIQUID CRYSTAL ELASTICITY

Because of the strong tendency of lipids to orient parallel any bending of a bilayer is associated with elastic energy. This corresponds to the splay elasticity of liquid crystals. The elastic energy is proportional to the square of the membrane curvature. An asymmetry in the structure of the two opposing monolayers - as is always the case in natural membranes - will cause a spontaneous curvature. Membranes are of course also elastic with respect to compression in the lateral (or normal) direction. In summary the elastic energy per unit area may be expressed as [4,5]

$$g_{el} = 1/2 \, K \, (c_1 + c_2 - c_0)^2 + 1/2 \, \varkappa^{-1} \, (\Delta A/A)^2 . \quad (1)$$

c_1 and c_2 are the principal and c_0 the spontaneous curvature, K is the splay elastic constant and $\varkappa$ the lateral compressibility. For fluid bilayers K is relatively small (K $\approx 10^{-12}$ erg) and the curvature elastic energy is of the order of kT. Membrane elasticity may be a transient phenomenon since an elastic strain may be relaxed by molecular exchange between the bilayer and the aqueous environment.

V LATERAL PHASE SEPARATION, SPONTANEOUS CURVATURE AND MEMBRANE STABILITY

As ordinary alloys (e.g.,of metals) lipid mixtures decompose into different phases under the condition that the pure components (1) form phases of different symmetry or (2) have strongly different molecular structures. In general, the phase separation occurs laterally, that is within the plane of the bilayer. A typical phase diagram of such a two-dimensional mixture is shown in Fig.6 for a case of two lipids which differ in chain length by four CH_2 groups. Note that the solid phases become miscible if the chain length difference is only two CH_2 groups.

Lateral phase separation will in general lead to an asymmetry of the bilayer and thus cause a local spontaneous curvature. The local variation in curvature may stabilize a mosaic-like organization of the components [2]. However, if the two-dimensional precipitate grows above a certain size it will be split off the bilayer as a small vesicle. Such processes play an extremely important role as transport mechanisms across membranes such as phagocytosis and endocytosis (cf. Fig.1).

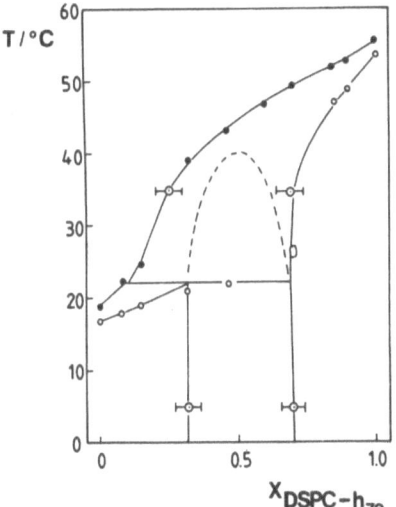

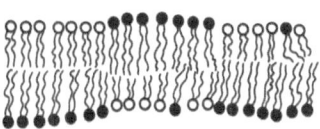

Fig. 6:

Typical phase diagram of mixture of dimyristoylphosphatidylcholine (DMPC) with 14 CH_2 groups and distearoylphosphatidylcholine (DSPC) with 18 CH_2 groups. As a consequence of the solid/solid miscibility gap the system exhibits critical demixing above the liquidus line as indicated on the right side

The transition temperatures of a number of naturally occurring lipids are well within the region of the physiological temperature. It is therefore very likely that lateral separation plays a role in biological membrane processes. Thus it appears that (1) abrupt changes in the activation energy of membrane-bound enzymes [9] and (2) in the lateral mobility observed at characteristic temperatures [6] are the consequences of such a localized segregation in fluid and rigid domains.

VI MECHANISMS OF LIPID/PROTEIN INTERACTION

Proteins may couple to lipid bilayers in a variety of ways. The major types of lipid-protein interaction are summarized in Fig.7. Extrinsic proteins (left side) can bind to membranes by Van der Waals interaction or - in the presence of charged lipid - via electrostatic forces. Integral proteins (right side) are more strongly bound due to the hydrophobic effect. However, the mechanisms mentioned in connection with the extrinsic proteins also play an important role for the interaction of the polar head groups of the integral proteins and the lipid molecules. Because of the latter, proteins with large head groups such as glycophorin - the major glycoprotein of erythrocyte plasma membranes - may interact with some hundred lipid molecules [7].

extrinsic	integral
a) electrostatic binding 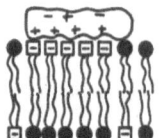 example: cytochrome C and polylysine presence of charged lipid	c) hydrophobic interaction  example: Gramicidin dimer
b) Van der Waals interaction 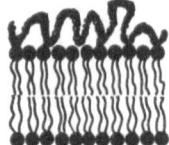 example: spectrin (adsorbed to inner monolayer of ery- throcytes)	d) hydrophobic/hydrophilic interaction 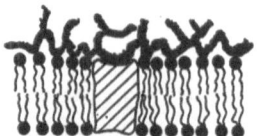 example: glycophorin and hormone receptors
	e) hydrophobic/electrostatic interaction 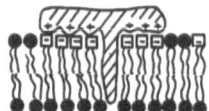 example: polymyxine in phosphatidylserine/ lecithin mixture

Fig. 7:

Types of lipid/protein interaction and possible examples:Left side: Binding mechanisms of extrinsic proteins.Right side: Mechanisms of integral proteins

VII PERTURBATION OF THE BILAYER BY PROTEIN BINDING

Fluid lipid bilayers are two-dimensional liquid crystals and the average orientation of the lipid chains may be characterized in terms of the well-known order parameter $S = 1/2 \langle 3\cos^2 \vartheta -1 \rangle$, where ϑ is the angle between the long axis of the hydrocarbon chain and the direction of average chain orientation and where $\langle \rangle$ denotes the average over all chain configurations.

Incorporation of a macromolecule will disturb the lipid chain organization in a number of ways. Some examples are shown in Fig.8. First the chain configuration of the lipids in the neighborhood of the hydrophobic core of the protein will be changed. This is a short-range effect because only the lipid in contact with the protein will be affected. Secondly, proteins with noncylindrical hydrophobic core or which penetrate only partially into the bilayer will tilt the lipid molecules as indicated in Fig.8b. Finally proteins may cause a compression (or dilatation) of the lipid bilayer if the lengths of the hydrophobic parts of the protein and of the lipids disagree. The latter two types of perturbations are long-range effects [2].

74

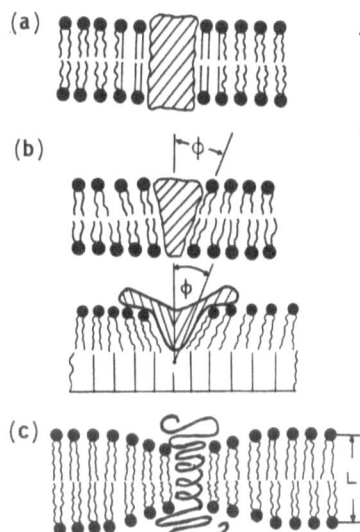

(a)

(b)

(c)

Fig. 8:

Perturbation of the lipid bilayer by the incorporation of proteins. a) Change in chain configuration or local order parameter (short-range effect). b) Splay deformation due to tilting of lipid molecules by protein. c) Compression of bilayer because hydrophobic core of protein is too short

VIII SELECTIVE LIPID-PROTEIN INTERACTION AND STRUCTURE FORMATION

In view of the complex lipid composition of membranes the question arises whether the activity of membrane-bound enzymes depends on their lipid environment. This is still a matter of controversy and is closely related to the problem of the selectivity of lipid/protein interaction. According to Fig.7 and 8 two mechanisms are conceivable: (1) electrostatic and (2) steric-elastic selectivity. Charged proteins - such as cytochrome C - which possess an excess positive charge bind preferentially to charged lipids such as phosphatidylserine or phosphatidylglycerol [2]. The second mechanism is suggested by Fig.8c. A protein penetrating the bilayer will surround itself preferentially with lipid molecules the lengths of which are best fitted to the length of its hydrophobic core. In a mixture this process will be accompanied by lipid phase separation as shown in Fig.9. The same holds if charged proteins bind to bilayers containing charged lipid components.

Clearly, a specific lipid-protein interaction together with lateral phase separation would provide a powerful mechanism for the formation of functional entities in membranes. In particular it would lead to the stabilization of concentration fluctuations in mixtures exhibiting critical demiscibility (cf. Figs.6 and 10).

IX LIPID MEDIATED PROTEIN-PROTEIN INTERACTION MECHANISMS. RANGE AND SPECIFICITY OF INTERACTION

If the incorporation of a foreign molecule into a condensed phase disturbs the structure of the latter a solvent mediated solute-solute interaction arises. A prominent example of this well-known principle of the physics of condensed matter is the Cooper pair formation in superconductors. This principle applies also to membranes. The distortion of the bilayer by the protein leads to indirect lipid mediated protein-protein interactions. Depending on the type of perturbation one may have short- and long-range forces.

1) The disturbance of the lipid order (Fig.8a) leads to a short-range force. The interaction energy decays exponentially with the distance between the proteins [2, 8].

2) The enforced inclination of the lipids (Fig.8b) and the deformation of the bilayer in the normal direction (Fig.8c) are elastic deformations and therefore

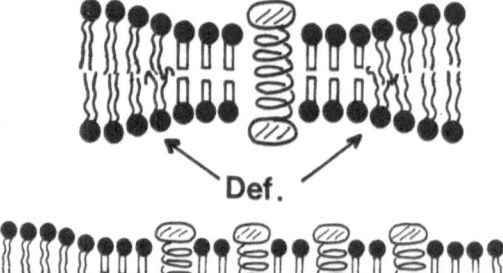

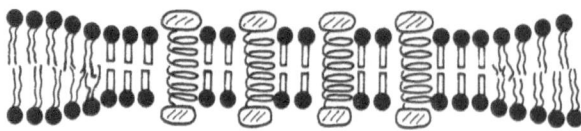

Fig. 9:

Example of protein-induced lipid segre-gation. Due to the preference of the protein for the shorter chain lipid, the elastic energy associated with the bilayer compression is reduced if the lipid/protein aggregates segregate in-to clusters

lead to long-range forces [2]. In the first case the force decays algebraically with the inter-protein distance R . The range of the forces is of the order of 10 nm .

Another type of long-range force arises in lipid mixtures exhibiting critical demixing in combination with a selective lipid/protein interaction. This is illustrated in Fig.10 for binary mixture. Because of the large correlation length of critical concentration fluctuations in two-dimensional systems, a long-range force results, which will decay slowly with the distance R between the proteins (force $\propto R^{-1/4}$).

All types of lipid mediated protein-protein interaction mechanisms discussed above can be either attractive or repulsive. In the case of Fig.8a for instance, the force is attractive if two (eventually different) proteins change the order parameter in the same direction. In the case of Fig.10 the a-b interaction would be attractive if the two proteins would prefer the same lipid. It thus follows that the lipid mediated protein-protein interaction can be selective.

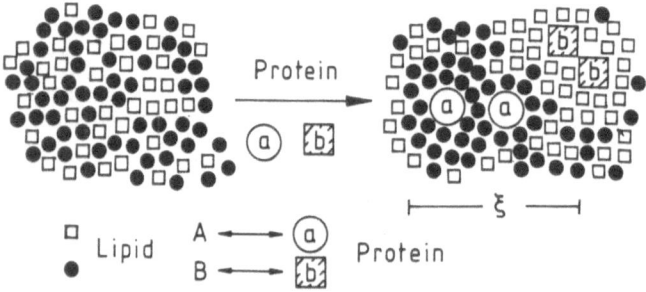

Fig. 10:
Illustration of long-range protein-protein interaction mechanism in mixture of two lipids A (▢) and B (●) exhibiting critical demixing. Two types of protein a and b are assumed to interact preferentially with lipid A and B, respectively. a and b will accumulate in clusters of lipid A and B, respectively. There results an attractive a-a and b-b and a repulsive a-b interaction

X LOCAL CURVATURE AND STABILIZATION OF STRUCTURES

Bilayer curvature could provide another possibility for structure formation. A local bilayer curvature has two interesting effects.
  1) Since each lipid molecule carries an electric dipole moment of some Debyes, a high local bilayer curvature will result in a spontaneous electric polarization [2,5] or space charge formation. The polarization leads to an additional electric membrane potential of about 10 mV.

2) Since the cross sections of the head group (A_h) and the hydrophobic part (A_c) of lipids (Fig.3a) and amphiphatic proteins are different, these molecules will redistribute within a strongly curved region of a bilayer in such a way that the curvature elastic energy is minimized.

Obviously a local curvature may cause a complete redistribution of the molecules which would simultaneously stabilize this protrusion. The spontaneous electric polarization could further lead to the adsorption of charged external protein molecules followed by an additional stabilization effect. Curvature could thus play a role for the long time stabilization of specialized regions in membranes. Indeed, specialized regions such as the "coated pits" in plasma membranes or the chromatophores of photosynthetic bacteria are located in regions of high membrane curvature.

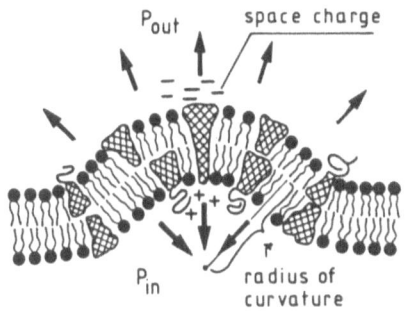

Fig. 11:

Illustration of redistribution of membrane components in region of high local curvature. Molecules with high A_h/A_c ratio for instance will accumulate in the outer monolayer at the top of the curved region and in the inner one at the transition from the flat to the curved area. The reverse holds for molecules with small A_h/A_c ratios. In addition, the strongly curved region exhibits a spontaneous electric polarization which could lead to the adsorption of external proteins

XI PROTEIN DECOUPLING BY SMALL MOLECULES: A POSSIBLE MECHANISM OF ACTION OF DRUGS AND LOCAL ANESTHETICS

As noted in section VIII the elastic distortion of bilayers may result in a protein aggregation. It is clear that the elastic strain caused by the protein could also be relaxed by the incorporation of small molecules. This holds both for the splay deformation of Fig.8b and the compression of Fig.8c. This could just be a possible effect of drugs or local anesthetics. Moreover, the elastic distortion could also lead to the accumulation of membrane-bound substrate molecules in the environment of enzymes. There is a remarkable analogy to the accumulation of solutes (Cottrell clouds) in the neighbourhood of dislocations [2]. An example illustrated in Fig.12 is the formation of pores by the aggregation of protomers. This could be prevented by clouds of small molecules formed in the environment of the proteins. The interesting aspect is that very small amounts of drug molecules

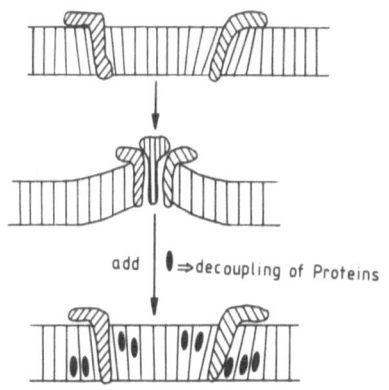

Fig. 12:

Decoupling of proteins by incorporation of small solute molecules. The case of a protein-induced tilt of the lipid is shown. Without the solute, the proteins aggregate to minimize the elastic splay energy and a pore is formed. Note that membrane escapes into third dimension. In presence of solute the splay deformation is relaxed without aggregation

can lead to large effects. A decoupling of proteins by local anesthetics has
indeed been observed in model membranes [7].

XII POSSIBLE BIOLOGICAL EXAMPLES

a) Coupling of proteins triggered by substrate binding. In smooth ER of liver
cells there exists a two-component electron transfer system composed of so-called
cytochrome P450 (a heme protein) and P450 reductase (a flavo protein) the task of
which is to activate drugs or to hydrolyse toxic substances. The reaction sequence
is (cf. Fig.13): (1) binding of substrate (S) to the P450; (2) binding of O_2; (3)
transfer of two electrons from the reductase to this complex; (4) decay of complex
in hydroxylated substrate, cytochrome P450 and H_2O. The membrane of the ER is
highly fluid and at $37^{\circ}C$ the enzymes are certainly decoupled. Recent experiments
by Greinert et al.[10] provide evidence that the interaction between P450 and the
reductase is mediated by a conformational change of the P450 upon substrate
binding. A reduction in the rotational mobility of P450 strongly suggests that the
protein penetrates more deeply into the bilayer.This could then lead to an elastic
deformation of the bilayer and a selective coupling of P450 and the reductase as
illustrated in Fig.13.

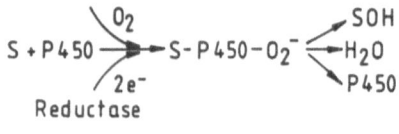

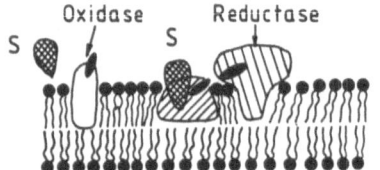

Fig. 13:

Example of selective protein-protein interaction induced by substrate binding. a) Sequence
of reactions for hydroxylation of substrate by
electron transport chain of cytochrome P450
and P450 reductase. b) Binding of substrate
causes conformational change of cytochrome
P450. Its penetration into membrane causes
attraction of P450 and reductase (after
Greinert et al. [10])

b) Variation of lipid composition may activate membrane processes. There are
many reports about changes in the lipid composition of cell membranes after
binding of hormones or of chemotactic factors [11]. Thus the ratio of
ethanolamine to phosphatidylcholine is changed. It appears that this so-called
lipid turnover is somehow involved in the amplification of the hormone signal and
in the activation of (possibly membrane-bound) enzymes the activity of which is
triggered by the c-AMP. The changes are very small, that is only about 1 per mille
of the phosphatidylcholine content is changed. Nevertheless, in combination with
local lateral phase separation, specific lipid/protein or
protein/protein interaction mechanisms, such small changes in composition could
well have large effects on the activity of enzyme complexes.

SAMPLE REFERENCES

1   Saffmann and M. Delbrück: Proc. Natl. Acad. Sci. 72,3111 (1975)

2   E. Sackmann: Biophysics (eds. Hoppe et al.) Springer Verlag 1983

3   M. Höchli and C.R. Hackenbrock: Proc.Natl. Acad.Sci.76, 1236 (1979)

4   W. Helfrich: Z. Naturf. 28C, 693 (1973)

5   A.G.Petrov, S.A.Seleznev and A.Derzhanski: Acta Phys.Pol.A55 385 (1979)

6   H.J. Kapitza and E. Sackmann: Biochim. Biophys. Acta 595, 56 (1980)

7   E. Sackmann, R. Kotulla and F.-J. Heiszler: Can. J. Biochem., to appear 1983

8   S. Marcelja: Biochim. Biophys. Acta 455, 1 (1976)

9   A. Stier and E. Sackmann: Biochim. Biophys. Acta 311, 400 (1973)

10 R. Greinert, S. Finch and A. Stier: Xenobiotica 12, 717 (1982)

11 F. Hirata and J. Axelrod: Science 209, 1082 (1980)

Generalized Moment Description of Brownian Dynamics in Biological Systems

K. Schulten, A. Brünger, W. Nadler, and Z. Schulten

Physik Department, Technische Universität München
D-8046 Garching, Fed. Rep. of Germany

1. Introduction

Many biological processes are controlled by the time that the participating bio-molecules need to diffuse around and encounter each other. Nature has devised, therefore, a variety of ways to shorten this time by guiding biomolecules into lower dimensional spaces, e.g., into the plane of membranes. Still the time spent on Brownian motion before the actual molecular reactions is exceedingly long compared to the time scale of the single diffusive displacements. Brownian transport processes also play a role in elementary biological reactions lasting as short as 10^{-12}s since even on this time scale the motion of molecules and molecular fragments in the dense biological media at physiological temperatures is of a Brownian nature. One may envisage that in such situations the fastest Brownian relaxation processes govern the reaction dynamics. This is not the case as is shown by a typical biochemical reaction proceeding along a reaction coordinate with a potential barrier. Diffusive barrier crossing is a very slow process and occurs only in the long time tail of Brownian relaxation. In recent years it has become apparent that proteins, the main carriers of biological function, exhibit an intrinsic dynamic disorder. The disorder originates from a local Brownian motion of the constituent atoms[1]. Fluctuations of the protein conformations which contribute to the protein function again occur very slowly compared to the time scale of local Brownian motion. All the processes described which last as long as 1 min and as short as 10^{-12}s require a description which accounts properly for the long time behavior of the intrinsic stochastic dynamics. In this article we will provide such a mathematical description. The mathematical derivation is based on 2-sided Padé approximants which reproduce the long-time behavior as well as the short-time behavior, i.e., the initial state. The description is applied to three sample situations: linear diffusive transport, lateral diffusion in membranes, and protein dynamics.

2. Observables

In this Section we define formally a class of observables which allow the monitoring of stochastic processes as they occur in biological systems. In the following section we will provide examples of such observables.

A particle under the influence of a potential $U(x)$ and a random force $F(t)$ and friction γ is described by the Langevin equation (for the general theory see Ref. 2,3)

$$m\ddot{x} = -\partial_x U(x) - \gamma\dot{x} + F(t) \quad . \tag{1}$$

In the case of white noise and strong friction conditions which apply in condensed biological media at physiological temperatures, after times of about 10^{-12}s the distribution $p(x,t|x_0)$ for an ensemble of biological particles initially ($t = 0$) at position x_0 satisfies the Fokker-Planck equation ($\beta = 1/kT$)

$$\partial_t p(x,t|x_0) = L(x)p(x,t|x_0) \tag{2a}$$

$$p(x,t=0|x_o) = \delta(x-x_o) \tag{2b}$$

$$L(x) = \partial_x D(x)\{\partial_x + \beta[\partial_x U(x)]\} - k(x) . \tag{2c}$$

These equations have to be complemented by two spatial boundary conditions at $x = x_i$, $i = 1,2$

$$D(x)\{\partial_x + \beta[\partial_x U(x)]\}p(x,t|x_o) = \kappa_i\, p(x,t|x_o) \tag{2d}$$

in case the diffusion space is confined to the interval $[x_1,x_2]$.[4]

In Eq.(2c) the first term describes the diffusive displacements of particles, the second term the drift in the force field $-\partial_x U(x)$, and the last term the occurrence of a possible first-order reaction with a rate constant $k(x)$. The Fokker-Planck equation (2) assumes a one-dimensional stochastic motion. The following theory is not restricted to this dimensionality.

We will consider the following class of observations: Particles are initially prepared in a distribution $v(x_o)$, most commonly a Boltzmann distribution. The particles propagate then according to Eq.(2) and are observed at time t at position x with a weight $w(x)$. The resulting observable is

$$J(t) = \int dx \int dx_o\, w(x)\, p(x,t|x_o)v(x_o) . \tag{3}$$

This can also be expressed in terms of the Fokker-Planck operator $L(x)$. It is actually most convenient to consider in the following the Laplace transform of (3) expressed by $L(x)$

$$\hat{J}(\omega) = \int dx\, w(x)\{[\omega - L(x)]^{-1}\}_b\, v(x) . \tag{4}$$

Here $\{....\}_b$ denotes that the operator .... is restricted to a function space in which all functions obey (2d).

In some instances an expression in terms of the adjoint Fokker-Planck operator $L^+(x)$

$$L^+(x) = \partial_x D(x)\partial_x - \beta[\partial_x U(x)]\partial_x - k(x) \tag{5}$$

may be preferred

$$\hat{J}(\omega) = \int dx\, v(x)\{[\omega - L^+(x)]^{-1}\}_{b^+}\, w(x) . \tag{6}$$

In this expression $\{...\}_{b^+}$ denotes that the function space is restricted to functions which obey the boundary condition adjoint to (2d)

$$D(x)\partial_x p(x,t|x_o) = \kappa_i\, p(x,t|x_o) . \tag{7}$$

The derivation of this condition involves a generalization of Green's theorem and the requirement that the concomitant of $L(x)$ and $L^+(x)$ vanishes on the boundary.

Equations (4) and (6) provide the most concise representation of the observables of interest and will be employed to construct proper approximations. $\hat{J}(\omega)$ as given by (4) and (6) are commonly referred to as correlation functions.

3. Examples

In our first example of observables of the type (4) or (6) we consider particles which diffuse freely $[U(x)\equiv 0]$ in the interval $[x_1,x_2]$. The diffusion is a prototype transport process which occurs intermediately between biochemical reactions. Of interest is the time which a biomolecule needs to arrive at its target. In our example the particles are initially positioned at $x = x_1$ and become absorbed if they reach the target positioned at $x = x_2$. The process is characterized by $L(x) = D\partial_x^2$ and by boundary conditions with $\kappa_1 = 0$ and $\kappa_2 \to \infty$. The latter corresponds to

$$p(x,t \mid x_o) = 0 \qquad \text{at } x = x_2 \qquad . \tag{8}$$

The observable of interest is the rate $r(t)$ of absorption by the target and can be expressed as the time derivative of the number of particles $N(t \mid x_o)$ not absorbed yet at time t. $N(t \mid x_o)$ is of the type (4) with $v(x) = \delta(x-x_o)$ and $w(x) \equiv 1$. The rate can be evaluated exactly by means of a spectral expansion

$$r(t) = \sum_{n=0}^{\infty} (-1)^n [(2n+1) \pi D/a^2] \exp[-(2n+1)^2 \pi^2 Dt/4a^2] . \tag{9}$$

This result furnishes a test of the approximation developed in Section 4. It should be pointed out, however, that the approximation to be developed can deal with more general situations than those considered in this example. For example, it can deal with a diffusion process with potential barriers lying between the initial position x_1 and the target position x_2.

The second example concerns an experimental method to measure the lateral diffusion in biological membranes, the "continuous fluorescence microphotolysis" method[5]. In an observation a laser beam with profile $k(r)$ is focused through a microscope to irradiate a small spot on a biological membrane with diameter of about $1\mu m$. The membrane constituents that one wishes to study, i.e.,proteins or lipids, are labeled with dye molecules which are partially damaged by the irradiation. The undamaged dyes fluoresce the incident light. The fluorescence is observed and provides information on the mobility of the dye-labeled membrane constituents. If one assumes a rotationally symmetric laser profile $k(r)$, the relevant radial Fokker-Planck operator for the distribution of the dye-labeled molecules is for flat membranes

$$L(r) = (D/r)\partial_r r \partial_r - k(r) \tag{10}$$

where the second term accounts for the photoreaction (damage) of the dyes, a process which is assumed to be first order. The observed fluorescence intensity is then described by an observable as given by Eq.(4) with $v(r) = 1$. Since the fluorescence of the dyes is proportional to the laser intensity, the weight function is given by the laser profile, i.e.,$w(r) = k(r)$, except for an overall unimportant factor. The observable can be written

$$\hat{J}(\omega) = \int d^2r \, k(r)\{[\omega - L(r)]^{-1}\}_b 1 \qquad .$$

The third example concerns the dynamics of proteins as observed through Mößbauer spectra of ^{57}Fe. Fe is a constituent of heme groups and iron-sulfur redox centers in proteins, and hence, the observation of the dynamics of this atom is of obvious interest. Mößbauer spectra result from a resonant scattering of γ quants involving a suitable metastable state of the Mößbauer atom. The spectrum entails information about the motion of the atom during the lifetime of the metastable state. In the case of ^{57}Fe the spectrum is sensitive to the atomic motion in the time window 1ns - 100ns[6].

The observable, the spectral line shape function $I(\omega)$, is given by the expression[7]

$$I(\omega) \sim \text{Re } \hat{J}(i\omega) \tag{11a}$$

where

$$\hat{J}(i\omega) = \int dx \, \exp(i\kappa x)[i\omega - L(x)]^{-1} \exp(-i\kappa x) \, p_o(x) \qquad . \tag{11b}$$

In this expression $\kappa = 7.3 \text{ Å}^{-1}$ is the momentum of the γ quantum, $p_o(x)$ the Boltzmann distribution of ^{57}Fe in a potential $U(x)$ and $L(x)$ as given in Eq.(2c) with $k(x) = \Gamma/2$ where $\Gamma = 7 \cdot 10^6 s^{-1}$ is the natural linewidth of the metastable state. Obviously the shape of the Mößbauer spectrum is related to a correlation function of the type (4).

4. Approximation

The starting point for an evaluation of the observables discussed in Section 3 are Eqs.(4) and (6). These expressions are expanded for low and high frequencies

$$\hat{J}(\omega) \underset{\omega \to 0}{\sim} \sum_{n=0}^{\infty} \mu_{-(n+1)} (-\omega)^n \tag{12a}$$

$$\hat{J}(\omega) \underset{\omega \to \infty}{\sim} - \sum_{n=0}^{\infty} \mu_n (-\frac{1}{\omega})^{n+1} \qquad . \tag{12b}$$

The expansion coefficients are in the case of (4)

$$\mu_n = (-1)^n \int dx \, w(x) \{L^n(x)\}_b \, v(x) \tag{13a}$$

and in the case of (6)

$$\mu_n = (-1)^n \int dx \, v(x) \{[L^+(x)]^n\}_{b^+} \, w(x) \qquad . \tag{13b}$$

These coefficients, the so-called generalized moments, can be constructed recursively in the order μ_0, μ_1, μ_2, ... and μ_{-1}, μ_{-2}, ... The determination of the moments with positive index is trivial. The coefficients with negative indices can be expressed by simple quadratures or evaluated numerically. For this purpose we consider the function $f_{-1}(x) = \{ [L^+(x)]^{-1} \}_{b^+} w(x)$ which is defined equivalently by

$$L^+(x) \, f_{-1}(x) = w(x) \tag{14}$$

complemented with the boundary conditions specified by $\{ \quad \}_{b^+}$. If an algorithm to obtain $f_{-1}(x)$ for arbitrary $w(x)$ exists, one can evaluate μ_{-1} in (6) and also recursively μ_{-2}, μ_{-3} ... .

The latter require the intermediate solution of

$$L^+(x) \, f_{-2}(x) = f_{-1}(x) \quad , \text{ etc.}, \tag{14'}$$

(complemented again with the proper boundary conditions) which is equivalent to (14).

The solution of Eq.(14) can be expressed by a quadrature in the case that L^+ is given by (5) with $k(x) \equiv 0$. The solution depends on the boundary condition. In case $x_1 < x_2$ and $\kappa_1 = 0$, $\kappa_2 \to \infty$ which corresponds to example 1 of Section 3, the solution is

$$f_{-1}(x) = -\int_x^{x_2} dy \, [D(y)p_0(y)]^{-1} \int_{x_1}^{y} dz \, p_0(z)w(z) \qquad . \tag{15}$$

Solutions for other boundary conditions and for moments expressed by (13a) can be constructed by means of the identities

$$L^+(x) = \exp[\beta U(x)] \, \partial_x \, \exp[-\beta U(x)]D(x) \, \partial_x \tag{16a}$$

$$L(x) = \partial_x \, \exp[-\beta U(x)]D(x) \, \partial_x \, \exp[\beta U(x)] \qquad . \tag{16b}$$

The reader may also consult Gardiner[3] who solves this problem in the framework of the mean first passage time approximation.

In the case $k(x) \neq 0$ one can solve (14) by employing a discretization scheme for the Fokker-Planck operator which results in a tridiagonal matrix for L or L^+. For the inverse of a tridiagonal matrix one can apply the well-known Gaussian elimination procedure[8] and thereby determine $f_{-1}(x)$. The computational effort of the Gaussian elimination grows only linearly with the matrix dimension and, therefore, can be applied to dimensions of a few thousand. Below a certain mesh size the moments have been found to be independent of the discretization scheme. Hence, the moments can be constructed to include any desired features of the model potential surfaces.

We will now construct an approximate observable $\hat{j}(\omega)$ which reproduces the N_ℓ leading terms of the low-frequency expansion (12a) and the N_h leading terms of the high-frequency expansion (12b) where $N_\ell + N_h = 2N$ is even. The functional form of $\hat{j}(\omega)$ should be such that the corresponding time-dependent function $j(t)$ entails a series of exponentials. This implies that $\hat{j}(\omega)$ can be expressed by the $(N+1,N)$ Padé approximants, i.e.,fractions of polynomials in ω of degree N+1 (denominator) and N (numerator). Such Padé approximants have been constructed before[9] for one-sided conditions only, namely to reproduce the leading terms of the high-frequency expansion in ω^{-1}. However, so-called 2-sided Padé approximants can be constructed[10] which reproduce a desired behavior in ω and ω^{-1}. We have derived a representation best suited for numerical applications which reproduces the desired terms of both (12a) and (12b). The approximant, later on referred to as (N_h,N_ℓ), is

where

$$\hat{j}(\omega) = \sum_{n=0}^{N-1} (\underline{b}^n \ \underline{A} \ \underline{b}^n)^{-1} (\hat{\underline{a}} \cdot \underline{b}^n)(\underline{b}^n \cdot \underline{a})[\ \omega + \lambda_n]^{-1} \tag{17}$$

$$(\underline{a}^j)_i = \mu_{-N_\ell+i+j} \qquad i=0,\ldots,N-1; \ j=0,\ldots,N$$

$$\underline{a} = \hat{\underline{a}} = \underline{a}^{N_\ell/2} \qquad\qquad (\ N_\ell \text{ even })$$

$$\underline{a} = \underline{a}^{(N_\ell-1)/2}; \ \hat{\underline{a}} = \underline{a}^{(N_\ell+1)/2} \qquad (\ N_\ell \text{ odd }) \tag{18}$$

$$\underline{A} = (\underline{a}^0,\ldots,\underline{a}^{N-1}) \qquad\qquad (\ N \times N \text{ matrix })$$

and where λ_n and b_n are the eigenvalues and eigenvectors, respectively, of the Frobenius matrix

$$\underline{F} = \begin{pmatrix} 0 & 0 & 0 & \cdots & \gamma_0 \\ 1 & 0 & 0 & \cdots & \gamma_1 \\ 0 & 1 & 0 & \cdots & \gamma_2 \\ \vdots & & & & \vdots \\ 0 & & \cdots & 1 & \gamma_{N-1} \end{pmatrix} \tag{19}$$

Here, the vector γ is the solution of

$$\underline{A} \ \underline{\gamma} = \underline{a} . \tag{20}$$

The eigenvectors may be obtained recursively by means of

$$(\underline{b}^n)_0 = \gamma_0/\lambda_n; \ (\underline{b}^n)_i = [(\underline{b}^n)_{i-1}+\gamma_i]/\lambda_n . \tag{21}$$

An equivalent representation of $\hat{j}(\omega)$ is furnished by expressing the $(N+1,N)$ Padé approximant in terms of a partial fraction expansion

$$\hat{j}(\omega) = \sum_{n=0}^{N-1} f_n/(\omega + \alpha_n) . \tag{22a}$$

The amplitudes f_n and relaxation constants α_n must obey

$$\sum_{n=0}^{N-1} f_n \alpha_n^m = \mu_m , \qquad m = -N_\ell, \ -N_\ell+1, \ \ldots\ldots, N_h-1 . \tag{22b}$$

The algebraic solution of (22) is only feasible for N = 1,2. For larger N one should solve for f_n and α_n from Eqs.(17)-(21).

The algorithm presented is closely related to the well-known moment expansion of correlation functions[11] which reproduces systematically only the high-frequency dependence (12b). The approximant is commonly provided in terms of a continued fraction expansion which, however, is equivalent to a representation by means of the Padé approximant given here. The low-frequency dependence is accounted for by a "memory kernel" for which no systematic representation exists. In our description the "memory kernel" is disposed of in favor of 2-sided conditions enforcing the correct low-frequency behavior (12a). The reason why this route had not been tried before is connected with the need for the moments μ_{-1}, μ_{-2}, ... It had probably not been realized that algorithms for these moments do exist at least for problems which are essentially one-dimensional, e.g.,three-dimensional transport with spherical symmetry. We have succeeded in generalizing the construction of the moments μ_{-1}, μ_{-2}, ... also to situations which deviate from a one-dimensional linear structure of L(x). Representing a tridiagonal L by a linear graph ————, the generalization applies also to nontridiagonal L corresponding to graphs with a finite number of structures ——————∠ and ———◯——— and reactions at the graph end points. The moments μ_{-1}, μ_{-2}, ... are related to mean first passage times as shown in the following section. One can therefore expect that the current work on mean first passage times for higher dimensional problems will contribute also to the evaluation of μ_n's for a wider class of problems.

5. Relationship to the 'Mean First Passage Time' Approximation

The mean first passage time $\tau(x_0)$ describes the mean time which particles, starting at some position $x=x_0$, need to pass the position $x=x_2$ for the first time. It is assumed that the particles are described by Eqs.(2a)-(2c). Particles which arrive at $x=x_2$ are taken out to achieve measurement of first passage. This corresponds to the boundary condition (2d) with $\kappa_2 \to \infty$, i.e.,(8). The second spatial boundary condition at $x=x_1$ is dictated by particle number conservation and requires $\kappa_1=0$ in (2d). However, there are also more general boundary conditions possible. The mean first passage time $\tau(x_0)$ is accessible through the particle number correlation function

$$N(t|x_0) = \int dx \, p(x,t|x_0) \qquad (23)$$

by means of the exponential approximation

$$N(t|x_0) \simeq N_0 \exp[-t/\tau(x_0)] . \qquad (24)$$

This approximation implies the identity

$$N_0 \tau(x_0) = \int_0^\infty dt \, N(t|x_0) . \qquad (25)$$

In order to determine $\tau(x_0)$ one starts from the adjoint equation to (2a)

$$\partial_t p(x,t|x_0) = L^+(x_0) p(x,t|x_0) . \qquad (26)$$

Equations (23), (25) and the appropriate boundary conditions yield

$$L^+(x_0) \tau(x_0) = -1 \qquad (27)$$

$$\partial_x \tau(x) = 0 \quad \text{at} \quad x=x_1 \; ; \quad \tau(x_2) = 0 .$$

This is the well-known differential equation for the mean first passage time[3,4].

In order to derive the relationship to the approximation in Section 4 we apply the algorithm derived to an observable with $v(x)=\delta(x-x_o)$, $w(x)\equiv 1$ and $N_\ell=1$, $N_h=1$. The leading terms of the low-frequency and high-frequency expansion are

$$\hat{N}(\omega|x_o) \sim N_o\int dx\,\delta(x-x_o)\,\{(L^+)^{-1}\}_{b^+}(-1) = \mu_{-1}(x_o) \qquad (28a)$$

and

$$\hat{N}(\omega|x_o) \sim N_o\int dx\,\delta(x-x_o)\,(1/\omega)\cdot 1 = N_o/\omega = \mu_o/\omega \qquad (28b)$$

where we have introduced the above definition of the moments μ_o, μ_{-1}. If we now set $f(x)=N_o\{(L^+)^{-1}\}_{b^+}(-1)$ or rather

$$L^+(x)\,f(x) = -N_o$$

applying the adjoint boundary conditions, a comparison with (27) shows $f(x)=\tau(x)N_o$. It follows $\mu_o=N_o$ and $\mu_{-1}=\tau(x_o)N_o$. If one now applies the algorithm of Section 4 one obtains according to Eq. (22b) $f_o\alpha_o^0=\mu_o$, $f_o\alpha_o^{-1}=\mu_{-1}$ and, hence, $f_o=\mu_o=N_o$, $\alpha_o^{-1}=\tau(x_o)$. Altogether, the algorithm yields

$$\hat{N}(\omega|x_o) \simeq N_o/[\omega+\tau^{-1}(x_o)]$$

or rather Eq. (24).

In Ref. 12 the mean first passage time monoexponential approximation had been generalized to a biexponential approximation reproducing correctly μ_1, μ_o, μ_{-1} and μ_{-2} of a particle number correlation function. In this article we have generalized the theory to arbitrary correlation functions and to an arbitrary number of exponential contributions.

6. Results

Figure 1 represents the observable of the first example of Section 3, the rate $\dot{N}(t)$ of absorbance of a particle diffusing freely from x_1 to x_2. The figure compares the exact rate with the approximants (N_h,N_ℓ) reproducing the N_h (N_ℓ) leading terms of the high (low)-frequency expansion. In this example the moments μ_{-1}, μ_{-2}, ... which correspond to $\dot{N}(0)$, $\ddot{N}(0)$, ... all vanish. The rate exhibits a typical threshold behaviour, i.e., the rate is completely flat and vanishes at short times and assumes nonzero values rather suddenly at later times. Such behavior is difficult to reproduce by a series of exponentials $\exp(\alpha_n t)$ and leads to spurious oscillations.

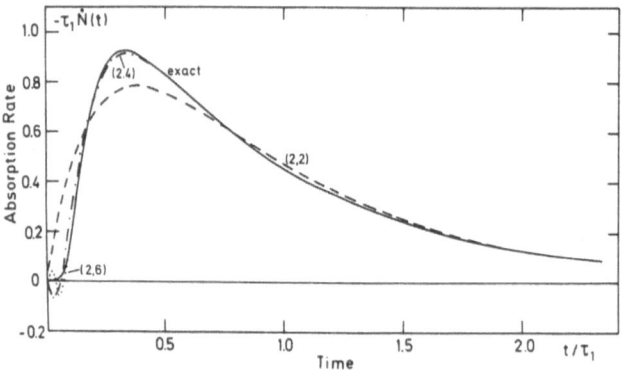

Fig. 1: Comparision of exact and approximate absorption rates for a particle diffusing linearly from x_1 to x_2 $[\tau_1 = (x_2 - x_1)^2/2D]$

Figure 1 shows that the most simple nontrivial approximation (2,2) which entails two exponential contributions describes rather correctly the overall increase, the maximum and the decay of the rate. Including more exponentials by way of (2,4) and (2,6) approximants leads to an improvement of the short-time behavior, albeit at the cost of the aforementioned spurious oscillations. The oscillations are connected with complex eigenvalues of the Frobenius matrix (19). In this respect this seemingly simple example actually constitutes a particularly difficult case, for in many other applications one can prove that oscillatory contributions do not arise.

The second example in Section 3 is concerned with the decay of fluorescence intensity as observed in a fluorescence microphotolysis experiment[5]. Figure 2 presents the fluorescence signal as observed in an actual experiment. The figure also demonstrates that the observed signal is described well by the evaluated fluorescence decay, proving thereby the validity of the theoretical description. The matching of the observed and calculated fluorescence signal yields the diffusion coefficient of the dye-labelled membrane constituents involved in the system investigated. In order to analyze the observations routinely a fast numerical procedure for the theoretical signal is desirable. The algorithm of Section 4 provides such procedure.

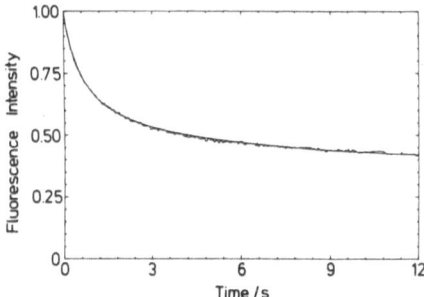

Fig. 2: Decay of the fluorescence intensity as observed in a Continuous Fluorescence Microphotolysis experiment and as calculated numerically (for details see Ref. 5)

Figure 3 compares the exact observable with the (8,0) and (0,8) approximants. The latter approximant yields by far the better description proving that the low-frequency expansion (12a) accounts for most of the fluorescence decay. Figure 3 also shows that a (2,6) approximant reproduces the fluorescence signal essentially exactly.

The third example considers the Mößbauer spectrum $I(\omega)$ of ^{57}Fe in proteins[13,14]. The accuracy of the observed spectrum is significant only in the central, low-frequency part of $I(\omega)$ and covers a time window of 1ns to 100ns. On this time scale the motion of a single atom in a protein is actually part of a concerted motion involving a larger protein fragment and, therefore, a large effective mass. One can safely describe this motion by the Fokker-Planck equation (2). In order to demonstrate the value of the algorithm developed here we consider the model of an ^{57}Fe atom diffusing in a harmonic potential.

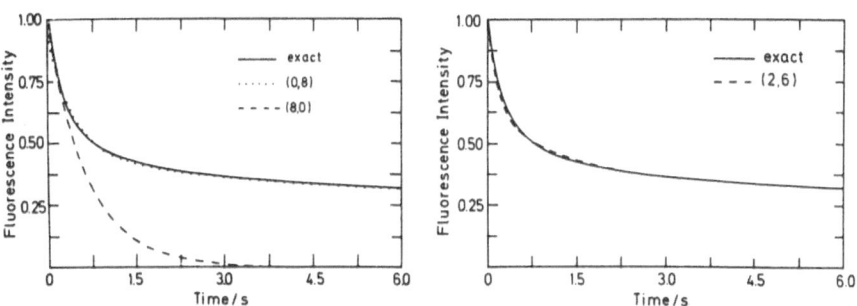

Fig. 3: Comparision of the exact and approximant fluorescence intensities

Figure 4 presents the exact spectrum obtained from the spectral expansion

$$J(\omega) = \exp(-a) \sum_{n=o}^{\infty} a^n / n! [i\omega + nD\kappa^2/a^2 + \Gamma/2]$$ (29)

where $a = \kappa^2 <x^2>_T$, $<>_T$ denoting the thermal average. This exact result is compared with various approximations, the moments reproduced in each case being indicated in Fig. 4. The comparision shows that the approximants reproducing 6 moments provide a satisfactory description of the Mößbauer spectrum. The central part of the spectrum, however, is described best if one chooses to reproduce the moments μ_{-5} - μ_0. The algorithm can be applied to essentially arbitrary potentials.

Studying a variety of potential shapes we have found that the observed Mößbauer spectra of proteins, in particular their temperature dependence, can be explained well with potentials exhibiting many minima. These minima correspond to metastable conformations of the proteins. The existence of such conformations had been previously suggested by Frauenfelder et al. The dynamics of proteins as monitored through Mößbauer spectra appears to originate from fluctuations between such metastable conformations.

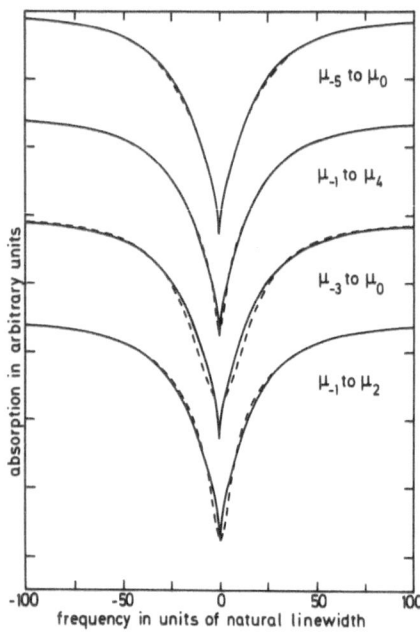

Fig. 4: Comparison of Mößbauer line shapes for Brownian motion in a harmonic oscillator ($D = 3 \cdot 10^6$ Å^2 s^{-1}, $<x^2>_T = 0.1$ Å^2) resulting from Eq.(29) (———) and from approximations reproducing the moments as indicated (----)

Acknowledgements

The authors would like to thank A. Szabo for important suggestions during the initial phase of this work. This work has been supported by the Deutsche Forschungsgemeinschaft (SFB-143 C1).

References

1. H. Frauenfelder, G.A. Petsko, and D. Tsernoglu, Nature 280, 558-563 (1979).

2. N.G. van Kampen, "Stochastic Processes in Physics and Chemistry", (North Holland Publ. Comp., Amsterdam, 1981).

3. C.W. Gardiner, "Handbook of Stochastic Methods" (Springer, Berlin, 1983).

4. A. Szabo, K. Schulten, and Z. Schulten, J. Chem. Phys. 72. 4350-4357 (1980).

5. R. Peters, A. Brünger and K. Schulten, Proc. Natl. Acad. Sci. USA 78, 962-966 (1981).

6. See for example F. Parak, E.N. Frolov, R.L. Mößbauer and V.I. Goldanskii, J. Mol. Biol. 145, 825-833 (1981).

7. This can be readily derived from Eq.(5) in K.S. Singwi and A. Sjölander, Phys. Rev. 120, 1093 (1960).

8. See for example B. Noble, "Applied Linear Algebra" (Prentice Hall, Englewood Cliffs, N.J., 1969).

9. O. Goscinski and E. Brändas, Int. J. Quant. Chem. 5, 131-156 (1971).

10. W.B. Jons, W.J. Thron and H. Waadeland, Trans. Am. Math. Soc., 261, 503-529 (1980); A. Ag. Németh and Gy. Paris, J. Math. Phys. 22, 1192-1195 (1981).

11. See for example S. Grossmann, Phys. Rev. A. 17, 1123-1132 (1978) and references therein.

12. K. Schulten, Z. Schulten and A. Szabo, J. Chem. Phys. 74, 4426-4432 (1981).

13. W. Nadler and K. Schulten, Phys. Rev. Lett. 51, 1712-1715 (1983).

14. W. Nadler and K. Schulten (to be submitted).

Nonequilibrium Current Noise Generated by Ordered Ion Transport Processes in Biological Membranes

Eckart Frehland

Fakultäten für Biologie und Physik, Universität Konstanz, Postfach 5560
D-7750 Konstanz, Fed. Rep. of Germany

1. Introduction

Electric phenomena, such as the directed (vectorial) transport of charges across
biological membranes, play an important role in a number of biological processes,
e.g. the ion transport through hydrophilic pathways (called channels or pores) is
the molecular mechanism responsible for the nerve excitation phenomena. Photosynthesis, the subject of the contribution of H.T. Witt, is also connected with the
vectorial transfer of charges across the membranes.

During the last years we have developed a theoretical concept by which generally
for those mechanisms the stochastic current *fluctuations* (current *noise*) can be
analyzed. Naturally, the interesting states to be analyzed in these systems are
nonequilibrium states. It turned out that theoretical formalisms, so far developed
mainly by physicists (e.g., [1,2]), are not applicable to these nonequilibrium *vectorial* processes.

At equilibrium the noise can be analyzed with the use of the famous Nyquist relation [3-5], or fluctuation-dissipation theorem, relating the *microscopic* fluctuations
to the *macroscopic* admittance. But we could show that generalizations of the Nyquist
relation to nonequilibrium states being proposed by several authors for *scalar* processes are not possible in the case of *vectorial* processes.

Before I shall discuss the influence of ordering mechanisms on the molecular level
on the intensity of the current fluctuations, I want to give a brief description
of the concept of transport in discrete systems and the treatment of stochastic
fluctuations.

2. Discrete Models for Ion Transport Through Biological Membranes

First I want to describe the essential ideas for a discrete modelling of ion transport through membranes. Two basic concepts are the concepts of transport mediated
by carrier molecules and of facilitated ion transport through hydrophilic pathways
(pores or channels).

In the first case the transport of the ions through the membranes is managed by
binding to a carrier protein. A most simple but very successful mechanism has been
proposed by Läuger and Stark [6] and Stark et al. [7]. It represents an important
example showing the applicability and usefulness of a discrete transport concept.
It is assumed that the single carrier molecules act independently. In Fig.1 the
state diagram for one carrier is shown.

The transport takes place in four steps: a) recombination of the ion M^+ and a neutral carrier at the left-hand interface ('), b) translocation of the complex to the
right-hand interface ("), c) dissociation of the complex and release of the ion into the solution and d) back transport of the free carrier.

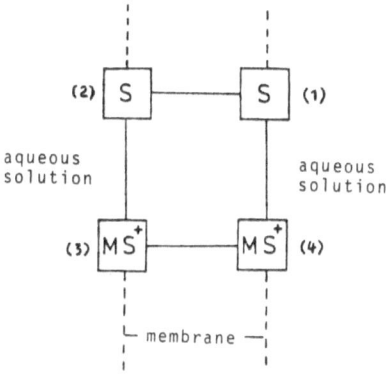

Fig.1. State diagram for the simplest model of carrier-mediated ion transport

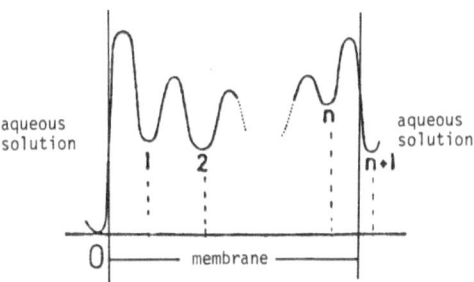

Fig.2. Potential profile in ionic pores with n-binding sites within the pore, 0 and (n+1) denoting the left-, right-hand reservoirs respectively

The second mechanism is the facilitated transport through hydrophilic pathways usually called *pores* or *channels*. The transport system consists of a membrane separating two ionic solutions and containing a great number of identical pores [8,9]. Often only one ion species may penetrate the channels. The concentrations of the ions at the pore mouths in the solutions are assumed to be held constant and act as reservoirs. The pores are considered to be a sequence of (n+1) activation barriers separated by n energy minima (binding sites) (see Fig.2). The movement of ions from binding site to binding site is jump diffusion.

More than 25 years ago the so-called single-file mechanism for ion transport through narrow channels in biological membranes had been proposed by Hodgkin and Keynes [10] to explain unexpected behavior of the unidirectional fluxes through the pores. I shall come back to this problem.

The essential property of single-file diffusion is that the movement of the ions is constrained to one dimension and the ions cannot overtake each other. The single-file mechanism includes interactions between the transported ions in so far as each binding site can be occupied by only one particle and the rate constants for jumps of ions over the energy barriers are dependent on the occupation state of the pore [11,12]. The individual channel or pore can be found in a certain number of different states depending on whether the binding sites are occupied by an ion or not. The case of a single-file pore with two binding sites is shown in Fig.3. An individual pore can be found in four different states.

00 : the empty pore, 10 , 01 : the left, right binding site occupied by an ion, 11 : both binding sites occupied.

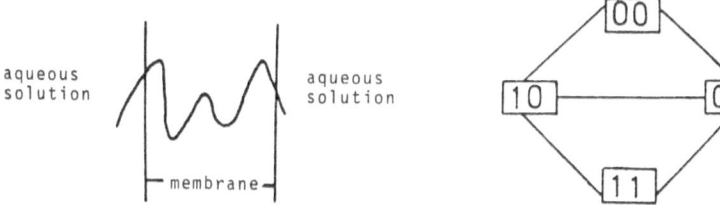

Fig.3. Potential profile and the corresponding state diagram for a rigid pore with two binding sites

Transport of one ion through the pore from left to right is given, e.g., by transitions through that closed cycle where no double occupancy occurs. Generally each transition is connected with a jump of one ion. According to Fig.3 jumps of two ions at the same time are excluded.

Ionic channels which may assume different conductance states (e.g., open and closed) play an important role for nerve excitation. Also such situations can be modelled by appropriate graphs (state diagrams) [13]. As before, the single channel in a special state is considered as a sequence of binding sites. The individual channel can be in a certain number of different states which generally depends a) on the number ion species which may penetrate the channel, b) on the number of binding sites within the channel and c) on the special structure of channel kinetics, e.g., the number of different conductance states. In order to illustrate how to get state diagrams for special models I have shown in Fig.4 the state diagrams for transport of one ion species through channels with one and two binding sites and two-state (open-closed) channel kinetics.

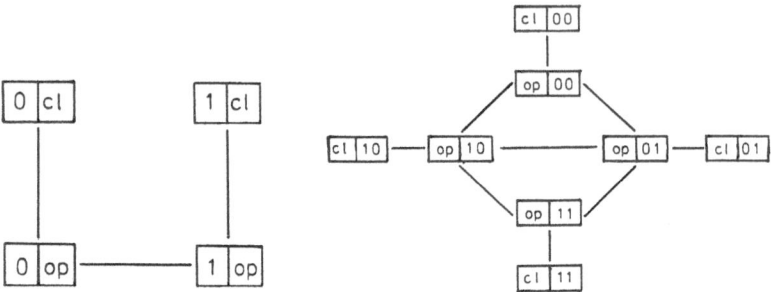

Fig.4. State diagrams for channels with one and two binding sites and two-state (open-closed) kinetics. op: open, cl: closed, 0: binding site unoccupied, 1: binding sites occupied

In the case of channels with one binding site four states are distincted. Recently, for acpical sodium channels in frog skin, when blocked by amiloride, we have proposed [14] a three-state model in order to account for new experimental current noise results of Hoshiko and van Driessche: open, sodium-occupied/open, unoccupied/amiloride-blocked. Even more complex situations can be modelled by discrete graphs, e.g., channels with more than one open conductance state [15], with the underlying idea that an ion entering a channel polarizes its surrounding thus changing the barrier structure of the channel. This is a simple example showing that memory effects can be taken into account by increasing the number of states of the system.

3. Transport in Discrete Systems

If we want to analyze the stochastic properties of transport phenomena in these systems, we have to develop a stochastic approach to the *transitions* between different states of the systems. Transport of charges is connected with changes of the states. The analysis must be done for nonequilibrium states, i.e., with a netto transport across the membranes.

For characterization of states μ of a discrete system the variable N_μ is introduced. N_μ is a step function:

$$N_\mu = \begin{cases} 1 & \text{if the system is in state } \mu \\ 0 & \text{if the system is } not \text{ in state } \mu \ . \end{cases} \tag{1}$$

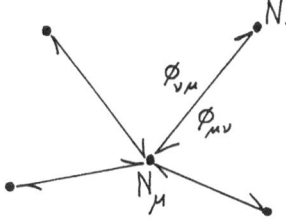

Fig.5. State variables and vectorial fluxes in discrete systems

The time derivative dN_μ/dt is a sequence of δ pulses with alternating sign determined by the transitions from states ν into state μ (positive sign) and from state μ to a state ν (negative sign). If at times $t_{i,\kappa,\rho}$ the transitions $\rho \rightarrow \kappa$ occur we can introduce the fluxes $\phi_{\kappa\rho}$ through

$$\phi_{\kappa\rho} = \sum_i \delta(t - t_{i,\kappa,\rho}) \ . \tag{2}$$

Then for dN_μ/dt holds the balance relation

$$\frac{dN_\mu}{dt} = \sum_\nu (\phi_{\mu\nu} - \phi_{\nu\mu}) \ . \tag{3}$$

We remark that in analogy to continuum systems the fluxes $\phi_{\mu\nu}$ are *vectorial* quantities, while the time derivates dN_μ/dt are *scalar* fluxes.

The transitions between different states are assumed to be governed by probalistic laws. Hence the N_μ are random variables. Introducing the probability P_μ for the system to be in state μ, we get with (1) the averaged or expectation value $<N_\mu>$

$$<N_\mu> = P_\mu \ . \tag{4}$$

And with (2) the expected flux $<\phi_{\kappa\rho}>$ is the mean transition rate, which under the assumption that the N_μ are Markovian variables is the transition probability (transition moment) $M_{\kappa\rho}$ per unit time multiplied by the probability P_ρ to find the system in state ρ:

$$<\phi_{\kappa\rho}> = M_{\kappa\rho} P_\rho \ . \tag{5}$$

With (4,5) we get by averaging the balance relation (3) as fundamental stochastic equation the master equation

$$\frac{dP_\mu}{dt} = \sum_{\substack{\nu \\ \nu \neq \mu}} (M_{\mu\nu} P_\nu - M_{\nu\mu} P_\mu) \ . \tag{6}$$

The steady state of the system, given by the steady-state solution P_μ^S of (6), is called an *equilibrium* state if the steady-state fluxes

$$<\phi_{\mu\nu}^S> = M_{\mu\nu} P_\nu^S \ , \qquad \mu \neq \nu \tag{7}$$

satisfy the detailed balance

$$<\phi_{\mu\nu}^S> = <\phi_{\nu\mu}^S> \qquad \text{for all } \mu,\nu \text{ with } \mu \neq \nu \tag{8}$$

Otherwise the steady state is a *nonequilibrium* state.

With the idea, motivated by the examples in the preceding section, that a transition $\nu \rightarrow \mu$ of the system yields a well-defined contribution to transport we introduce a *transport observable* J as a linear mapping of the fluxes:

$$J = \sum_{\substack{\mu,\nu \\ \mu \neq \nu}} \gamma_{\mu\nu} \phi_{\mu\nu} \quad . \tag{9}$$

In the following J will be the electric current.

As an example we consider the ion carrier (Fig.1). Only the transitions of the charged carrier-ion complex MS^+ of charge q between states 3 and 4 contribute to J. Hence

$$\gamma_{43} = -\gamma_{34} = q, \ \gamma_{\mu\nu} = 0 \ \text{otherwise} \quad . \tag{10}$$

4. Transport Noise in Discrete Systems

The analysis of steady-state current noise is usually done with the autocorrelation function or the spectral density of the fluctuations. For derivation of general expressions for these quantities within the described concept of transport we first studied the time correlations between the fluxes [16,17]:

$$<\phi_{\mu\nu}(0)\phi_{\kappa\rho}(t)> \ = \ <\phi_{\mu\nu}^S> \{\delta_{\mu\nu,\kappa\rho}\delta(t) + M_{\kappa\rho}\Omega_{\rho\mu}(t) + <\phi_{\kappa\rho}^S>\}$$

$$\delta_{\mu\nu,\kappa\rho} = \begin{cases} 1 & \text{for } \mu,\nu = \kappa,\rho \\ 0 & \text{else} \end{cases}$$

$$\Omega_{\rho\mu}(t) = (P_\rho - P_\rho^S)|_{P_\rho(t=0)=\delta_{\rho\mu}} \quad . \tag{11}$$

From (11), because current J is linearly dependent on the fluxes, follows the autocorrelation function $C_{\Delta J}(t)$

$$C_{\Delta J}(t) = \sum_{\substack{\mu,\nu,\kappa,\rho \\ \mu \neq \nu, \kappa \neq \rho}} \gamma_{\mu\nu}\gamma_{\kappa\rho}\{<\phi_{\mu\nu}(0)\phi_{\kappa\rho}(t)> - <\phi_{\mu\nu}^S><\phi_{\kappa\rho}^S>\} \quad . \tag{12}$$

According to the Wiener-Khintchine relations [18] one gets from (12) by Fourier transformation the spectral density $G_{\Delta J}$ as a function of circular frequency ω

$$G_{\Delta J}(\omega) = 2\sum_{\substack{\mu,\nu \\ \mu \neq \nu}} \gamma_{\mu\nu}^2 <\phi_{\mu\nu}^S> + 4\sum_{\substack{\mu,\nu,\kappa,\rho \\ \mu \neq \nu, \kappa \neq \rho}} \gamma_{\mu\nu}\gamma_{\kappa\rho} <\phi_{\mu\nu}^S>M_{\kappa\rho}\int_0^\infty \Omega_{\rho\mu}(t)\cos\omega t \ dt \quad . \tag{13}$$

With the expressions (11-13) we are in a position to analyze the nonequilibrium fluctuations of vectorial processes in discrete systems. The fluctuations of scalar quantities based on the N_μ have been treated with the so-called master equation approach [1,2,19]. While for scalar quantities possible generalizations of the fluctuation-dissipation theorem (Nyquist relation) to nonequilibrium steady states have been proposed (e.g., [1,19,20]), such generalizations are not possible for vectorial quantities [17,21].

We emphasize that via the balance relation (3) the fluctuations of scalar fluxes can be derived as a special case of our general approach to vectorial processes from the time correlations (11). Indeed we recently could explicitly show [21] that the resulting expressions agree with the results from the master equation approach.

5. Reduction of Noise as Consequence of Ordering

As described in Sect.2, in the case of single file diffusion the ionic motion is not free, because the ions cannot overtake each other within the pores and each 'place' (ionic binding site) can be occupied by only one ion. Obviously these ionic interactions lead to a mutual ordering of the movement of the ions: we expect a positive flux coupling. Indeed, historically the single-file concept has been proposed by Hodgkin and Keynes [10], who observed a positive flux coupling for K^+ transport in deviation from the predicted behavior under the assumption of free diffusion.

In the case of free movement the unidirectional fluxes $\overrightarrow{\phi}$ and $\overleftarrow{\phi}$ (measurable by tracers) of an ion obey the so-called Ussing criterion

$$\frac{\overleftarrow{\phi}}{\overrightarrow{\phi}} = \frac{[S]r}{[S]l} \exp\left(\frac{qv}{k_BT}\right) \tag{14}$$

$[S]r$, $[S]l$: ionic concentrations on the right, left reservoirs respectively, q: charge of the ion, V: applied voltage, k_B: Boltzmann constant, T: absolute temperature. Hodgkin and Keynes found that the Ussing criterion is *not* obeyed by the K^+ fluxes through nerve axon membranes. They observed

$$\frac{\overleftarrow{\phi}}{\overrightarrow{\phi}} = \frac{[S]r}{[S]l} \exp\left(\frac{qv}{k_BT}\right)^{n'} \tag{15}$$

with $n' = 2.5$. Hence for applied voltage V there is a positive flux coupling, i.e., an ordering of ionic motion. For explanation Hodgkin and Keynes proposed the single-file mechanism of transport and argued that the exponent n' is a measure for the number of ions (multiple occupancy) in the pores. The steady-state theory of single-file transport has been extensively developed by Heckmann [23] and by Hille and Schwarz [11]. More recently W. Stephan and I have discussed the time-dependent properties of single-file transport in rigid pores [12], which are oscillatory as consequence of the ionic interactions, and have investigated (theoretically) the resulting current noise [24].

Because the single-file mechanism leads to an ordering of transport on the microscopic level, this should have consequences for the stochastic current fluctuations: we expect them to be reduced compared with fluctuations generated by free-diffusion mechanisms. There is a limited situation where a comparison with current fluctuations from free ionic movement (no interactions) is especially simple. In the case of high applied voltage, where ionic motion is formed to take place only in one direction, for low frequencies the current noise spectrum is determined by the discrete nature of current carriers and is usually called 'shot noise' [25]:

$$G_{\Delta J}(\omega \to 0) = 2qJ^s \quad . \tag{16}$$

The essential condition for derivation of the shot noise relation is that the current pulses generated by the single charges obey Poisson statistics, i.e., are mutually independent. In case interactions essentially influence the transport process this relation can no longer be expected to be valid. We have applied the described approach for investigating this problem. First we regard pores with regular barrier structure where the pores are to be occupied by not more than one ion (one ion case). The corresponding state diagram for a single pore with n binding sites is shown in Fig.6. From (13) the current noise intensity is calculated to be [17]

$$G_{\Delta J} = \frac{1}{n+1} 2qJ^s \quad . \tag{17}$$

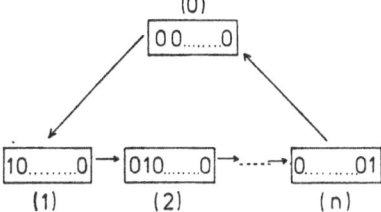

Fig.6. One-ion state diagram for pores with n binding sites in the case of high applied voltage

95

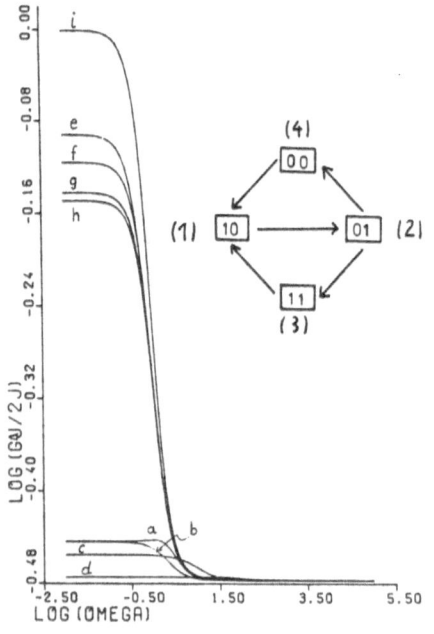

Fig.7. Single-file current noise for high applied voltage in channels with two binding sites. All $\gamma_{\mu\nu}$ are set equal to 1/3. The rate constants $M_{\mu\nu}$ ($\nu \neq \mu$) for the different curves are zero with the exceptions

a) $M_{14} = 1$, $M_{21} = 1$, $M_{42} = 1$, $M_{13} = 1$, $M_{32} = 1$; b) as a) except $M_{32} = 1/2$, $M_{13} = 2$;
c) as a) except $M_{32} = 1/10$, $M_{13} = 10$; d) as a) except $M_{32} = 1/100$, $M_{13} = 100$ (one-ion case); e) $M_{14} = 1/10$, $M_{32} = 1/10$, $M_{13} = M_{21} = M_{42} = 1$; f) as e) except $M_{13} = 2$, $M_{32} = 1/20$;
g) as e) except $M_{13} = 10$, $M_{32} = 1/100$; h) as e) except $M_{13} = 100$, $M_{32} = 1/1000$;
i) as h) except $M_{14} = 1/1000$ (no-interaction case)

The result clearly shows that even in the simple one-ion case the spectral density is below the shot noise level by the factor $1/n + 1$. For one binding site it is the half of the shot noise intensity. With increasing number of binding sites it is further reduced.

In Fig.7 are presented numerical calculations of current noise in the complete single-file model for pores with two binding sites. As to be seen from the legend the degree of ionic repulsion is varied by variation of the corresponding transition rate constants. The noninteraction case is also approximated in the examples. By division through the shot noise intensity the spectral density is normalized. This makes possible a sensible comparison of noise intensities. In the noninteraction cases the low-frequency limit approximates 1, i.e., shot noise. A reduction of noise is to be seen in those cases where ionic interactions are regarded. Corresponding numerical calculations for pores with three binding sites have been done with similar results. The single ion carrier acts as well as an ordering mechanism generating a controlled movement of material across the membrane. As long as the carrier-ion complex exists no further ion can be bound and transported across the membrane. The situation is similar to pores with one binding site. Indeed, theory predicts that in the high field limit, the low-frequency noise is below the shot noise level. In Fig.8 showing the measured current noise results for a voltage of 100 mV, I have marked the shot noise level, which indeed is above the measured low-frequency noise level by a factor of nearly 2. As far as I know this is the first experimental evidence for a reduction of current noise by an ordering process on the molecular level.

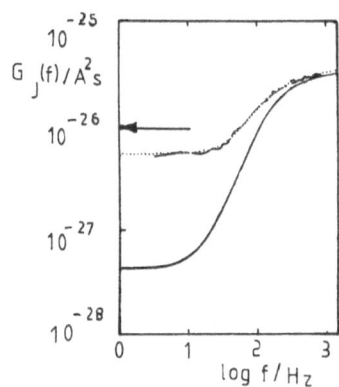

Fig.8. Spectral density of current noise from a lipid bilayer membrane in the presence of 10^{-7} M tetranactin at an applied voltage of 100 mV. The dotted line is a fit to the experimental curve with the use of our theoretical approach to current noise. The lower curve is 4 k_BT Re $Y(\omega)$, which would be predicted as spectral density, if the fluctuation dissipation theorem were valid also at nonequilibrium. The experiments have been carried out by H.A. Kolb [26]

←——— shot noise level

We are planning to investigate further and to search for experimental confirmation of these predictions, which are occurring in the low-frequency region and should be measurable also for pores, though the transport processes in pores are extremely fast.

6. Conclusions

The reception, processing and transport of information in biology is connected with vectorial transport of electric charges in biological membranes. The same is true, e.g., for the molecular process of conversion of solar energy (photosynthesis). For optimization of such processes it should have been an advantage to reduce the noise. One can speculate if and how biological nonequilibrium transport processes (systems) during evolution have minimized the size of stochastic fluctuations connected with transport.

The experimental measurement of low-frequency current noise intensities is a possibility for obtaining information about the transport parameters from noise analysis though the characteristic transport times are too fast and the corresponding dispersive frequency regions too high to be directly measurable.

With very simple examples for vectorial ion transport processes at nonequilibrium I have shown some effects and consequences of interaction of subunits (ion) on the molecular level on the fluctuation properties. The presented examples are still simple and we are at the very beginning of analyzing more complex systems. But we believe that it might be a fruitful way to characterize synergetic effects in vectorial nonequilibrium processes with the use of their fluctuation properties, using quantities as 'autocorrelation' and 'spectral density' which are available to experimental measurement.

7. Acknowledgments

I wish to thank the Deutsche Forschungsgemeinschaft for financial support (Heisenberg fellowship) and The Institute for Advanced Study Berlin for its great hospitality during my one year stay as a fellow.

References

1. M. Lax: Rev. Mod. Phys. **32**, 25 (1960)
2. K.M. van Vliet, J.R. Fassett: In *Fluctuation Phenomena in Solids*, ed. by R. Burgess (Academic, New York 1965)
3. H. Nyquist: Phys. Rev. **32**, 110 (1928)

4. H.B. Callen, T.A. Welton: Phys. Rev. **83**, 34 (1951)
5. H.B. Callen, R.F. Greene: Phys. Rev. **86**, 702 (1952)
6. P. Läuger, G. Stark: Biochim. Biophys. Acta **211**, 458 (1970)
7. G. Stark, B. Ketterer, R. Benz, P. Läuger: Biophys. J. **11**, 981 (1971)
8. P. Läuger: In *Membrane Transport Processes*, Vol.3, ed. by C.F. Stevens and R.W. Tsien (Raven, New York 1979)
9. B. Hille: In *Membrane Transport Processes*, Vol.3, ed. by C.F. Stevens and R.W. Tsien (Raven, New York 1979)
10. A.L. Hodgkin, R.D. Keynes: J. Physiol. **128**, 61 (1955)
11. B. Hille, W. Schwarz: J. Gen. Physiol. **72**, 409 (1978)
12. E. Frehland, W. Stephan: J. theor. Biol. (1983, in press);
    W. Stephan, E. Frehland: J. theor. Biol. (1983, in press);
    W. Stephan, B. Kleutsch, E. Frehland: J. theor. Biol. (1983, in press)
13. E. Frehland: Biophys. Struct. Mech. **5**, 91 (1979)
14. E. Frehland, T. Hoshiko, S. Machlup: Biochim. Biophys. Acta (1983, in press)
15. P. Läuger, W. Stephan, E. Frehland: Biochim. Biophys. Acta **602** (1980)
16. E. Frehland: Biophys. Chem. **8**, 255 (1978); **10**, 128 (1979)
17. E. Frehland: Stochastic Transport Processes in Discrete Biological Systems. Lecture Notes in Biomathematics, Vol.47 (Berlin, Heidelberg, New York 1982)
18. N. Wiener: Acta Math. **55**, 117 (1930)
19. V.D. Chen: Advances in Chemical Physics **37**, 67 (1978)
20. F. Jähnig, P.H. Richter: J. Chem. Phys. **64**, 4645 (1976)
21. E. Frehland: To be published
22. E. Frehland: Biophys. Chem. **12**, 63 (1980)
23. K. Heckmann: Biomembranes 3, 127 (1972)
24. E. Frehland, W. Stephan: Biochem. Biophys. Acta **553**, 326 (1979)
25. W. Schottky: Ann. Physik **57**, 541 (1918)
26. H.A. Kolb, E. Frehland: Biophys. Chem. **12**, 21 (1980)

The Molecular Machine of Photosynthesis – Physico-Chemical Aspects

H.T. Witt

Max-Volmer-Institut für Biophysikalische und Physikalische Chemie
Technische Universität Berlin, Strasse des 17. Juni 135
D-1000 Berlin 12, Fed. Rep. of Germany

A biological molecule is a molecule with the best possible structure
for its biological function. In this sense, chlorophyll is the best
molecule for photosynthesis, rhodopsin the best for vision, cytochromes the best for respiration, etc. However, it is evident that
such a biological molecule can achieve its function only when it is
incorporated in a highly specific organization. For instance, chlorophyll per se behaves with respect to its possible reaction sequences,
as far as we know, like any other dye molecule. However, when chlorophyll is embedded in a specific organization, excited chlorophyll
creates reaction patterns of extraordinary attributes, specificities
and efficiencies. Several different biological molecules use for
various functions a common principle of organization, the vesicle.
A vesicle consists of a closed membrane of lipoprotein. Some of its
characteristics can be recognized when we discuss the presumable
pathway of its evolution (see Fig. 1).

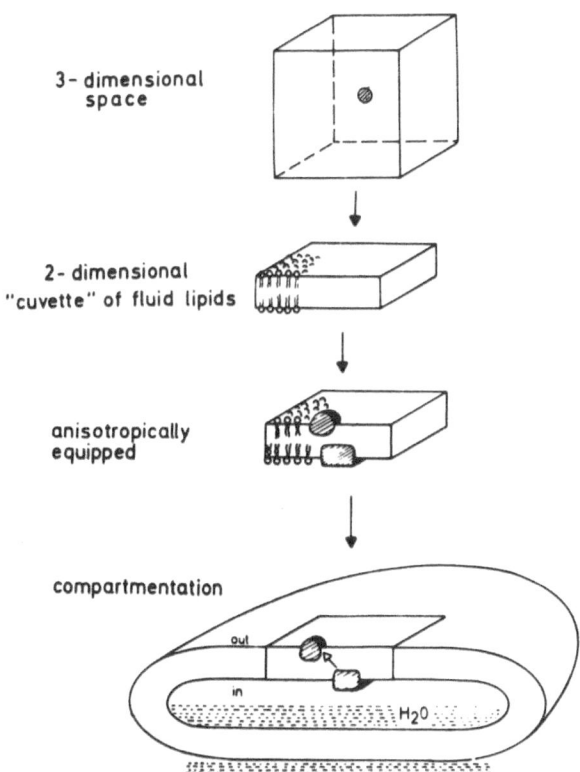

3- dimensional
space

2- dimensional
"cuvette" of fluid lipids

anisotropically
equipped

compartmentation

out

in

H_2O

Fig. 1 Development of compartmentation

Billions of years ago chemistry occurred in a 3-dimensional space in
a gaseous atmosphere and later in water, respectively. In a 3-
dimensional space the reactants are relatively diluted, intermediates,
e.g., triplet states, with long lifetimes are useful and the reactions
can be relatively slow. Later, with the "intervention" of membranes
that consist of fluid bilayers of lipids a "2-dimensional-like reaction vessel" was made available for chemistry. This has the advantage that the reactive substances can be much more concentrated.
Intermediates, e.g., singlet states, with short lifetimes are useful
and the reactions can be relatively fast. Furthermore, the membranes
can be equipped anisotropically with biological molecules in such a
way that the excited biological molecules can react only in one
specific way. Thereby, excited biological molecules are prevented
from wasting their energy in meaningless side reactions. For instance,
an excited dye molecule can react in 3-dimensional space by many
different pathways. The excited state can be used for photooxida-
tions, dissociation, protolytic reactions, rearrangements, fluorescence, phosphorescence, etc. By the specific arrangement in a lipid
membrane, excited chlorophyll can be forced, however, to react with
high efficiency in only one way so that, for example, it is only
photooxidized. This is the paradoxical reason why in some respects
studies of the photochemistry of chlorophyll in vivo are much simpler
than its analysis in vitro. In a further step of evolution,
membranes were used for compartmentation to realize a closed vesicle.
Under these conditions of an interior and exterior system, the ani-
sotropical arrangement in the membrane can induce vectorial reactions in the inner space. In this way, the system can have the
additional function of a storage device, and new types of reactions
are possible. Such "interior-exterior" systems are used as an
operational board for biological molecules in photosynthesis, vision,
respiration, etc. In photosynthesis the vesicles are called
thylakoids.

In the overall reaction of photosynthesis, absorbed light performs
a transfer of an electron from H_2O to the terminal electron
acceptor, $NADP^+$. The electron transport is coupled with the formation of ATP from ADP + P. With reduced NADPH and ATP, absorbed CO_2
can be reduced to sugar and other energy-rich organic compounds
(Calvin). About 200 of such electron transport chains are embedded
in one thylakoid membrane. Each chain is surrounded by about 500
light-energy conducting pigment molecules called antennae pigments
(chlorophyll-a and -b and carotenoids).

In photosynthesis the molecular machinery must be adaptable to dim
light as well as to bright light. At low intensity the very few
quanta nevertheless excite the small number of reaction centers with
a high probability because singlet energy transfer and migration in
the antennae pigments channel the quanta via the singlet states from
Car to Chl and from Chl to Chl (Emerson, Arnold, Gaffron, Wohl).
In this way, also the photoactive Chl centers are "visited"; whereby the energy is trapped (see Fig. 2, top). At high intensities
superfluous energy stays in the excited Chl states. The energy
must be dissipated quickly, because it is known that excited Chl
is irreversibly destroyed by photooxidation with the omnipresent O_2.
Energy dissipation may occur by reemission as fluorescence or by nonradiative transitions, but these two ways cannot be utilized
efficiently. An additional "valve reaction" is energy transfer
and migration after intersystem crossing via triplet states in the
opposite direction from Chl to Car (Wolff and Witt)(see Fig. 2).

Two different pigment systems were tentatively postulated by Emerson
in 1958 from the O_2 evolution as a function of the wavelength of

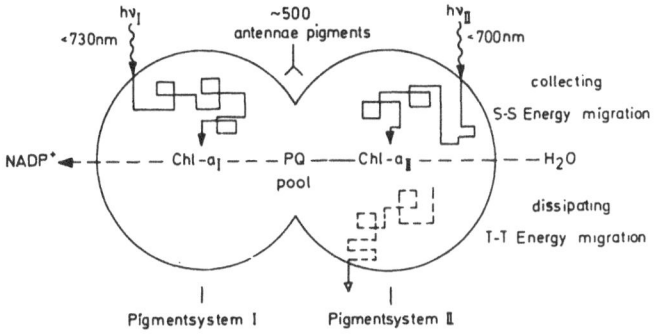

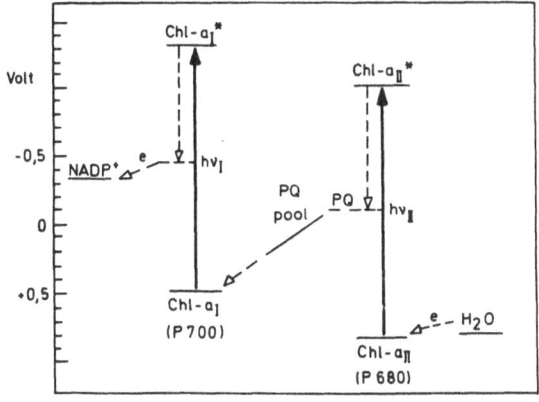

Fig. 2 Top: Electron transfer within an electron transport chain; bottom: energy
diagram

light, a system excitable at long wavelengths (< 730 nm) and a second
system excitable at shorter wavelengths (< 700 nm). Hill and Bendall
postulated in 1960 that the two suggested pigment systems may be
coupled in series and that the coupling is effected by cytochromes.
First experimental evidence for a coupling of two systems was given
independently by three spectroscopic phenomena observed by Kok,
Duysens and Witt and their co-workers. In all three experiments it
was shown that an intermediate reaction of photosynthesis, followed
by absorption changes, is "switched" from one direction to the
opposite course of reaction when photosynthesis is triggered alterna-
tively with < 700 nm and 730 nm light.

The photoactive molecules within the two coupled systems have been
observed directly by absorption changes. The spectrum of the absorption change of the center in system I has been discovered by
Kok and indicates the photooxidation of a special chlorophyll-a
called $Chl-a_I$ or P700. The difference spectrum of the center in
system II was observed by Döring, Witt and co-workers and indicates
the photooxidation of another special chlorophyll-a called $Chl-a_{II}$
or P680. As a link between both chlorophylls, a pool of plastoquinone (5 molecules) has been identified (Stiehl and Witt).

On the basis of the redox potentials of the components there results
an energy diagram which is depicted in Fig. 2, bottom. Excited
$Chl-a_{II}$ lifts the electron from H_2O up to PQ. Excited $Chl-a_I$ lifts
the electron a second time to $NADP^+$. It is obvious that at $Chl-a_I$
and $Chl-a_{II}$ much more energy is absorbed than is used for the electron transfer. It was shown that part of this energy is used for a
vectorial electron transfer which needs additional energy for charging the membrane (see Fig. 3).

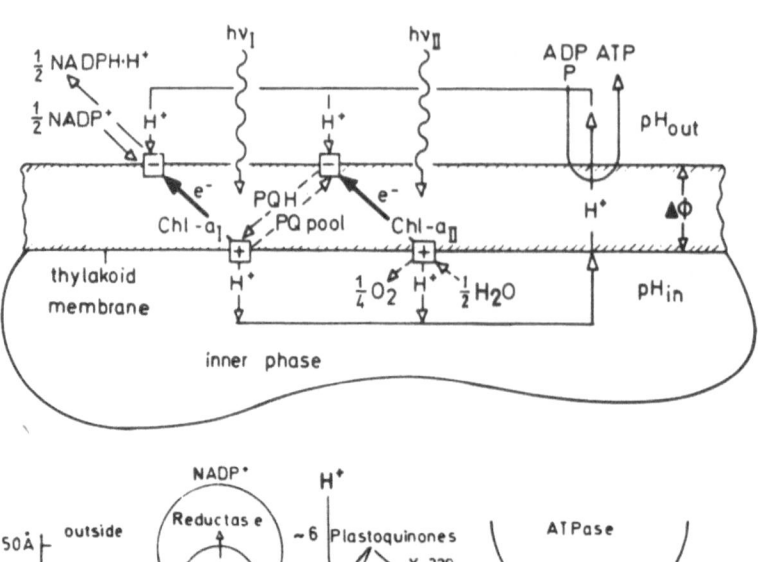

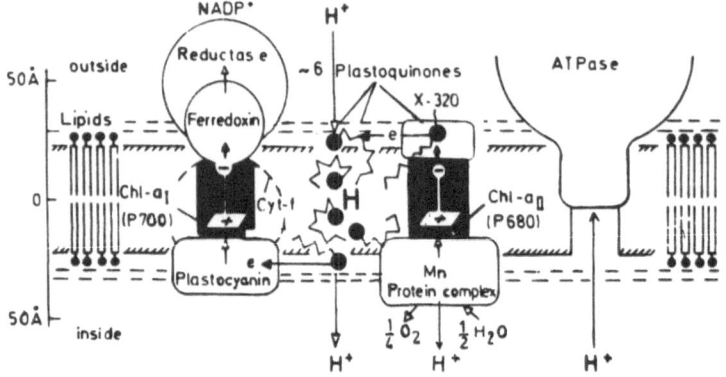

Fig. 3   Top: vectorial pathways within the electron
         transport chain; Bottom: preliminary topo-
         graphy of the molecular machine

A vectorial electron displacement must create an electric field
across the membrane. This has been measured as follows (Junge and
Witt). If during photosynthesis a potential of say 100 mV is set
up across the thylakoid membrane (about 50 Å thickness), this
corresponds to 10^5 V/cm. In such high fields the absorption bands
of all pigments should be shifted by the order of one Å, inducing
field-indicating absorption changes (Stark effect). This technique
has been used for the analysis of the electrification of the membrane. The extent of the changes indicated the transmembrane voltage (~ 100 mV) and the slope of the time of course the changes of the
transmembrane current. The polarity of the voltage indicates the
direction of the electron shift and the location of the electron
acceptors and donors, respectively. It was shown that the voltage
setup is caused by a shift of one electron across the membrane at
$Chl-a_I$ as well as at $Chl-a_{II}$ from the inside of the membrane to the
outside. The decay of the absorption changes and current, resp., is
caused by H^+ efflux from the inner to the outer space of the thylakoid
membrane.

The first stable electron acceptor of $Chl-a_{II}$ is a special plastoquinone, X-320 (Stiehl and Witt). A transient intermediate between
$Chl-a_{II}$ and X-320 is probably a pheophytin molecule (Klimov). With

the acceptance of an electron by PQ, a proton uptake is coupled. This is also true for the ultimate electron acceptor of $Chl-a_I$ if instead of $NADP^+$, e.g., the artificial acceptor benzylviologen is used. The positive hole produced by $Chl-a_{II}$ at the inside of the membrane (on the unknown electron donor of $Chl-a_{II}$) is finally reduced by H_2O releasing $1/4 \ O_2$ and $1 \ H^+$ to the inside. The positive hole produced by $Chl-a_I$ at the inside of the membrane (on plasto-cyanine) is reduced by reduced quinone releasing also one H^+ to the inside. In this way, in a single turnover, at each light reaction one proton should be translocated into the inner space of the thyla-koid. This type of H^+ translocation was outlined in a hypothesis by Mitchell. This proton translocation has been proven by measure-ments of absorption changes of added artificial pH indicators. The results indicate that when in a single turnover flash one electron is transferred through the electron chain, an uptake of $2 \ H^+$ at the outside as well as a release of $2 \ H^+$ at the inside is observed (Junge, Ausländer, Gräber, Witt).

In continuous light, large amounts of H^+ ions are translocated into the inner thylakoid space by the extrusion of an equivalent number of counter ions (e.g., K^+). Thereby, the pH_{in} value decreases and electrical energy is in part transformed into a pH gradient. A maximum pH gradient of ~ 3.0 was observed across the membrane (Jagen-dorf, Rumberg).

The ultimate donor of $Chl-a_{II}^+$ is H_2O. With the oxidation of H_2O, liberation of O_2 and H^+ takes place. The oxidation is catalyzed by an enzyme system, S. It has been shown by Kok and Joliot that S has to accumulate stepwise four oxidizing equivalents ($S_0 - S_4$) by four turnovers of $Chl-a_{II}$ before two H_2O molecules are split into one O_2, four electrons and four protons. The nature of the S states is un-known except for the involvement of 2-4 manganese molecules. The individual donation times of the four electrons out of the four different S states have been analyzed down to the nanosecond range (Brettel, Schlodder, Witt).

One can draw conclusions as to the topography of the membrane from the functional molecular concept developed above (see Fig. 3). For example, the vectorial transmembrane translocation of an electron + H^+ (i.e., H translocation) via the plastoquinone pool makes it necess-ary for the pool to be located from the outside through the membrane to the inside. Furthermore, the observed photoinduced negative charging of the membrane at the outside and positive charging at the inside, at both light reaction centers, and the fact that both $Chl-a_I$ and $Chl-a_{II}$ are photooxidized simultaneously, lead to the conclusion that the porphyrin rings of both photoactive chlorophylls must be located at the inside of the membrane and the primary electron accep-tors, ferredoxin and a special plastoquinone, X-320, at the outside. The cleavage of H_2O was established to take place at the membrane in-side. A preliminary model of the topography is depicted in Fig. 4. The indicated ATPase is the center at which proton efflux and phos-phorylation are coupled.

According to a hypothesis of Mitchell, the energy that is released during the discharging of the electrically energized membrane and the breakdown of the transmembrane pH gradient by the efflux of protons should be coupled with the generation of ATP. Such a model has been proven in detail, especially by the observed correlation between the electric events (voltage, current) and ionic events (H^+, pH) on one side and the ATP formation (kinetics and amount) on the other side (Gräber, Schlodder, Jagendorf, Rumberg, Witt, et al.). It was also possible to demonstrate that the electrification of the thylakoids with an external electric field, i.e., in the dark without any light, leads to ATP formation with practically the same amount as in light-induced thylakoids (Witt and Schlodder, et al.).

The extraordinary specificity of the reactions in photosynthesis outlined above is characterized by: (1) the regulation of the utilization of quanta via two types of energy migration; (2) the cooperation of two photoreaction centers, $Chl-a_I$ and $Chl-a_{II}$, for the transfer of one electron from H_2O to $NADP^+$; (3) the vectorial electron ejection out of the excited state of $Chl-a_I$ and $Chl-a_{II}$ and electric field generation, resp.; (4) the subsequent protolytic reactions, H^+ circulation and pumping of H; (5) the use of a proton potential as energy source for ATP formation, and (6) the use of H_2O as electron source.

Each of these events per se is a unique "invention" by nature and is, in part, applied to molecular machineries of other life cycles.

Although progress has been made, it is evident that important questions are open. For instance: (1) The mechanistic details of the vectorial electron ejection across the membrane and the stabilization of the primary charge separation are not understood in detail. (2) The molecular reactions that lead to the cleavage of H_2O are still unknown. (3) The mechanism by which ATP is synthesized by the H^+ flux through the ATPase enzyme must also be left to future work.

(For details and literature see Witt, H.T. (1979) Biochim. Biophys. Acta 505, 355-427.)

Part III

Evolution of Structures

Polynucleotide Replication and Biological Evolution

Peter Schuster

Institut für Theoretische Chemie und Strahlenchemie der Universität Wien
Währingerstraße 17, A-1090 Wien, Austria

Abstract

Polynucleotide replication is considered as an example of primitive
biological evolution which can be tested experimentally. We distinguish deterministic selection which is based on differences in the
rate constants of replication and degradation and neutral selection,
a stochastic phenomenon in finite populations which is a result of
the nature of the replication process only. Depending on the structural
details of the fitness surface and on the distance from the temporal
evolutionary optimum the changes in the population are predominantly
due to adaptive or neutral mutations: far from the optimum almost
every advantageous mutation will be fixed by adaptive selection
whereas fixation of neutral mutation dominates close to the optimum.

1. Mutations and Population Dynamics

Biological evolution can be visualized as a succession of transients
and metastable states. A state of the population under consideration
is characterized by a distribution of frequencies for different
types of individuals. We denote these types by

$$I_1, I_2, \ldots \ldots, I_n$$

and the corresponding variables describing the time dependence of
their frequencies by

$$x_1(t), \ x_2(t), \ldots., x_n(t).$$

In addition we assume the variables to be normalized

$$\sum_{i=1}^{n} x_i(t) = 1. \tag{1}$$

Hence, every state of the population is represented by a point or a
vector $\vec{x}$ on the unit simplex

$$S_n = x = \{(\vec{x}_1, \ldots, x_n) \in \mathbb{R}^n : x_i \geqslant 0, \ \sum_{i=1}^{n} x_i = 1\}. \tag{2}$$

The distribution of the frequencies $x_i(t)$ changes during a transient state and becomes constant when the population reaches a metastable state. Metastability is understood here as a potential instability against some microscopic fluctuations which may lead to a new - again metastable - state. These microscopic fluctuations are mutations which appear in the population at first as single modified individuals. When such a fluctuation is amplified and reaches the macroscopic level - this is the case when its frequency increases to values as large as an appreciable fraction of unity - we speak of fixation of a mutation. In general, a previously predominant type will be replaced by the new variant during the transient state.

The changes in the distribution of frequencies in a population are consequences of replication processes. Individuals replicate and in addition they have a finite lifetime. They are removed by degradation. A metastable state is in dynamical equilibrium, the number of individuals of a given type which are born through replication compensates losses through degradation or dilution fluxes. In this contribution we restrict ourselves to very simple mechanisms of replication and degradation:

$$(A) + I_i \quad \xrightarrow{\quad} \quad
\begin{cases}
\xrightarrow{\;f_i Q_{1i}\;} I_1 + I_i & (3b) \\[4pt]
\xrightarrow{\;f_i Q_{2i}\;} I_2 + I_i & (3b) \\[4pt]
\xrightarrow{\;f_i Q_{ii}\;} 2I_i \qquad ;i,j=1,\ldots,n & (3a) \\[4pt]
\xrightarrow{\;f_i Q_{ni}\;} I_n + I_i & (3b)
\end{cases}$$

and

$$I_i \xrightarrow{\quad d_i \quad} (B); \quad i=1,\ldots,n . \tag{3c}$$

A and B stand here symbolically for the material from which new individuals are built and for the degradation products respectively. Here, we assume the existence of a large reservoir of A. Then, the amount of A available for the processes (3a) and (3b) is practically unchanged, hence, set constant. Because of the irreversibility of the degradation process B does not enter as a variable into the equations for population dynamics either.

The rate constants for replication are factorized into two contributions: the gross replication rate constant f_i which gives the total number of descendants of I_i per unit time, and the mutation frequency factor Q_{ji} which represents the frequency at which I_j is obtained as a mutant of I_i. Hence, Q_{ii} is the frequency of correct replication. It reflects the accuracy of the replication process. There is a conservation law which is easily verified:

$$\sum_{j=1}^{n} Q_{ji} = 1 .$$ (4)

Accordingly, $1-Q_{ii}$ is a measure of the total frequency of mutations and Q_{ii} a measure of inheritance in our system. The rate constant of the degradation process of I_i is denoted by d_i.

Population dynamics in the simple example chosen here is described in the deterministic approach by the following system of differential equations:

$$\dot{x}_i = x_i (w_{ii} - \bar{E}) + \sum_{j=i} w_{ij} x_j ; \quad i,j=1,\ldots,n .$$ (5)

The elements of the matrix W are given by

$$W = \left\{ \begin{matrix} w_{ii} = f_i Q_{ii} - d_i \\ w_{ij} = f_j Q_{ij} ; \quad i=j \end{matrix} \right\} .$$ (6)

The term containing the mean excess production

$$\bar{E} = \sum_{i=1}^{n} x_i (f_i - d_i)$$ (7)

compensates the net growth in the population. Equation (5) has been studied extensively in the literature [1-5]. Solution curves $x_i(t)$ are obtainable in terms of the eigenvalues and eigenvectors of an n x n eigenvalue problem.

The deterministic equation, in principle, describes the dynamics of infinite populations. Effects of finite population size are negligible only if two conditions are fulfilled:
(1) There is no kinetic degeneracy and all rate constants are sufficiently different from each other. We call systems in which the rate constants for different types are equal, $f_1=f_2=\ldots$ and $d_1=d_2=\ldots$, kinetically degenerate.
(2) The replication process is accurate enough to sustain inheritance. There is a sharply defined error threshold [1,4,5]; see also section 3 which determines a minimum accuracy of the replication process as a

function of the differences in rate constants. If the accuracy of replication is below threshold the hereditary mechanism breaks down and, within the limits of the deterministic approach, mutants are present in the amounts determined by the statistics of mutations. In the case the probabilities for forward and backward mutations are equal, $Q_{ij}=Q_{ji}$, we obtain equipartition of types in the deterministic limit.

If the population under consideration is finite - as is always the case in reality - and if conditions (1) and (2) are not fulfilled, we have to go beyond the deterministic description. Then, only statistical results on the evolution of the distribution of types in the population can be obtained and the dynamics is described by a stochastic process. In continuous time the mechanism (3) can be translated into a master equation of the following form [6] (for related attempts to study mechanism (3) by stochastic methods see [7-9]):

$$\frac{dP(x_1,\ldots,x_n,t)}{dt} = \sum_k \{ \left[(x_k-1)f_kQ_{kk}+\sum_{j\neq k} w_{kj}x_j \right] P(x_1,\ldots,x_k-1,\ldots,x_n,t) +$$

$$+ (x_k+1)\left[d_k+\bar{E}(x_1,\ldots,x_k+1,\ldots,x_n) \right] P(x_1,\ldots,x_k+1,\ldots,x_n,t) -$$

$$- \left[f_kQ_{kk}x_k + \sum_{j\neq k} w_{kj}x_j+d_kx_k+\bar{E}(x_1,\ldots,x_n) \right] P(x_1,\ldots,x_n,t) \} . \qquad (8)$$

Herein, we use the same rate constants as in equation (5). $\bar{E}$ is the expectation value of f_i-d_i as in (7) defined for a specified set $(x_1,\ldots,x_n)$ and

$$P(x_1,\ldots,x_n,t) = Prob\{X_1(t)=x_1,\ldots,X_n(t)=x_n\} . \qquad (9)$$

$X_1,\ldots,X_n$ are the discrete stochastic variables describing particle numbers. In general, it is very difficult to find solutions of equation (8). Some approximate results on expectation values and variances have been obtained [6]. We shall mention one particularly interesting example, the case of complete kinetic degeneracy in section 4 [10,11].

The simple mechanism studied here serves well as a model for test tube evolution of polynucleotides and eventually also for the evolution of prokaryote populations. In case of higher organisms we have sexual replication which requires a different mechanism of gene distribution in the sense of Mendelian genetics. The problems of mutant fixation in populations and the evolution of gene distributions in

Mendelian systems has been treated extensively by the schools of
population genetics. For two recent introductory texts   to this field
and the mathematical structure of its theory see [12,13].

2. Some Experimental Data on Polynucleotide Replication

Polynucleotides, RNA or DNA, are the only known molecules which have
an intrinsic capability of self-replication. This property is a direct
consequence of their molecular structure which is characterized by
three basic and indispensible features:

(1) polynucleotides consist of a repetitive backbone with digits or
"bases" occupying structurally equivalent positions

(2) these bases are complementary in the Watson-Crick pairs, A=U(T)
and G=C, and

(3) polynucleotides form double helices with a uniform structure.
Every base in this structure is uniquely defined by that in the
opposite or vis-à-vis position. Thus, there is a uniquely defined
negative for every given polynucleotide. The double-helical structure
of polynucleotides is used in all known biological copying mechanisms.

Polynucleotides, in particular DNA and RNA in exceptional cases only,
are the hardware of molecular genetics and biological evolution. In-
heritable variations of the phenotype have their ultimate causes in
changes of the base sequence in the DNA. In principle, it should be
possible therefore to trace back evolution to the level of DNA and
its replication dynamics. In practice, however, the unfolding of the
genotype of higher, multicellular organisms is an extremely complicated and not yet understood process. One basic problem concerns the
processing of genetic signals during morphogenesis. The evolution of
prokaryotes, accordingly, is a much simpler problem and many more
details are known. But still, one of the key processes, DNA replication, is such an enormously complicated many-step reaction [14],
that no detailed kinetic studies have been successful so far. The replication of RNA from simple bacteriophages by means of virus-specific
enzymes has been studied in great detail. A cyclic polymerization
mechanism shown in figure 1 was found to agree with all the available
kinetic data. Despite the highly complex molecular mechanism RNA
replication obeys the law of simple, first-order, autocatalysis provided energy-rich monomers (GTP, ATP, CTP and UTP) as well as the
enzymes are present in excess to polynucleotides. This fact can be
used and systems for studies on test tube evolution of RNA molecules
can be built up in the laboratory.

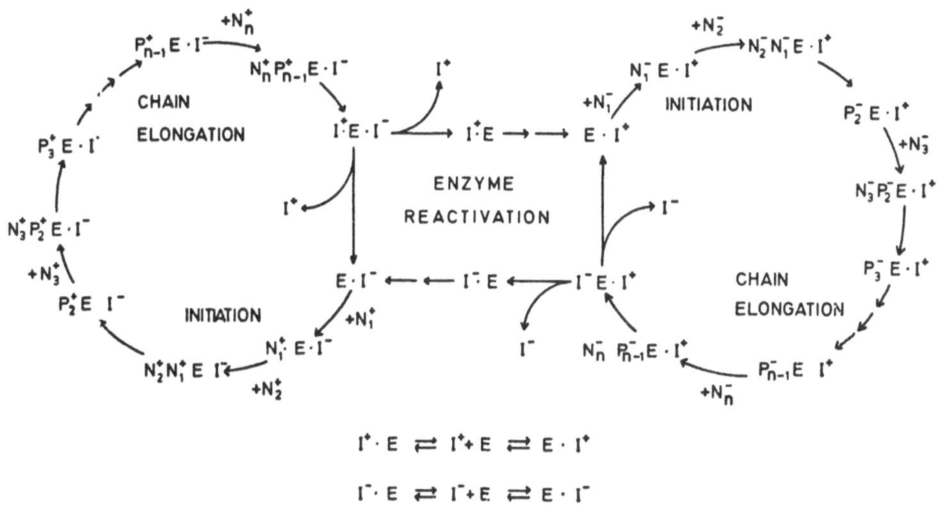

$$I^+ \cdot E \rightleftharpoons I^+ + E \rightleftharpoons E \cdot I^+$$

$$I^- \cdot E \rightleftharpoons I^- + E \rightleftharpoons E \cdot I^-$$

Figure 1. A cyclic mechanism for bacteriophage RNA replication in vitro as used by Biebricher et al. [15,17] to interpret the kinetic results obtained with the Qβ system. E, I^+ and I^- represent free Qβ-polymerase, and the plus and minus strand of the RNA to be replicated. $N_1^+, N_2^+, \ldots$, N_n^+ and $N_1^-, N_2^-, \ldots, N_n^-$ are the nucleoside triphosphates in a sequence as they appear in the plus and minus strand respectively. The chain of the newly synthesized molecule always grows from the 5' to the 3' end. $P_2^+, P_3^+, \ldots, P_{n-1}^+$ and $P_2^-, P_3^-, \ldots, P_{n-1}^-$ are used as symbols for the growing chains. A nucleation length of two bases is assumed. Enzyme reactivation is necessary because the two polynucleotides are not in suitable positions to restart polymerization: the newly synthesized strand is bound with the correct end (3') but in the wrong site (synthetic site) whereas the old strand is sitting in the correct (reading) site but bound with the wrong end (5'). The many-step mechanism thus consists of three distinguishable processes: initiation of RNA synthesis, chain propagation and reactivation of the enzyme

The first studies on selection and evolution in an ensemble of RNA molecules in the test tube were performed by Spiegelman and co-workers (for a review of the early works see [18]). RNA of the simple bacteriophage Qβ was transferred into a medium which is most favorable for RNA synthesis. This medium, mentioned already in the preceding section, consists of a solution of nucleoside triphosphates and of the specific polymerase of Qβ, both in excess. In this solution RNA synthesis starts instantaneously after template has been added. In the test tube experiments an open system is created by the serial transfer technique (figure 3): after a certain period of replication a small sample containing RNA from one test tube is transferred into the next one containing fresh solution. Thereby, the material consumed by the replication process is renewed. The whole procedure is repeated many times. The most interesting result of these experiments is a spontaneous increase in the rate of RNA

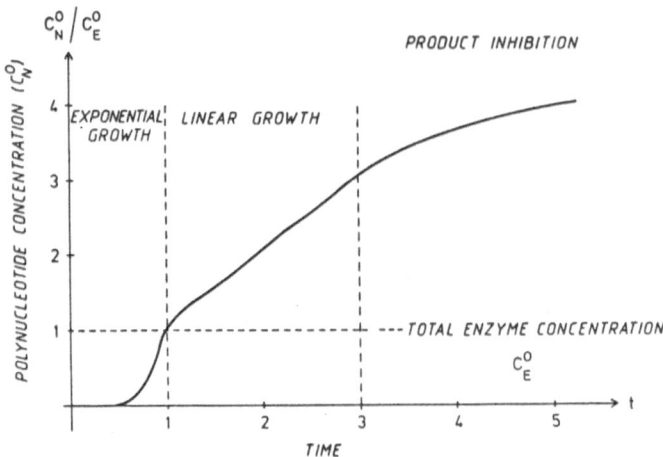

Figure 2. RNA synthesis in the test tube. A characteristic experimentally recorded curve shows three phases of growth: exponential
growth at low polynucleotide concentration, linear growth above
saturation of the enzyme by polynucleotides and further levelling
off at still higher concentrations of template when enzyme reactivation becomes the rate-determining step

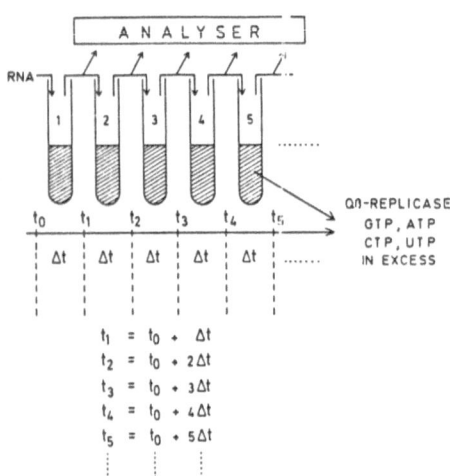

Figure 3. The technique of serial
transfer experiments [18]. RNA, in
particular RNA of the bacteriophage
Qβ, grows in a medium which contains
the enzyme Qβ replicase as well as
the four nucleoside triphosphates
ATP, UTP, GTP and CTP in excess.
After a time Δt a sample is taken
out of the test tube. Part of it is
analyzed, part of it is transferred
into a new medium. The procedure is
repeated after time intervals of Δt.
The conditions are chosen such that
the RNA is growing in the exponential phase (see figure 2)

synthesis which occurs several times before an optimum rate is
attained (figure 4). The increase in the replication rate is accompanied by a decrease in the molecular weight of the RNA.

More recently, it was shown in the laboratory of Eigen [15,19] that an
analogous process can be carried out from lower to higher molecular
weights of the RNA as well. Starting from highly purified Qβ replicase,
i.e. an enzyme sample without any detectable impurity of polynucleotides, and an excess of triphosphates GTP, ATP, CTP and UTP, RNA is
synthesized "de novo". In serial transfer experiments these "de novo"

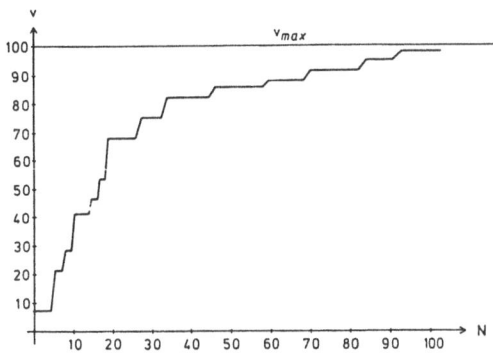

Figure 4. Spontaneous increase in the rate of RNA synthesis (schematically) as observed in serial transfer experiments (the rate is given in percent of the optimum value, v_{max}; N is the number of transfer)

products show increasing rates of RNA synthesis. Thereby, the increase in rate is accompanied by an increase in molecular weight. The optimum rates of RNA synthesis and the molecular weights ultimately attained in these experiments are very close to those of serial transfer experiments starting from high molecular weight RNA samples.

RNA replication under the conditions of serial transfer experiments, thus, has a defined optimum rate which depends on temperature, ionic strength and other experimental parameters. What one observes is optimization of the rate of RNA replication in the sense of Darwin's theory. Accidental replication errors leading to RNA which replicates faster are amplified through selection. Most replication errors will lead to less efficient RNA molecules and, hence, are instantly discarded. For more details on test tube evolution the interested reader is referred to the recent review by Biebricher [20].

3. Adaptive Selection

At first we consider the simple case of almost perfect replication. Mutation is an extremely rare event and we may write

$$Q_{ij} = \delta_{ij}$$

and hence $w_{ii} = f_i - d_i = w_i$ and $w_{ij} = 0$ if $i \neq j$. Then, equation (5) can be integrated and yields

$$x_i(t) = \frac{x_i(0)\exp\{w_i t\}}{\sum_j x_j(0)\exp\{w_j t\}}; \quad i=1,\ldots,n . \tag{10}$$

In absence of kinetic degeneracy equation (10) describes adaptive selection. The population becomes homogeneous after long enough time; only one type survives:

$$\lim_{t\to\infty} x_m(t) = 1 \quad \text{and} \quad \lim_{t\to\infty} x_j(t) = 0, \quad j \neq m \tag{11a}$$

113

$$w_m = \max\{w_i; \; i=1,\ldots,n\}. \qquad\qquad (11b)$$

The selection process is global, i.e. the long-time results do not depend on initial conditions, provided $x_m(0)>0$. The selection criterion is the excess production $w_i=f_i-d_i$. It is easy to show that the mean excess production or the total rate of RNA synthesis is optimized during the selection process:

$$\frac{d\bar{E}}{dt} = \sum_i w_i \dot{x}_i = \sum_i w_i^2 x_i - (\sum_i w_i x_i)^2 = \overline{w^2} - (\bar{E})^2 = \overline{(w-\bar{E})^2} \geqslant 0. \qquad (12)$$

Figure 5 presents a numerical example. Advantageous mutants which appear once in a while in the population are amplified. The previously selected type is replaced by the more efficient variant. The population passes through transients which separate the metastable states.

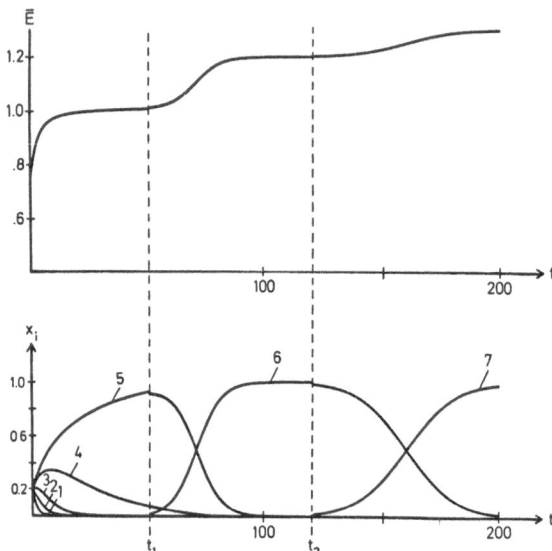

Figure 5. Selection and optimization in an ensemble of polynucleotides in the evolution reactor. We present solution curves of the differential equation (5) with $n=7$ and $w_{ij}=0$ iff $i \neq j$. The values for w_{ii} were chosen in the sequence 0.3, 0.65, 0.82, 0.96, 1.01, 1.2 and 1.3 for $i=1,2,\ldots,7$. The initial conditions are: $x_1(0)=x_2(0)=\ldots=x_5(0)=0.2$ and $x_6(0)=x_7(0)=0$. Thus, the mean excess production starts from an initial value $\bar{E}=0.748$ and increases steadily as the population becomes inhomogeneous through selection: $\bar{E} \to 1.01$; $x_1,x_2,x_3,x_4 \to 0$ as $x_5 \to 1$. We recognize easily the case for the increase of $\bar{E}$: the less efficiently growing polynucleotides are eliminated. They disappear in the same sequence as their excess production increases, namely $1,2,\ldots,5$. At $t=t_1=50$ we observe a fluctuation $x_6=0 \to x_6=\delta$. The appearance of I_6 as a favourable mutant leads to a further increase in $\bar{E}$, which approaches the value $\bar{E}=1.2$ as I_5 disappears now and the population becomes homogeneous again: x_1,x_2,x_3,x_4, $x_5 \to 0$ as $x_6 \to 1$. The same story happens when a more efficient mutant appears at $t=t_2=120$. I_6 is replaced by I_7 and $\bar{E}$ approaches the temporary optimum $\bar{E}=1.3$. This example illustrates the nature of the selection process and the role of the mean excess production $\bar{E}$ as the quantity which is optimized

114

The succession of metastable states is accompanied by a monotonous increase of the total rate of RNA synthesis, precisely as in the experimental system we have described in the last section.

In our final discussion we shall refer to the time of fixation of an advantageous mutant in a population of n individuals. We start at time t=0 from n-1 molecules of type I_1 and one molecule of type I_2. At time $t=\Delta t_f$ (which stands for "time of fixation") we have only one molecule of type I_1 left and n-1 molecules of type I_2. We can calculate a very rough estimate for Δt_f from equation (10) when we ignore stochastic effects. For the sake of simplicity we assume $d_1=d_2$ and put $f_2-f_1=\Delta f>0$. Then

$$t_f = \frac{2\ln(n-1)}{f} .$$
(13)

At higher mutation frequencies ($Q_{ij}>0$) similar results can be obtained although the mathematical treatment is somewhat more involved. The details have been presented in the literature [1-5] and we need not repeat them here. Selection, in this case, does not lead to a homogeneous distribution of a single type only, but to a distribution of types called "quasispecies". The quasispecies consists of a master sequence I_m and its most probable mutants. It represents the stable stationary state of equation (5). In precise mathematical terms a quasispecies is the "largest" eigenvector, the eigenvector corresponding to the largest eigenvalue of the matrix W. According to a well-known theorem of Frobenius [21] all components of this eigenvector are positive. Despite the presence of mutation terms ($Q_{ij}>0$) it is possible to visualize selection as an optimization process. The function which is optimized, however, is somewhat more complicated than $\bar{E}(t)$, but it converges to $\bar{E}(t)$ near the steady state. For further details see the derivation by Jones [22].

In figure 6 we show the quasispecies as a function of the accuracy of replication. The quasispecies represents a non-trivial mutant distribution only if the replication process of the master sequence I_m is accurate enough: $1>Q_{mm}>Q_{min}$. Below this minimum accuracy we have equipartition of types: every mutant, including the master sequence, is present in equal amount, provided the matrix Q is symmetric ($Q_{ij}=Q_{ji}$). Otherwise we find a mutant distribution different from equipartition but still exclusively determined by the structure of Q. Then, the kinetic parameters (f_i, d_i) have no influence on the stationary mutant distribution. Therefore, replication at low accuracy, $Q<Q_{min}$, has been called "stochastic or random replication"

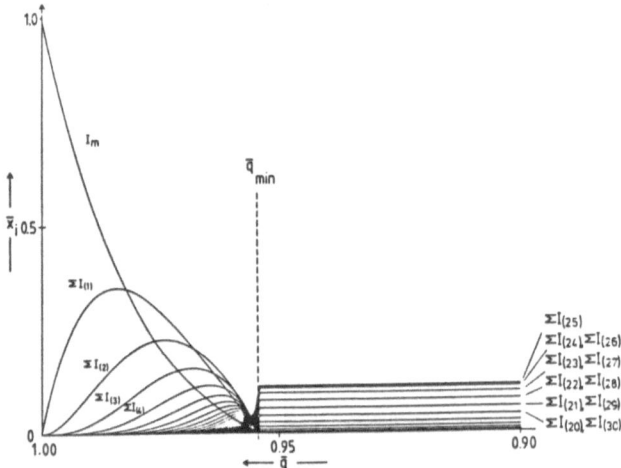

Figure 6. Stationary mutant distribution (quasispecies) in an replication ensemble of polynucleotides. The curves represent the relative
concentrations of the master sequence (I_m), the sum of the relative
concentrations of all one error mutants ($\Sigma I_{(1)}$), of all two error mutants ($\Sigma I_{(2)}$), etc., as functions of the mean single-digit replication
accuracy ($\bar{q}$). The single-digit accuracy is related to the total accuracy by the relation $Q=\bar{q}^{\nu}$. In this example the length of the polynucleotides is $\nu=50$ bases. The formation rate constant for the master sequence is chosen to be $f_m=10$, for all other sequences $f_1=f_2=...=f_n=1$.
All decomposition rate constants are equal, $d_m=d_1=d_2=...=d_n$, and hence
do not enter the differential equation (5). From these parameter values
we calculate $\sigma_m=10$ and a critical single-digit accuracy $\bar{q}_{min}=0.945$.
With decreasing replication accuracy a pronounced decrease in the relative concentration of the master sequence is observed. One-error
mutants, then two-error mutants dominate the polynucleotide distribution. Below the critical accuracy $\bar{q}_{min}$ the concentrations are exclusively determined by the statistical weight of the corresponding
sequences. Hence, the sum of the concentrations of 25-error mutants
($\Sigma I_{(25)}$) is largest, followed by 24- and 26-error mutants ($\Sigma I_{(24)}$,
$\Sigma I_{(26)}$), etc. For further details see [5]

[5]. It is interesting that the transition between adaptive selection
and random replication is very sharp already at small chain lengths
of polynucleotides (figure 6; we have chosen $\nu=50$ here). The transition becomes still sharper at larger chain lengths.

Random replication has another important aspect: the number of possible
polynucleotide sequences (4^{ν}) is "hyperastronomic" already for polynucleotides of moderate chain lengths. Thus, the number of possible
sequences exceeds by far the number of individuals in any realizable
population and we expect a severe effect of finite population size
which makes the deterministic description inadequate. We really cannot have less than a single copy of a given sequence in the volume
under consideration and all stationary concentrations will be far less

than this critical limit. Hence, we are dealing with a set of sequences which changes from generation to generation. New sequences appear due to copying errors and a certain percentage of the old sequences disappears as a consequence of degradation and dilution. The notion of "presence in equal amounts" can be replaced at best by "equal probability of realization" in a long-term experiment.

4. Random Selection

Selection can occur also by a completely different mechanism which is not based on differences in kinetic parameters. Random selection has its origin exclusively in the stochasticity of replication in finite populations. It plays an important role in the "neutral theory" of evolution of higher organisms [23] which is strongly supported by data from molecular evolution. Neutral mutations do also occur in polynucleotide replication [20]. We shall illustrate random selection by means of a simple example derived from replicating polynucleotides.

The starting point of our consideration is equation (8). In order to be able to perform complete mathematical analysis we have to introduce several simplifications:

(1) We neglect mutation terms: $Q_{ij} = \delta_{ij}$ and release the constraint on the population size which is expressed by the flux terms involving $\bar{E}$.

(2) We assume complete kinetic degeneracy: $f_1 = f_2 = \ldots = f_n = f$ and $d_1 = d_2 = \ldots = d_n = d$. Now, we can compensate the neglect of regulation of the population size somewhat by putting $d=f$.

Assumption (1) is essential for our treatment and more serious with respect to its consequences. It allows to factorize the probability density:

$$P(x_1, \ldots, x_n, t) = P(x_1, t) \cdot P(x_2, t) \cdot \ldots \cdot P(x_n, t) \tag{14a}$$

wherein

$$P(x_i, t) = Prob\{X_i(t) = x_i\}. \tag{14b}$$

The individual probability densities according to assumption (2) fulfil the master equation

$$\frac{dP(x,t)}{dt} = f\{(x-1)P(x-1,t) + (x+1)P(x+1,t) - 2xP(x,t)\} \tag{15}$$
$$x = 0,1,2,\ldots.$$

Solutions of equation (15) have been derived by Bartholomay [24]. For our treatment we need only the probability of extinction:

$$X(O) = m \longrightarrow P(O,t) = \{1+(ft)^{-1}\}^{-m}. \tag{16}$$

In order to describe random selection appropriately we choose the following initial conditions (t=0): n different polynucleotide sequences, each one present in a single copy (m=1). The event when its polynucleotide sequence becomes extinct is considered to be a random variable T_k [10]. The set of random variables (k=0,1,2,...,n) has been called characteristically "sequential extinction times". The index k gives the number of different polynucleotide sequences which are present just after the event T_k. Thus, we have exactly n different polynucleotide sequences between the events T_n and T_{n-1}, n-1 sequences between T_{n-1} and T_{n-2}, etc. The variables T_k form a time-ordered set:

$$T_n < T_{n-1} < T_{n-2} < \ldots < T_2 < T_1 < T_0 .$$

From equation (16) we derive expectation values for the sequential extinction times in a straightforward calculation [10]:

$$\overline{T}_k = \frac{n-k}{k} \cdot \frac{1}{f} . \tag{17}$$

In figure 7 we present a concrete numerical example (n=20). Most of the sequences become extinct within the initial period. Then, the number of different sequences decreases more slowly until we have finally a single sequence left which is then present in many copies. The expectation value $\overline{T}_o$ diverges although we have certain extinction times according to equation (16):

$$\lim_{t \to \infty} P(O,t) = 1 . \tag{18}$$

From this probabilistic result we conclude that the homogeneous state after T_1, in general, is metastable but long-lived.

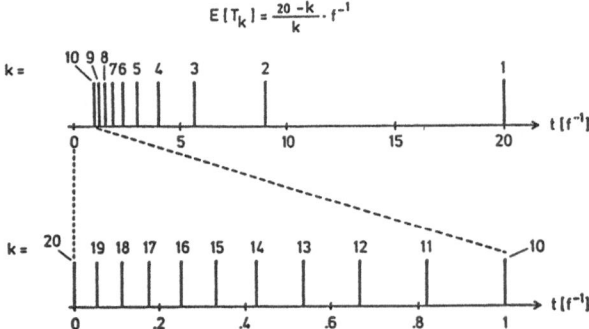

Figure 7. The distribution of the expectation values for the sequential extinction times, $\overline{T}_k$ with n=20

The process of selection described here is completely random with respect to the particular sequence which ultimately survives. In case we start from the same number of initial copies for all sequences as we did here the initial chances to survive by random selection are equal for all sequences, namely 1/n.

Let us now consider the spreading of a mutant in a population. The mathematical analysis centres around two stochastic variables:
(1) the time of fixation (Δt_f) of a selectively neutral mutant present in a single copy at time t=0 in a population of n individuals, and
(2) the time of replacement (Δt_r) of one type by another type in a population of n individuals. We calculate both quantities from our simple model. Any mutation starts from a single copy. Since individual replication steps are independent we can mimic the process of mutant fixation by the initial condition of n-1 replicating molecules of type I_1 and one of type I_2. Provided I_2 reaches fixation at all the time Δt_f will be identical with the sequential extinction time T_1:

$$\overline{\Delta t_f} = \overline{T}_1 = (n-1) \frac{1}{f} \simeq \frac{n}{f} \quad \text{for large n}. \tag{19}$$

The expectation value of the time of replacement can be obtained from the mutation rate f.v=f(1-Q) - the latter expression holds in case practically all mutants are selectively neutral - the probability of fixation (1/n) and the population size (n):

$$\Delta t_r = \frac{1}{f \cdot v} \cdot \frac{1}{n} \cdot n = \frac{1}{f \cdot v} . \tag{20}$$

Note that the factors which contain the population size cancel here and hence, the mean time of replacement is independent of n.

5. Conclusion

In the last two sections we discussed two different mechanisms leading to uniform populations: adaptive and random selection. Both processes are nothing but the two limiting cases of replication in finite populations which is always of stochastic nature and has a deterministic as well as a random component. In order to illustrate this we compare adaptive and random selection. A rough estimate of the prevailing mechanism consists in a comparison of the time of fixation in the deterministic and the neutral model. The condition that adaptive selection is faster than the stochastic process can be formulated as

$$\Delta t_f^{DET} < \overline{\Delta t_f^{RAN}} \; .$$

From equations (13) and (19) we derive

$$\frac{\Delta f}{f} > \frac{2\ln(n-1)}{n-1} \simeq \frac{2\ln n}{n} \quad \text{for large } n \; . \tag{21}$$

In the table we illustrate the function given above.

Table: Comparison of the rates of adaptive and random selection. The deterministic process is faster if $\Delta f/f > (\Delta f/f)_{min}$.

Population size n	Minimum relative difference in replication rates $(\Delta f/f)_{min}$
10	0.46
100	0.09
10^3	$1.4 \cdot 10^{-2}$
10^4	$1.8 \cdot 10^{-3}$
10^6	$2.8 \cdot 10^{-5}$
10^{10}	$4.6 \cdot 10^{-9}$
10^{15}	$6.9 \cdot 10^{-14}$
10^{20}	$9.2 \cdot 10^{-19}$

In large populations the random process becomes very slow and minute differences in rate constants are sufficient to make adaptive selection faster than random selection.

Let us now consider a population far below the replication optimum. Many mutations will be advantageous and provided the population is large enough, fixation of the advantageous mutations is fast. Hence, adaptive selection prevails. In case we are already close to the optimum advantageous mutations are rare and the relative increases in replication rates are small. Hence, neutral mutations prevail and provided the population is not too large random selection leads to a succession of selectively neutral types.

The problems treated here are in close analogy to the concepts of adaptive and neutral evolution in population genetics although there they are complicated by the effects of recombination and sexual replication. For a comparison of results see [11].

Acknowledgements: The work reported here has been supported financially by the Austrian "Fonds zur Förderung der Wissenschaftlichen Forschung" Project no. 4506 and 5286. Technical assistance in preparing the manuscript by Mrs.J.Jakubetz and Mr.J.König is gratefully acknowledged.

References

1   M.Eigen: Naturwissenschaften $\underline{58}$, 465 (1971)

2   C.J.Thompson, J.L.McBride: Math.Bioscience $\underline{21}$, 127 (1974)

3   B.L.Jones, R.H.Enns, S.S.Ragnekar: Bull.Math.Biol. $\underline{38}$, 15 (1976)

4   M.Eigen, P.Schuster: The Hypercycle - A Principle of Natural Self-Organization (Springer, Berlin 1979)

5   J.Swetina, P.Schuster: Biophys.Chem. $\underline{16}$, 329 (1982)

6   B.L.Jones, H.K.Leung: Bull.Math.Biol. $\underline{43}$, 665 (1981)

7   W.Ebeling, R.Feistel: Ann.Phys. $\underline{34}$, 81 (1977)

8   H.Inagaki: Bull.Math.Biol. $\underline{44}$, 17 (1982)

9   R.Heinrich, I.Sonntag: J.Theor.Biol. $\underline{93}$, 325 (1981)

10 P.Schuster, K.Sigmund: Bull.Math.Biol. $\underline{45}$, in press (1983)

11 P.Schuster, K.Sigmund: "Random Selection and the Neutral Theory", in P.Schuster, ed.:Stochastic Phenomena and Chaotic Behaviour in Complex Systems (Springer, Berlin 1984)

12 J.F.Crow, M.Kimura: An Introduction to Population Genetics Theory (Harper and Row, New York 1970)

13 W.J.Ewens, Mathematical Population Genetics (Springer, Berlin 1979)

14 A.Kornberg: DNA-Replication, 2nd ed. (Freeman, San Francisco 1980)

15 C.K.Biebricher, M.Eigen, R.Luce: J.Mol.Biol. $\underline{148}$, 369 and 391 (1981)

16 C.K.Biebricher, S.Diekmann, R.Luce: J.Mol.Biol. $\underline{154}$, 629 (1982)

17 C.K.Biebricher, M.Eigen, W.C.Gardiner: Biochemistry $\underline{22}$, 2544 (1983)

18 S.Spiegelman: Quart.Rev.Biophys. $\underline{4}$, 213 (1971)

19 M.Sumper, R.Luce: Proc.Natl.Acad.Sci.(USA) $\underline{72}$, 162 (1975)

20 C.K.Biebricher: Evolutionary Biology $\underline{16}$, 1 (1983)

21 G.Frobenius:Sitz.Ber.Akad.Wiss., phys.-math.Klasse Berlin $\underline{417}$ (1908) and $\underline{412}$ (1912)

22 B.L.Jones: J.Math.Biol. $\underline{6}$, 169 (1978)

23 M.Kimura, ed.: Molecular Evolution, Protein Polymorphism and the Neutral Theory (Springer, Berlin 1982)

24 A.F.Bartholomay: Bull.Math.Biophys. $\underline{20}$, 97 (1958)

The Evolution Strategy. A Mathematical Model of Darwinian Evolution

Ingo Rechenberg

Technische Universität Berlin, Fachgebeit Bionik und Evolutionstechnik
Ackerstraße 71-76, D-1000 Berlin 65, Fed. Rep. of Germany

1. Introduction

Everybody is fascinated, realizing the technical performance of living systems. Thus we should remember that living beings are the result of a large-scale experiment on the Earth, called biological evolution. It may be worthy to reflect on this biological development. Over the past years it has become obvious that rules of biological evolution are the result of an evolution process itself. Suppose a population of organism with slightly modified hereditary rules compared to the existing norm. If these modifications will help the population to adapt faster to their particular environments, then this population will have a better chance to survive in future than a population with less effective hereditary rules. Therefore it should be assumed that evolution, during its action over more than a thousand million years, gave itself an optimal mode of operation. This hypothesis results to the following statements:

1. The imitation of rules of biological evolution should yield an excellent experimental method in engineering to design better technical apparatus.

2. The similarities between biological evolution and optimization give rise to the idea that common concepts in optimization theory may be used to describe Darwinian evolution.

2. Evolution Strategy as an Optimization Technique

It was in 1964 [1] when I started the first experiment to imitate the method of biological evolution in a laboratory of fluid mechanics. An aluminium plate flexible at five positions was mounted in an open wind tunnel (Fig. 1). The articulated plate can be altered stepwise. There are more than 345 million possible forms. The task is to find the shape with minimum drag. We all know that this is a flat plate, directed parallel to the air stream. But suppose we don't know that. Therefore the plate is set into a random starting configuration. To produce the random alternations of the five hinges (the mutations in biology) we used in our first experiment a mechanical apparatus (Fig. 2). Five balls, representing the five hinges of the plate, pass the pyramid of pins and land in the ground boxes. The box markings determine the alternations of angle.

Our experimental arrangement makes it possible to measure the fitness of the mutated shapes. The technical fitness is the drag of the plate, which has to become a minimum. At the beginning of the wind-tunnel experiment the plate was folded into a zig-zag shape of high drag. The experimental scheme to imitate rules of biological evolution will be discussed later in detail. The basic idea is to reject all mutations with increasing drag (decreasing fitness). But if a randomly generated form has a lower drag, then it becomes parental shape for the next mutations. Figure 3 shows the result of our first evolution-like experiment. The drag of the plate is plotted versus the number of generations. Below the diagramm the best of the plate after every ten generations is shown. We achieve the plain shape after 300 generations.

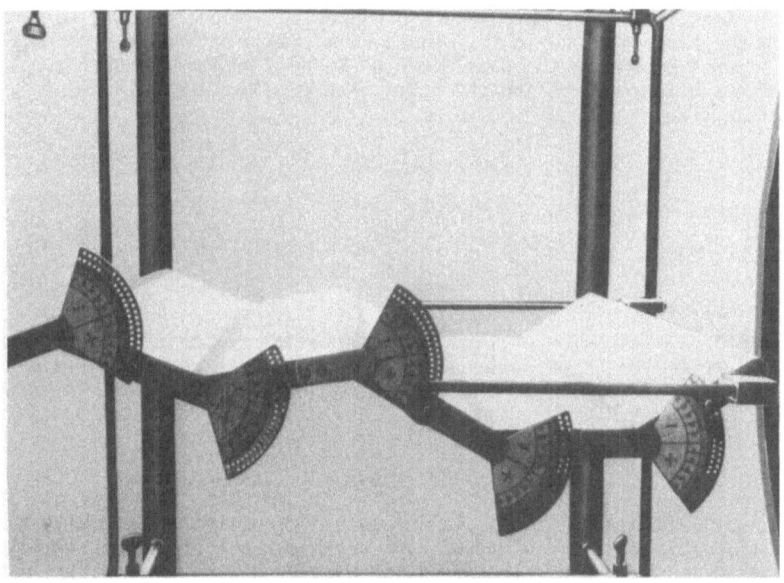

Fig. 1 The articulated plate for an experiment in the wind tunnel to imitate Darwinian evolution

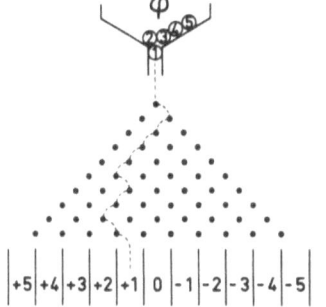

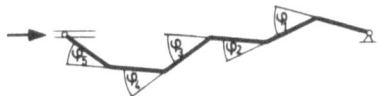

Fig. 2 Galton's pin board. Passing the pyramid the balls land in compartments. The labels indicate the alternations of angle

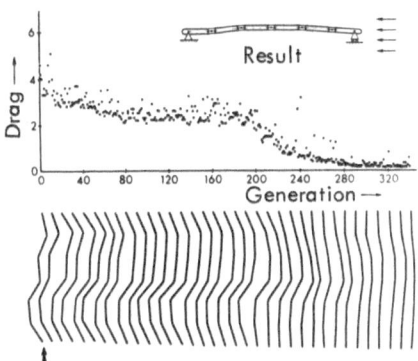

Fig. 3 The experimentum crucis. Starting the experiment the plate was set in a random configuration. Applying rules of biological evolution the plate develops to the form of minimum drag. The expected solution of a flat plate is found

For the next experiment we wished to change a boundary condition. Suppose we turn the wind tunnel. In the biological world this would be a change of the environment. Because it is too expensive to turn the complete wind tunnel, we turned the flat plate (Fig. 4). Now evolution goes on and the flat plate develops to an S-shaped curvature with minimum drag.

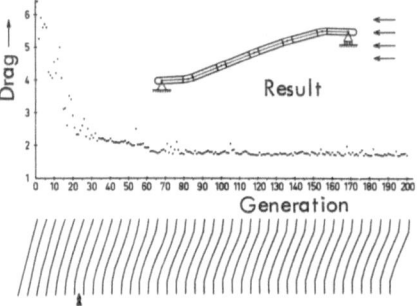

Fig. 4 Continuation of the experiment. The boundary conditions were changed by raising the leading edge of the plate. The new environmetal factor will cause the plate to evolve to an S-shaped curvature

Whilst the tests with the articulated plate are more of academic value, the following three experiments in the field of fluid dynamics are of practical interest [2]. The next task was to find the form of a right-angled pipe bend with minimum flow resistance. Figure 5 shows the experimental arrangement. Two flexible plastic hoses are held by adjustable bars in the deflection zone. The positions of the six bars of each pipe are the optimization variables. At the beginning of the experiment both pipe elbows were adjusted to a quarter of a circle. While one pipe bend was continuously varied according to the algorithm of evolution strategy, the second pipe remained unchanged as the reference system. Figure 6 shows the initial form and the optimum form one upon the other. The optimum pipe bend, having 10% less deflection losses, starts with a steadily increasing curvature (similar to Euler's spiral) and ends with a small reverse in curvature.

In another experiment made by SCHWEFEL[3] a two-phase supersonic flow nozzle was developed using evolution strategy. Heated water vapourises partly in the throat of a convergent-divergent nozzle. The expanding vapour then forms the propellent for the remaining liquid. It is impossible to calculate the shape of the nozzle for maximum thrust. For the experimental optimization the nozzle was made up of segments. A total of 400 segments with different conical borings were available. By continually

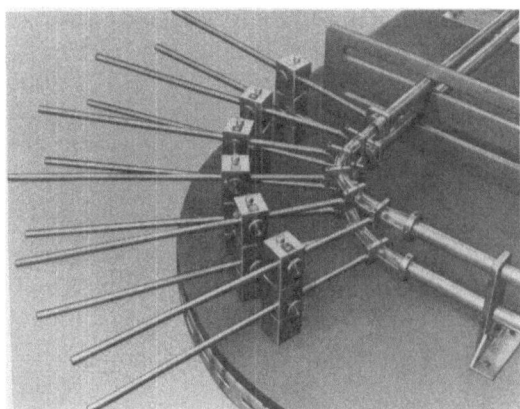

Fig. 5 Experimental arrangement for the evolution of a pipe elbow

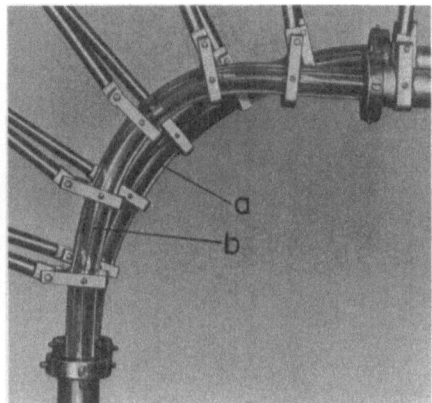

Fig. 6 Circular form (a) and optimum form (b) of the pipe elbow

124

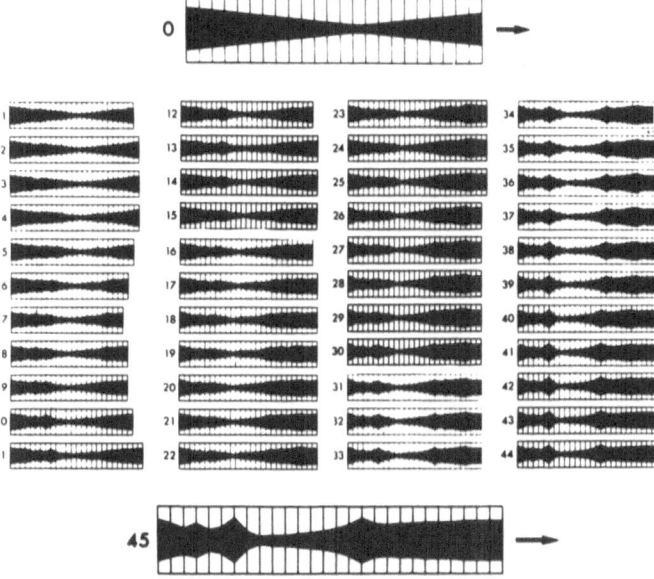

Fig. 7 Evolution of a hot-water flashing nozzle. A set of 400 segments, each coni-cally bored, made it possible to build up different nozzle configurations

exchanging segments in accordance with the rules of evolution strategy, an optimum nozzle form was found which looks like a modern vase (Fig. 7). In this experiment the evolution strategy has invented something new . That is a chamber in the diver-ging part of the nozzle, where mixing takes place between liquid and vapour. The en-ergetic efficiency of the initial conical nozzle, calculated on the basis of super-sonic theory, was 55%. The optimum nozzle has an efficiency of nearly 80%.

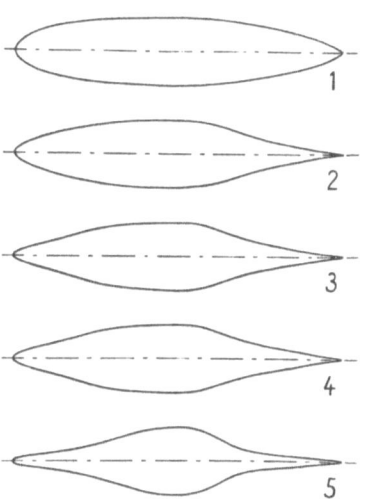

Fig. 8 Development of a body of revo-lution for minimum drag shown at inter-vals of 600 generations

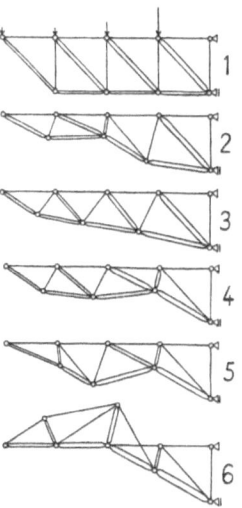

Fig. 9 Development of a lattice frame for minimum weight using evolution strategy

It is a classical problem of fluid dynamics to find the minimum drag profile of an axisymmetric body. Recently PINEBROOK solved this problem on a computer, combining boundary layer calculation techniques with the evolution strategy. Figure 8 gives an example from his work [4]. A body of revolution with a constant maximum diameter and a given length evolves to a dolphin-like form with minimum drag. The computer run required 2400 generations to fit the 21 intensities of the variational problem.

Finally an application of the evolution strategy is presented in the field of structural engineering. A lattice frame with given loads has to be constructed for minimum weight. The variables of the structure are the plain coordinates of the six joints. Figure 9 shows the development of the frame. The initial design as the result of a linear optimization procedure has a weight of 922 kg. The optimum solution, which looks like a crane jib, weights only 718 kg. This computer simulation was performed by HOEFLER [5].

3.   Sequence of Operations for the Evolution Strategy

So far the evolution strategy has proved to be an excellent optimization procedure. Now the time has come to elucidate the exact algorithm. It is useful to introduce at this point a nomenclature for the different modes of the evolution strategy [6], [7]. This will be cone by the formal abbreviation:

$$(\mu + \lambda) - ES .$$

We shall call μ the number of parents of a generation and λ the number of descendants (the children of the parents). We start with the simplest imitation of evolution. This is a (1+1)-ES. A game of cards (data cards) demonstrates how this algorithm works (Fig. 10a). A data card represents a genotype in biology. By card re-

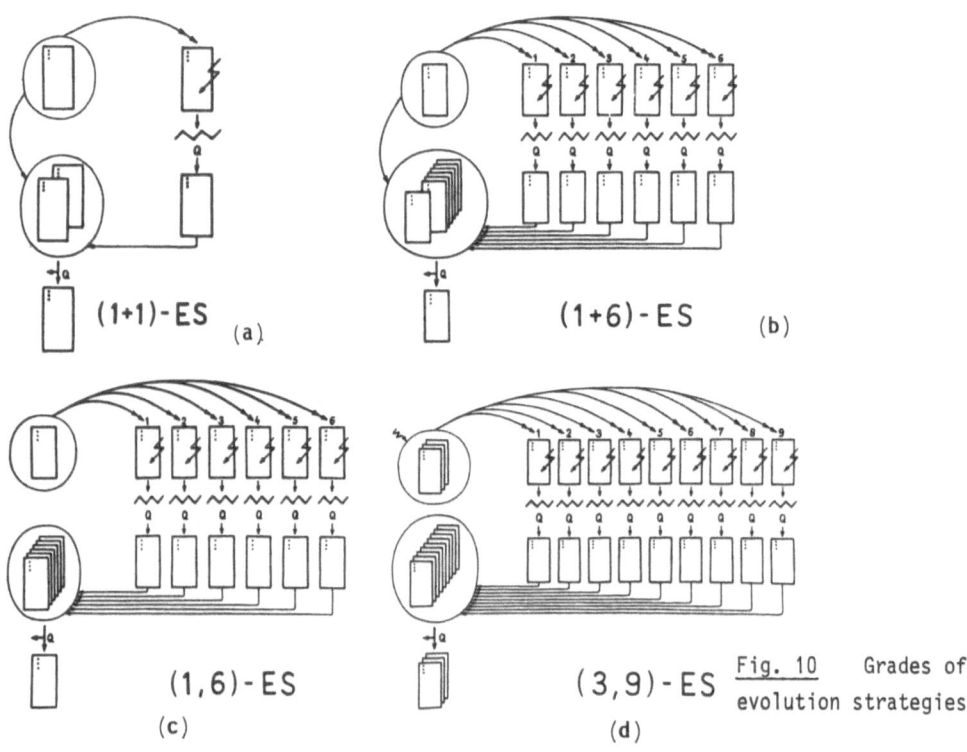

(1+1)-ES   (a)

(1+6)-ES   (b)

(1,6)-ES   (c)

(3,9)-ES   (d)

Fig. 10   Grades of evolution strategies

peating the parental information will be doubled. This is the DNA replication in biology. DNA replication errors are introducedin our game by using a random number generator. Then the randomly altered information of the offspring has to be realized. We get the phenotype, which is in our example a slightly changed form of the articulated plate. We measure the drag of the plate within the airstream of the windtunnel. The drag becomes the fitness of the offspring genotype. The parental card plus the offspring card are put into a ballot box. The best card holds the parental information for the next generation. This so-called two-membered evolution strategy has been applied in our first experiments.

A higher imitation level of biological evolution will be obtained with the algorithm of a (1+6)-ES. Now the parent will produce 6 offspring (Fig. 10b). Random alternations are introduced and the information is translated into the phenotypes. After the drag has been measured all 6 offspring cards plus the parental card are put into the ballot box. The best of them will become the parent for the next generation.

The next scheme takes into account that the parents have a limited lifespan (Fig. 10c). We set up a (1,6)-ES. The comma in this notation indicates that the parental card is not included in the selection. With the exception of this modification the operations are the same as in the preceding algorithm. This kind of evolution strategy is preferred in our mathematical treatment today.

Last we introduce a population of three parents. They produce in a random sequence 9 offspring. Because it is the comma version of the evolution strategy only the 9 offspring are put into the ballot box. The three best of them survive and become the parents for the next generation. This scheme will be named a (3,9)-ES.

What can we do with this formalism? Certainly we can apply the statistical operations to optimize an airfoil in a wind tunnel or even to find the solution of Rubik's cube. But we are far away from the formulation of an evolution equation. Actually I take a new course for the mathematical treatment of Darwinian evolution. The key idea is to interpret the method of biological evolution as a hill-climbing procedure. Hill climbing is a principle common to many optimization strategies. A hill-climbing strategy acts like gravity, forcing a ball to roll down the gradient of a hill, but it works in the opposite direction. The effect is that the vast space of possibilities is reduced to a narrow street, on which the optimum seeking takes place. Applying such a strategy you must make sure that a hill exists to climb up. This was the case for all our engineering experiments using the evolution strategy. I claim that there is no difference in biology. Piecewise smooth relationships between the fitness of an organism and the structure of the variables form a genetic landscape with hills to climb up.

To demonstrate the gradient climbing of a (1,10)-ES we look at an ordinary optimization problem in automobile engineering. The object is a carburettor with two adjusting screws. The contour lines in Fig. 11 represent screw settings of equal efficiency. To find the optimum setting we start with the parent at a random position (a). This parent will produce 10 children (b). Then the parent will die out (c). The offspring with the highest efficiency is declared to become the new parent (d). All other children die out (e). The new parent will produce the 10 children of the next generation (f). The setting point of the variables moves up the hill with a certain speed.

4.   The Discovery of the Evolution Window

We will now direct our attention on the calculation of the rate of progress. To do this we must know the local form of the fitness function. A smooth fitness function may be described by a general quadratic equation with the variables $x_1, x_2, \ldots, x_n$:

$$(1) \qquad Q = Q_o + \sum_{k=1}^{n} a_k x_k - \sum_{i=1}^{n} \sum_{k=1}^{n} b_{ik} x_i x_k .$$

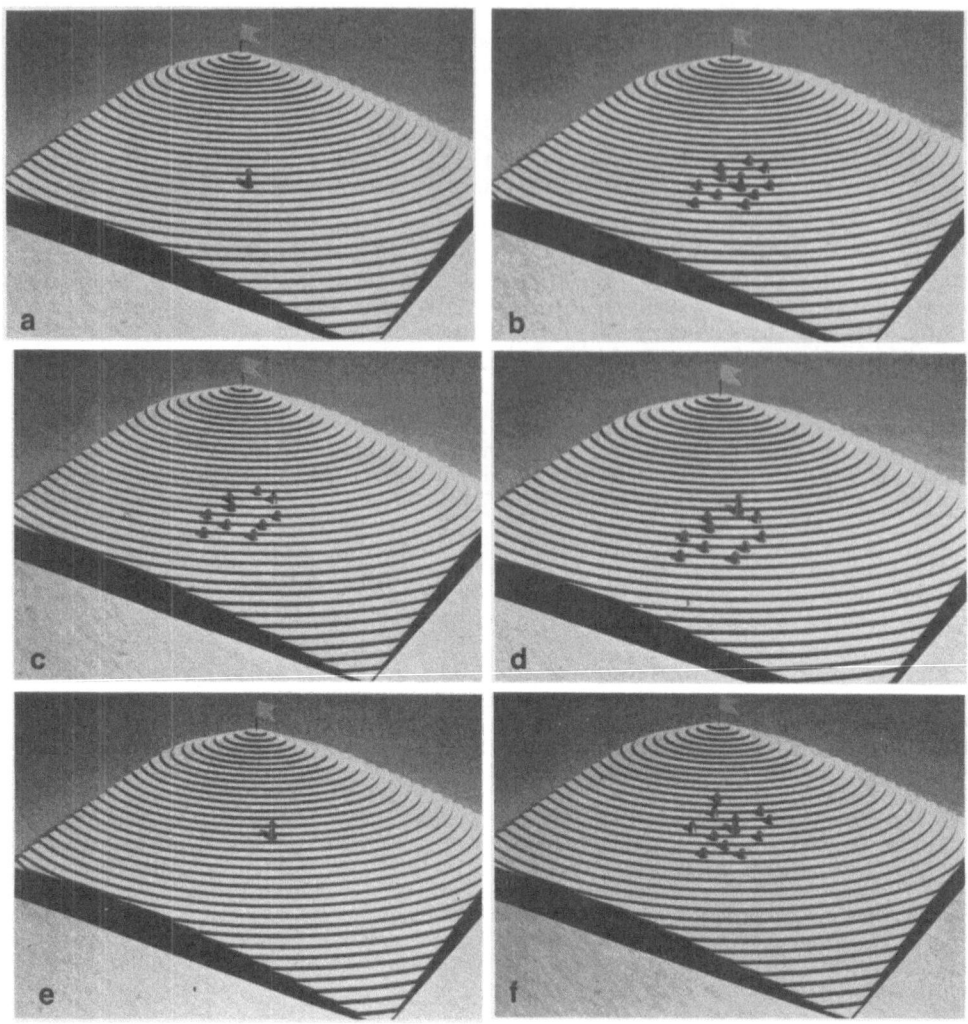

Fig. 11 The hill-climbing procedure of the multimembered evolution strategy

The variables are altered by random numbers satisfying a $(0,6)$ normal distribution. Because the mutations are distributed spherically symmetric around the parental point, a rotation of the coordinate axes will be allowed. The transformation of equation (1) to the principal axes gives the result:

$$(2) \qquad Q = Q_0 + \sum_{k=1}^{n} c_k y_k - \sum_{k=1}^{n} d_k y_k^2 \ .$$

For this general description of the local form of a fitness function I succeeded in calculating the rate of progress in the case of a (μ,λ)-ES [8]. Here we introduce a simplification. Assuming equal coefficients $d_k=d$ for each variable - we speak of an isotropic non-linear local behaviour of the fitness function - the rate of progress is given by the formula:

$$(3) \qquad \varphi_{\mu,\lambda} = \frac{c_{\mu,\lambda}\,\delta}{\sqrt{n}} - \frac{\delta^2}{2r}$$

$c_{\mu,\lambda}$ = progress coefficient
δ = mutation step length
n = number of variables
r = $\sqrt{\sum c_k^2}/2d$.

All difficulties are concentrated in the evaluation of the progress coefficient. In the case of a $(1,\lambda)$-ES one has to solve the integral:

$$(4) \qquad c_{1,\lambda} = \sqrt{\frac{2}{\pi}}\,\frac{\lambda}{2^{\lambda-1}} \int\limits_{z=-\infty}^{\infty} z\,e^{-z^2}\left[1+\mathrm{erf}(z)\right]\,dz .$$

Fortunately the progress coefficient will change only from 1,5 to 3,5 when λ is increased from 10 to 1000 [7].

Figure 12 shows the theoretical result in the form of a diagram . The climbing speed Φ (in a universal notation) is represented as a function of the mutation step size Δ (also in a universal notation). It is a very remarkable result that the evolution strategy works only within a small band of the mutation step size. I have named this band the "evolution window" (Evolutionsfenster). Out of this window no evolution occurs. This is a very exciting fact, because SCHWEFEL [6] has found in his doctor thesis that the main difference between the evolution strategy and other optimum seeking methods is the exponent by which the search effort will rise with increasing number of variables. This means that the result of the evolution window is of more general evidence. It is worthy to reflect on this fact for engineering optimization as well as for the development of economic and social systems. Right to the window you may locate the field of revolution with negative values of Φ . Left to the window you will have the region of conservatism with no progress. The logarithmic scale for Δ has been chosen, because this quantity (the mutation rate in biology) will change in a decimal power mode.

Evolutionsfenster

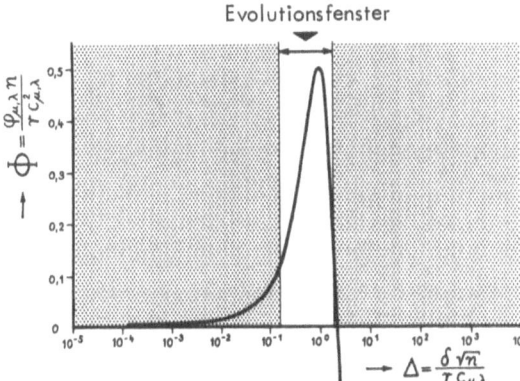

Fig. 12 The evolution window. Evolution strategy works only within a small band of the mutation step length

5.   The Twofold Algorithm of the Evolution Strategy

The existence of the narrow evolution window gives rise to a new problem. How can the evolution strategy find the window in order to be effective? The answer is: the multimembered evolution strategy is a twofold optimization procedure. This very important feature of the evolution strategy is illustrated in Fig. 13 for the example of a (2,10)-ES. This time we have two parents on our fitness hill (a). Each parent will produce five offspring (b). But one of the two parents will do it with a large

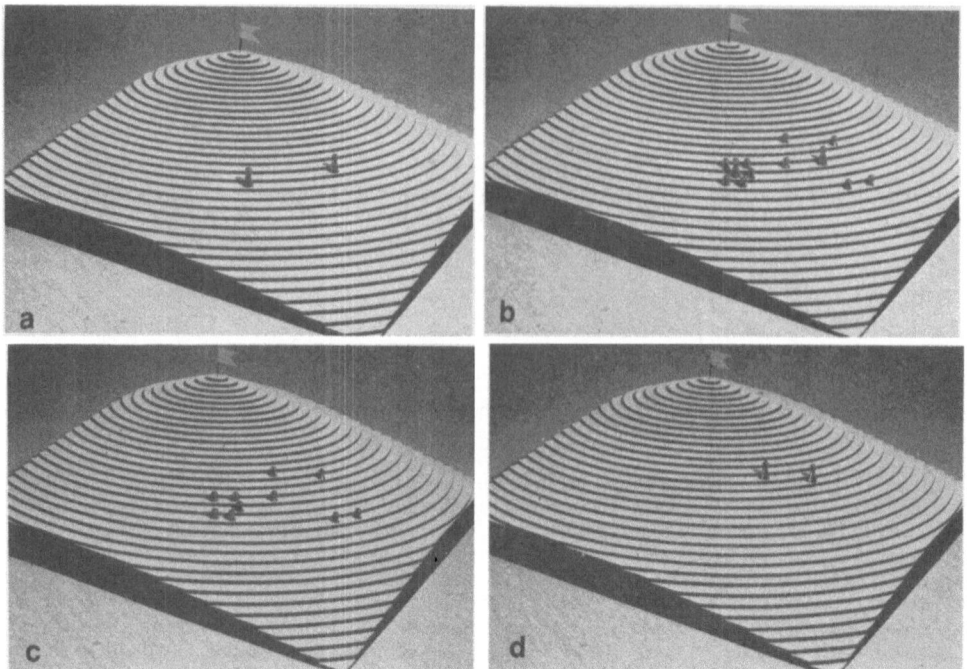

Fig. 13 Hill climbing and automatic adjustment of the mutation step length

step size, the other one with a small step size. In the language of biology: the two parents have different kinds of error correcting DNA polymerases. Next the parents will die out (c). The two offspring with the highest fitness are declared to become the parents of the next generation and all other children die out (d).

We come now to the main point: most probably these children will win having a step size to be located nearer to the centre of the evolution window. In biology, however, the mutation step size, or more precisely the mutation rate, is a hereditable character. That does mean that the step size of a surviving offspring has to be transferred to the corresponding parent of the next generation. Before creating a new set of children, however, we must not forget to mutate the step length of each parent. In this case the mutation step size adapts itself to the local topology of the fitness funtion to get maximum rate of progress.

We look at a similar situation in the life of man. Suppose you are an alpinist climbing a difficult mountain-side. You have picked out an appropriate climbing technique. You can't, however, decide if it is the best, because there is no standard of comparison. Next day you are climbing in a group. Each member of the group is using a slightly different climbing technique. After a short time it becomes evident which technique is the best and you may copy it.

The $(1, \lambda)$-ES, taking step length mutations into account, can be formalised as follows:

$$\delta_{N1}^{g} = \delta_{E}^{g} \cdot \mathcal{F}_1$$
$$\underline{X}_{N1}^{g} = \underline{X}_{E}^{g} + \delta_{N1}^{g} \cdot \underline{Z}_1$$

(offspring 1 of generation g)

$$\vdots$$

$$\delta_{N\lambda}^{g} = \delta_{E}^{g} \cdot \mathcal{F}_{\lambda}$$
$$\underline{X}_{N\lambda}^{g} = \underline{X}_{E}^{g} + \delta_{N\lambda}^{g} \cdot \underline{Z}_{\lambda}$$

(offspring λ of generation g)

Let NB denote the best offspring: $Q(\underline{x}_{NB}^g) = Max\{Q(\underline{x}_{N1}^g), \cdots Q(\underline{x}_{N\lambda}^g)\}$

$$\delta_E^{g+1} = \delta_{NB}^g$$
$$\underline{x}_E^{g+1} = \underline{x}_{NB}^g$$

(parent of generation g+1)

In this algorithm the random vectors z_i have normally distributed components. In accordance to the logarithmic scale of the mutation step length in the evolution window, the deviates $\mathcal{F}_i$ are obtained from log-normally distributed numbers.

6. Gradient Climbing in Darwinian Evolution

What is the optimum mutation step length of the evolution strategy adapting itself to maximum rate of progress? Our hill-climbing theory gives the analytical expression:

(5) $$\delta_{opt} = \frac{c_{\mu,\lambda} \cdot r}{\sqrt{n}} .$$

At this point it has to be stated that the formulas (2),(3) and (5) are the result of an asymptotic theory ($n \to \infty$). The general solution becomes more complicated. However, we are just interested in a complex system having many degrees of freedom (variables). Now, if the number of variables is increasing, the optimum step length to get maximum rate of progress will decrease. If it would be possible to look into a multidimensional Euclidian space designed by the n variables of the problem, then one would observe that an evolution process looks like a one-dimensional diffusion process winding up the gradient path of the fitness function. The picture of an evolution process to find the optimum is not that of a concentrating cloud of points. It is an elongating chain of points following the gradient line of the fitness function (Fig. 14). The gradient path acts as a guiding thread from the starting point to an optimum of the functional landscape. Evolution strategy will not scatter in the vast space of possibilities.

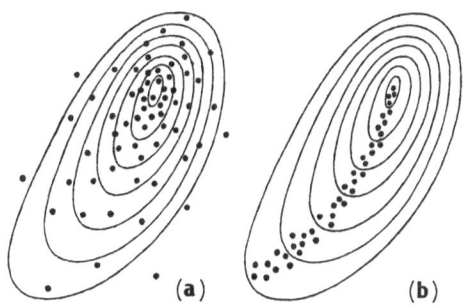

Fig. 14 Hill-climbing picture of the evolution strategy

a) False: shrinking cloud of random points

b) True: diffusion along the gradient line

Many people extrapolate the two-dimensional view of the problem into n dimensions. This turns out to be a fatal error. You can poke around in a two-dimensional manifold to find an optimum. Of course, applying pure random search you can jump from a local to a global optimum in two dimensions. However, you have no chance to repeat this in n dimensions. There is only one way to overcome the curse of dimensionality: to follow the gradient line. All optimum seeking methods act in this way. The evolution strategy, however, will do that with a minimum search effort.

References

1. I. Rechenberg: "Cybernetic Solution Path of an Experimental Problem", Royal Aircraft Establishment, Library Translation 1122, Farnborough (1965)

2. I. Rechenberg: Evolutionsstrategie: Optimierung technischer Systeme nach Prinzipien der biologischen Evolution (Frommann-Holzboog, Stuttgart 1973)

3. J. Klockgether, H.-P. Schwefel: "Two-Phase Nozzle and Hollow Core Jet Experiments", Eleventh Symposium on Engineering Aspects of Magnetohydrodynamics. Pasadena, California (1970)

4. W. E. Pinebrook: "Drag Minimization on a Body of Revolution", PhD Thesis, Faculty of the Department of Mechanical Engineering of the University of Houston (1982)

5. A. Höfler, U. Leyssner, J. Wiedemann: "Optimization of the Layout of Trusses Combining Strategies Based on Michell's Theorem and on the Biological Principles of Evolution", AGARD Conf. Proc. 123, Mailand (1973)

6. H.-P. Schwefel: Numerical Optimization of Computer Models (Wiley, Chichester 1981)

7. I. Rechenberg: "Evolutionsstrategien", in Medizinische Informatik und Statistik, Vol. 8, Ed. B. Schneider and U. Ranft, Springer, Berlin 1978

8. I. Rechenberg: Lectures on the Evolution-Strategy, Technische Universität Berlin (unpublished)

Multigene Families and their Implications for Evolutionary Theory

Tomoko Ohta

National Institute of Genetics, Mishima, 411, Japan

Summary   Multigene families of hemoglobins, immunoglobulins and histocompatibility antigens are evolving under continued occurrence of unequal crossing-over
and gene conversion. They provide a very different picture of evolution from the
conventional models of population genetics; nevertheless their polymorphisms and
gene diversity may be understood in terms of a common concept with the previous
theory, i. e. probability of gene identity.

The mechanisms of evolution will be ultimately understood in terms of change
of genetic material, and the new findings of molecular biology need to be incorporated. It is now known that genes of higher organisms are often split and
that many repeated gene families exist, and these facts have significant implications for our evolutionary thought. As has been shown by Kimura and his associates (KIMURA [1, 2]), gene substitutions at the molecular level in evolution
are mainly caused by random genetic drift rather than by natural selection.
While the primary structure of genes mainly evolves by random drift, organization
of genes, especially that of multigene families, has important bearings on progressive change of organisms (HOOD et al. [3]), and it is an urgent task of population geneticists to provide theoretical understanding of the evolutionary
change of multigene families. For the past several years, I have been working
on this problem (OHTA [4-7], KIMURA and OHTA [8]), and the purpose of this review
is to present facts and theories on three multigene families (of globins, immunoglobulins and histocompatibility antigens) that have been studied in greatest
detail and considered to be most important. The recent article of DOVER [9]
mainly treats evolutionary dynamics of gene families of ribosomal RNA and non-
coding sequences, whereas this review concentrates on 'informational multigene
families' that carry out extremely diverse function.

MULTIGENE FAMILY OF HEMOGLOBINS

Structure of gene families of hemoglobins has been clarified for man, mouse and
others (PROUDFOOT et al. [10]). The two families are on different chromosomes
and both have a few pseudogenes. All members have evolved by duplication of a
primordial gene, and are arranged according to the order of their expression.
From the standpoint of population genetics, it is interesting to note that two
very similar genes, such as $^G\gamma$-$^A\gamma$, $\alpha1$-$\alpha2$ and $\zeta1$-$\zeta2$, exist in tandem. The above
structure is known to be common to man, gorilla and orangutan, therefore gene
duplication must have preceded their speciation. Nevertheless the two tandem
genes are very similar to each other. This is a common feature observed in all
multigene families, and is called coincidental (HOOD et al. [3]), concerted
(ZIMMER et al. [11]) or horizontal (BROWN et al. [12]) evolution, and more recently molecular drive (DOVER [9]). By repeated unequal (but homologous)
crossing-over and gene conversion, the two gene members are thought to be kept
similar. The unequal crossover products are unbalanced with respect to the number
of genes and therefore have deleterious effects on the organisms. In fact, individuals carrying chromosomes with one copy or three copies of hemoglobin α gene
may show mild thalassemia (GOOSSENS et al. [13]). From the standpoint of popu-

lation genetics, it is reasonable to suppose that the unbalanced chromosomes are kept in the population in low frequencies by balance between unequal crossing-over and selective elimination (OHTA [14], TAKAHATA [15]). However, unequal crossing-over contributes to homogenization of the two genes. Figure 1 shows an example of concerted evolution by unequal crossing-over.

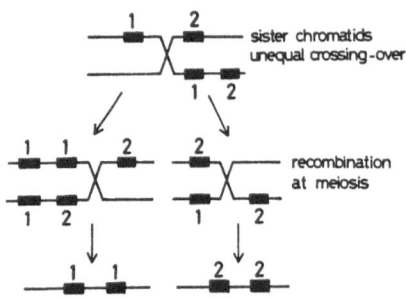 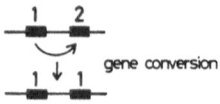

Figure 1. Diagram showing the models of concerted evolution by unequal crossing-over and gene conversion (from OHTA [26])

It also depicts a diagram of gene conversion, in which there is no change of the number of genes. At present, it is not known which one of the two mechanisms contributes more to concerted evolution. If unequal crossing-over is the main mechanism, it is estimated to occur at the rate of about 10^{-4} per two genes per generation and the selective disadvantage of unbalanced chromosomes is estimated to be roughly 10^{-2} (OHTA [14]). The above picture of evolution of globin gene families is remarkably different from the conventional neo-Darwinian view. Continued occurrence of duplication, deletion and gene conversion would provide various opportunities for the progressive evolution of gene families, and the creation of new genes such as the ζ gene from the α gene would be relatively easy under such a situation.

MULTIGENE FAMILY OF IMMUNOGLOBULINS

The gene family of immunoglobulins is known as the most complicated organization so far studied. There are three main classes, IgG, IgM and IgA; IgG being used in mature mammals. The molecule of this class consists of two identical light and two identical heavy chains. The light chain is made of two homologous domains, the constant and variable regions; whereas the heavy chain is made of a variable region and three constant regions with a small hinge region in the middle. The variable region is further classified into the framework and the hypervariable regions and the amino acid sequence of the latter is mainly responsible for the antigen specificity.

Thanks to the remarkable progress of molecular biology, the structure of this complicated gene family has been elucidated. See LEDER [16] for review, but the outline of gene organization is as follows. The variable region is split into two gene families (V and J) for the light chain, and three gene families (V, D and J) for the heavy chain. The V family usually contains a few hundred gene copies, and the J family, several copies. The V gene codes for about 90 amino acid sites, and J, roughly 15 amino acid sites. The D gene codes for several amino acid sites between V and J for the heavy chain and this family contains only several copies. An amazing discovery is that one gene each from the V, D, and J families are joined together during differentiation to make a complete gene, and enormous diversity is supplied by combinatorial use of split genes.

From the standpoint of evolutionary genetics, the present organization is the product of numerous trials and errors of duplication and deletion of gene segments. A most interesting phenomenon is again the concerted (coincidental) evolution of V, D and J families of each species. In other words, each family evolves as a set and species specific DNA (amino acid) sequence is observed, although consider-

134

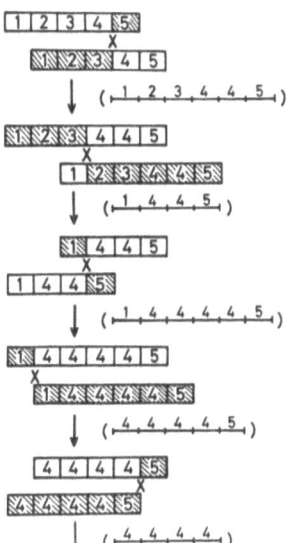

Figure 2. A model of gene fixation by unequal crossing-over (from OHTA [7], KIMURA and OHTA [8])

able variability exists among the gene copies. As before, unequal crossing-over and gene conversion are considered to be responsible for the concerted evolution. Figure 2 shows an example of spreading of one copy (gene 4) on the chromosome by continued unequal crossing-overs.

The figure is for a single chromosomal line, but in the actual process, genes are exchanged between the haploid sets, and the process is much more complicated. It is a two-fold process of random drift on the chromosome and in the population. See DOVER [9] for an interesting discussion of two-fold processes of drift of ribosomal DNA family. If drift is rapid, gene members are expected to be uniform as in rDNA family, whereas if it is slow, copy members contain genetic variation as in immunoglobulin gene families.

I have been working on the population genetics theory of this process, and it is interesting to apply the theory to amino acid diversity of immunoglobulins, of which much data are available. The results are summarized here; as to the details, see OHTA [7]. By using the data compiled by KABAT et al. [17], an amino acid identity coefficient either within or between species is calculated separately for both the hypervariable and the framework regions. The relationship between the identity coefficient of the hypervariable region and that of the framework region suggests that the rate of evolutionary accumulation of amino acid replacements in the hypervariable region is roughly three times higher than in the framework region. Based on the theory, the high variability within a species at the hypervariable region is just as predictable from this high substitution rate. In addition, this rate of evolution of the hypervariable region is roughly the same as that of fibrinopeptides, known to be the most rapidly evolving protein, i. e. it is close to the maximum predicted from the neutral theory of evolution (KIMURA [2]) when there is no functional or structural constraint on the molecule. The above results imply that antibody diversity is mainly produced by combinatorial use of randomly accumulated mutations (i. e. selectively neutral mutations) in redundant copies. It should be noted that these statistical analyses do not exclude the possibility that several nucleotide substitutions may occur somatically during differentiation. The above picture of the evolution of antibody genes is surprising to most geneticists, and has a significant meaning for the Darwinian theory of evolution. Darwinian natural selection increases genes in the population that are needed, whereas the combinatorial use of randomly accumulated mutations would imply that a species keeps a repertory which is more than required at one time.

OHNO et al. [18] argue that the evolution of the immune system is Promethean in that a species often possesses an antibody molecule capable of reacting to a pathogen that the species had never encountered before.

Another topic on the evolution of immunoglobulin gene family is the dynamic change of the constant region gene families of man and mouse. The various classes of constant region genes of the heavy chain are arranged on the chromosome according to the order of expression, and each gene contains several exons which corresponds to a protein domain. Thus I use here domain and exon interchangeably. Honjo, Miyata and others (MIYATA et al. [19], SLIGHTOM et al. [20], YAMAWAKI-KATAOKA et al. [21]) have found that homologous domains are often transferred from one locus to another in the course of evolution. In the case of constant region genes, domain transfer occurs between genes belonging to the same class when more than one copy is present. Figure 3 shows a hypothetical case of domain transfer.

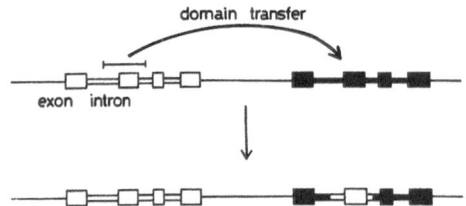

Figure 3. Diagram showing a hypothetical case of domain transfer (from OHTA [26])

As before, domain transfer is considered to be due to gene conversion or unequal crossing-over (in this case, double unequal crossing-over). In any case, it seems that whenever homologous genes that are split are tandemly arranged on the chromosome, dynamic evolutionary change is expected to occur. Frequent occurrence of domain transfer has an important bearing on the evolution of the gene family of histocompatibility antigen and other families.

MAJOR HISTOCOMPATIBILITY COMPLEX (MHC)

Major histocompatibility antigen was discovered as causing skin graft rejection, and has drawn much attention of evolutionary geneticists because of the extreme polymorphism. In the case of man, there are at least three linked loci: HLA-A, -B and -C, all of which are located in the chromosomal region called major histocompatibility complex (MHC). The region contains, other than genes of histocompatibility antigen, the genes of immune response and complement, and is called "supergene." H2 complex of mouse corresponds to HLA complex of man. According to Bodmer and his associates (PLOEGH et al. [23]), heterozygosity exceeds 90% at HLA-A or -B loci in the English population. Why are they so highly polymorphic? There has been much discussion on how diversifying selection may increase heterozygosity. However with the accumulation of data at the molecular level, it has become more and more difficult to explain the polymorphism by the conventional models of diversifying selection based on single locus. Comparisons of direct gene products (amino acid sequences) show that amino acid identity is about 90% when allelic products are compared whereas it is 85 ~ 90% when nonallelic products are compared (i. e. H2-K vs. -D) (PLOEGH et al. [23]). In other words, there is neither K-ness nor D-ness. On the other hand, it has been found that there are 30-50 homologous loci of the antigen in a mouse genome, and each gene has a split structure with 7-8 domains (STEINMETZ et al. [24]), i. e. they constitute a multigene family.

It is thus likely that high polymorphism is caused by multigenic structure and domain transfer as is reviewed in the previous section. By using the model of gene conversion, I here summarize the theory for evaluating probability of gene identity (identity coefficients) which is also applicable to hemoglobin and immunoglobulin gene families discussed in the previous sections. In the following, allelic and nonallelic identities of homologous units (amino acids or nucleotides) are examined. Various parameters are, N = effective population size, n = the number of genes in a family, v = mutation rate per gamete per generation and β = interchromosomal recombination rate per adjacent loci. Under the assumption that gene conversion is mainly asymmetric with no bias, let λ be the rate at which a domain is converted by one of the remaining (n - 1) homologous domains in one generation. Figure 4 depicts the domain transfer.

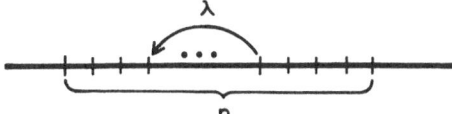

Figure 4. The model of gene conversion (from OHTA [25])

It is possible to formulate changes in the probabilities of allelic and nonallelic identities; let f be the average probability of allelic identity, C_1 be the average identity probability of genes at different loci of the family on one chromosome, and C_2 be that of two genes taken from different loci of two homologous chromosomes of the population. Figure 5 depicts these three identity coefficients.

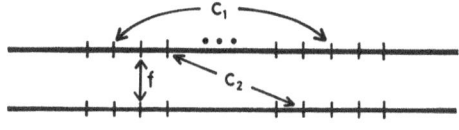

Figure 5. The meaning of three identity coefficients (from OHTA [25])

Then it can be shown that the vector $c = (f, C_1, C_2)$ changes according to the following equation in one generation, by assuming v, N^{-1}, λ and β << 1,

$$
\Delta c =
\begin{bmatrix}
-2v - \dfrac{1}{2N} - 2\lambda & 0 & 2\lambda \\[2ex]
0 & -2v - \dfrac{2\lambda}{n-1} - \dfrac{(n+1)}{3}\beta & \dfrac{(n+1)}{3}\beta \\[2ex]
\dfrac{2\lambda}{n-1} & \dfrac{1}{2N} & -2v - \dfrac{2\lambda}{n-1} - \dfrac{1}{2N}
\end{bmatrix}
c +
\begin{bmatrix}
\dfrac{1}{2N} \\[2ex]
\dfrac{2\lambda}{n-1} \\[2ex]
0
\end{bmatrix}
. \quad (1)
$$

For details of the derivation of equation (1), see OHTA [25] and [26].

The equilibrium values may be obtained by solving the above equation at $\Delta c = 0$. Table 1 gives the equilibrium values for some interesting cases. As can be seen from the table, the observed degree of allelic and nonallelic identity can be explained by assuming that a domain is converted at a rate of 10^{-5} - 10^{-6} per generation. It is also noted that this rate of domain transfer is compatible with the observed high mutation rate at marker loci of MHC detected by skin grafting (KLEIN [27]).

When the gene family size is kept fairly constant, the previous model of unequal crossing-over (OHTA [7]) becomes the same as the conversion model. However the former analysis is more approximate than the latter, since only two identity coefficients, C_{w1} and C_{w2} corresponding to C_1 and C_2, are formulated there. For details of the correspondence between the two models, see OHTA [26].

PROBLEM OF PROGRESSIVE EVOLUTION

The examples of evolving multigene families in the previous sections and many others suggest that minor DNA rearrangements, such as duplication deletion or

TABLE I. Equilibrium values of allelic and nonallelic identity of a multi-gene family

Parameter value		Allelic identity f	Nonallelic identity $c_1 \approx c_2$
λ	0.5	0.956	0.538
	5	0.932	0.867
(in units of 10^{-6})	50	0.942	0.937
v	0.1	0.992	0.977
	1	0.936	0.811
(in units of 10^{-8})	10	0.756	0.298
n	30	0.949	0.852
	50	0.923	0.774
N	1	0.986	0.855
	10	0.879	0.762
(in units of 10^4)	100	0.422	0.366

Parameters: $\lambda = 2.5 \times 10^{-6}$, $v = 10^{-8}$, $n = 40$, $N = 5 \times 10^4$, and $(n - 1)\beta = 10^{-3}$, unless otherwise specified.

domain transfer, are very important for the progressive evolution of higher organisms. The rate of unequal crossing-over or gene conversion is much higher than previously thought, and in any species these changes are constantly occurring. Selectively advantageous or neutral changes may survive.

In this regard, I mention two examples of quantitative characters that have been studied in some detail. One is viability polygenes by MUKAI [28, 29], and the other is abdominal bristle number by FRANKHAM et al. [30], both in *Drosophila*. Mukai accumulated mutations of viability polygenes in *Drosophila* experimental populations by minimizing competition (natural selection) and found that the mutation rate is about ten times higher than the ordinary mutation studies indicate. It is likely that most "mutations" of viability polygenes are duplication deletion, gene conversion or transfer of DNA segment, that is now under intense study (DOVER and FLAVELL [31]). The bristle number is known to be related to the ribosomal RNA gene family, and unequal crossing-over was considered to be a cause of unexpected response to artificial selection (FRANKHAM et al. [30]).

I would like to emphasize that evolutionary biologists have to seriously consider the implications of new discoveries of molecular genetics as reported here. As WRIGHT [32] remarked many years ago, Darwin's general contention that evolution is a process of statistical transformation of populations still holds, but now our view of evolution is enriched by our knowledge of additional mechanisms of intragenomic changes. For quantitative assessment of the dynamics of gene material, the concept of identity coefficients provides a useful tool and enables one to grasp the nature of genetic diversity of the gene pool of evolving species.

ACKNOWLEDGMENTS

I thank Dr. Motoo Kimura for his stimulating discussions and encouragement, and Dr. Kenichi Aoki for carefully going over the manuscript and making many valuable suggestions to improve the presentation.

REFERENCES

1 M. Kimura: Nature 217, 624-626 (1968)
2 M. Kimura: Sci. Amer. 241(5), 94-104 (1979)

3 L. Hood, J.H. Campbell and S.C.R. Elgin: Ann. Rev. Genetics 9, 305-353 (1975)
4 T. Ohta: Nature 263, 74-76 (1976)
5 T. Ohta: Genet. Res. Camb. 31, 13-28 (1978)
6 T. Ohta: Genetics 88, 845-861 (1978)
7 T. Ohta: Evolution and variation of multigene families, Lecture Notes in
   Biomathematics, Vol. 37 (Springer-Verlag, Berlin, New York 1980)
8 M. Kimura and T. Ohta: Proc. Natl. Acad. Sci. U.S.A. 76, 4001-4005 (1979)
9 G. Dover: Nature 299, 111-117 (1982)
10 N.J. Proudfoot, M.H.M. Shander, J.L. Manley, M.L. Gefter and T. Maniatis:
   Science 209, 1329-1336 (1980)
11 E.A. Zimmer, S.L. Martin, S.M. Beverley, Y.W. Kan and A.C. Wilson: Proc. Natl.
   Acad. Sci. U.S.A. 77, 2158-2162 (1980)
12 D.D. Brown, P.C. Wensink and E. Jordan: J. Mol. Biol. 63, 57-73 (1971)
13 M. Goossens, A.M. Dozy, S.H. Embury, Z. Zachariades, M.G. Hadjiminas,
   G. Stamatoyannopoulos and Y.W. Kan: Proc. Natl. Acad. Sci. U.S.A. 77,
   518-521 (1980)
14 T. Ohta: Genet. Res. Camb. 37, 133-149 (1981)
15 N. Takahata: Genet. Res. Camb. 38, 97-102 (1981)
16 P. Leder: Sci. Amer. 246(5), 72-83 (1982)
17 E.A. Kabat, T.T. Wu and H. Bilofsky: Variable regions of immunoglobulin
   chains, Medical Computer Systems (Bolt, Beranek and Newman, Cambridge,
   Mass. 1976)
18 S. Ohno, T. Matsunaga, J.T. Epplen and T. Hozumi: in *Immunology 80, Progress
   in Immunology IV*, edited by M. Fougereau and J. Dausset (Academic Press,
   London, 1980), pp. 577-598
19 T. Miyata, T. Yasunaga, Y. Yamawaki-Kataoka, M. Obata and T. Honjo: Proc.
   Natl. Acad. Sci. U.S.A. 77, 2143-2147 (1980)
20 J.L. Slightom, A.E. Blechl and O. Smithies: Cell 21, 627-638 (1980)
21 Y. Yamawaki-Kataoka, S. Nakai, T. Miyata and T. Honjo: Proc. Natl. Acad.
   Sci. U.S.A. 79, 2623-2627 (1982)
22 W.F. Bodmer: in *Human Genetics: possibilities and realisties*, Ciba
   Foundation Series 66 (Elsevier, North-Holland, 1979), pp. 205-229
23 H.L. Ploegh, H.T. Orr and J.L. Strominger: Cell 24, 287-299 (1981)
24 M. Steinmetz, A. Winoto, K. Minard and L. Hood: Cell 28, 489-498 (1982)
25 T. Ohta: Proc. Natl. Acad. Sci. U.S.A. 79, 3251-3254 (1982)
26 T. Ohta: Theor. Pop. Biol. 23, 216-240 (1983)
27 J. Klein: Adv. Immunol. 26, 55-146 (1978)
28 T. Mukai: Genetics 54, 1-19 (1964)
29 T. Mukai: Proc. XII Intern. Congr. Genetics 3, 293-308 (1969)
30 R. Frankham, D.A. Briscoe and R.K. Nurthen: Nature 272, 80-81 (1978)
31 G. Dover and R.B. Flavell (eds.): Genome Evolution (Academic Press, London,
   1982)
32 S. Wright: Bull. Amer. Math. Soc. 48, 223-246 (1942)

Evolution of Biothermodynamic Systems

J.U. Keller

Institut für Thermodynamik, TK 7, Technische Universität Berlin
Strasse des 17. Juni 135, D-1000 Berlin 12, Fed. Rep. of Germany

Abstract

The evolution of simple sets of biothermodynamic systems is considered within the framework of thermodynamics of irreversible processes. An integral equation for the time derivative of the total
number $\dot{n}_\alpha$ of systems of type α included in a set is derived. This
equation includes the constitutive laws which govern the exchange
of heat, work and mass of a single biothermodynamic system with its
environment, the state of the environmental system and the exchange
of work, heat and mass of the total system with its surroundings as
well. An example is given for the evolution of a set of systems
growing only by isothermal absorption processes. Given constant mass
supply, the total number of biothermodynamic systems in the set exhibits a relaxation type behaviour, i.e. starting from any initial
value n_o it approaches monotonously an asymptotic value $n_\infty > n_o$.

1. Populations of biothermodynamic systems

We consider a fluid thermodynamic system $\overset{\circ}{\Sigma}$, [1] which inclu-
des sets of biothermodynamic systems $\Sigma_{i_\alpha}^{(\alpha)}$, $i_\alpha = 1, 2, \ldots n^{(\alpha)}(t)$,
of various types $\alpha = 1, 2, 3, \ldots A$.

A biothermodynamic system (BTS) is defined as a certain
piece of matter which
i) may exchange heat, mechanical work and mass with its surroun-
    dings and
ii) may reproduce itself or another kind of systems.
Any BTS has to have clear defined boundaries dividing its interior
from its surroundings or environmental system Σ^*.
We assume a BTS $\Sigma_{i_\alpha}^{(\alpha)}$ to start its evolution in a certain, thermodynamically well-defined state at a certain time $\tau_i^{(\alpha)}$ prior to which
it is not considered to be an individual system but rather part

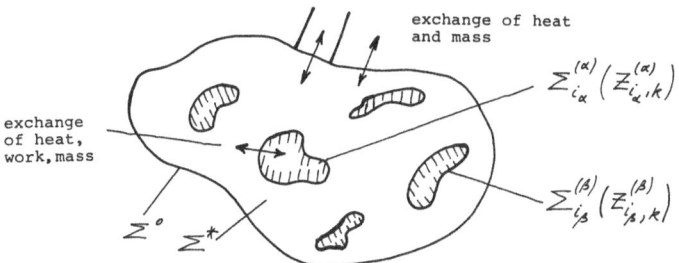

$$\sum_{i_\alpha}^{(\alpha)} \left(z_{i_\alpha , k}^{(\alpha)} \right)$$

exchange
of heat,
work, mass

$$\sum_{i_\beta}^{(\beta)} \left(z_{i_\beta , k}^{(\beta)} \right)$$

$$\sum^{0} \quad \sum^{*}$$

Fig.1. Thermodynamic system $\sum^{0}$ including sets of biothermodynamic systems $\left\{ \sum_{i_\alpha}^{(\alpha)} \right\}$

of another system or of the environmental system $\sum^{*}$.

The total number $n^{(\alpha)}(t)$ of BTSs of type α being present (in various states of their individual evolution) in $\sum^{0}$ at time t can obviously be written as

$$n^{(\alpha)}(t) = \int_{-\infty}^{t} \sum_{i}^{(\alpha)} \delta \left(\tau - \tau_i^{(\alpha)} \right) d\tau , \qquad (1.1)$$

or, using a continuous quasi-distribution function $\nu^{(\alpha)}(t)$, as

$$n^{(\alpha)}(t) = \int_{-\infty}^{t} \nu^{(\alpha)}(\tau) d\tau . \qquad (1.2)$$

We then have

$$\dot{n}^{(\alpha)}(t) = \nu^{(\alpha)}(t) \qquad (1.3)$$

and $\left[n^{(\alpha)}(t) \right]$, i. e. the largest capital number included in $n^{(\alpha)}(t)$, is the total number of BTSs of type α being present in $\sum^{0}$ at time t.

We now consider only BTSs which exclude each other, i.e. which are not part of any other BTS but rather are individual systems included in the overall system $\sum^{0}$.
We then have the relation (cp. Fig.1):

$$\sum^{0} = \int_{-\infty}^{t} \sum_{\tau}^{(\alpha)} \nu^{(\alpha)}(\tau) + \sum^{*\alpha} \qquad (1.4)$$

with the environmental system $\sum^{*\alpha}$ being defined as

$$\sum^{*\alpha} = \sum_{\beta \neq \alpha}^{A} \int_{-\infty}^{t} \sum_{\tau}^{(\beta)} \nu^{(\beta)}(\tau) d\tau + \sum^{*}, \quad \alpha = 1...A . \qquad (1.5)$$

These relations also hold in algebraic sense at any time t for
any extensive quantity Z_k like the volume, mass, internal energy
or entropy of the various systems:

$$Z_k^o(t) = \int_{-\infty}^{t} r^{(\alpha)}(\tau)\, Z_{\tau k}^{(\alpha)}(t)\, d\tau + Z_k^{*\alpha}(t), \tag{1.6}$$

$$k = 1, 2, 3 \ldots ,$$
$$\alpha = 1 \ldots A .$$

Here $Z_{\tau k}^{(\alpha)}(t)$ is the value of Z_k in a BTS of type α which started
its evolution at time τ and is considered at time t. The relation
(1.6) will be the starting point for the derivation of an integral
equation for the quasi-distribution function $r^{(\alpha)}(t)$. However, to
be well prepared we have to make a few remarks on thermodynamics
of processes occurring in the various systems.

2. Thermodynamics of processes

We consider a simple thermodynamic system [1,p.5,3,p.10], Σ with
extensive parameters

$$\left(Z_k, \; k = 1 \ldots N+2 \right) = \left(U, V, M_1 \ldots M_N \right) \tag{2.1}$$

denoting its internal energy, volume and masses of components
i=1,...N, respectively. The behaviour of the system against quasistatic exchange of heat, work and mass can be described by its
fundamental equation for the entropy [3,4]

$$S = S\left(Z_k, \; k = 1 \ldots N+2 \right) , \tag{2.2}$$

or Gibbs' equation

$$dS = \sum_{k=1}^{N+2} z_k \, dZ_k , \tag{2.3}$$

with z_k being so-called accompanying intensive parameters which
can be calculated from the (generalized) thermostatic equations
of state

$$z_k = \left(\frac{\partial S}{\partial Z_k} \right)_{Z_\ell, \; \ell \neq k} , \quad k = 1 \ldots N+2 . \tag{2.4}$$

To describe phenomenologically the behaviour of the system
during dynamic exchange processes of heat, work and mass with its
surroundings Σ^*, one has to consider the balance equation for the
extensive parameters:

142

$$\dot{Z}_k = J_k + P_k \,, \qquad k = 1 \dots N+2 \,. \qquad (2.5)$$

Here J_k denotes the (overall) flux and P_k the production term of Z_k.
These equations can be complemented by dynamic equations of state
or constitutive equations, a sound physical base of which is the
second law of thermodynamics which is due to R.Clausius [1,3,5] and
can be written as

$$S(t) - S(-\infty) \geq \int_{-\infty}^{t} \sum_k \mathcal{S}_k^* J_k \, dt \quad \dots \text{ all } t. \qquad (2.6)$$

Here $S(t)$ denotes the accompanying thermostatic entropy of the
system at time t, defined by (2.2) also during processes or in nonequilibrium states of Z. The quantities $\mathcal{S}_k^*$ denote well-known
functions of the (normally prescribed) intensive parameters of the
surrounding system Z^*! Equation (2.5.) combined with (2.6) yield
the passivity inequality of the system

$$\int_{-\infty}^{t} dt \sum_k \left[(Z_k - \mathcal{S}_k^*) J_k + Z_k P_k \right] \geq 0 \qquad (2.7)$$

from which the constitutive equations can be read as

$$J_k = \mathcal{F}_k \left\{ Z_\ell, \mathcal{S}_\ell^* \right\}, \qquad (2.8)$$

$$k = 1 \dots N+2,$$

$$P_k = \mathcal{T}_k \left\{ Z_\ell, \mathcal{S}_\ell^* \right\}. \qquad (2.9)$$

The symbols $\mathcal{F}$ and $\mathcal{T}$ denote (generally nonlinear) functionals of
their arguments, the mathematical structure of which however is due
to several conditions which cannot be discussed in detail here [5,6].
Inserting (2.8, 2.9) in the balance equation (2.5), we get generalized evolution equations for a simple thermodynamic system

$$\dot{Z}_k = \mathcal{F}_k \left\{ Z_\ell (Z_m), \mathcal{S}_\ell^* \right\} +$$

$$+ \mathcal{T}_k \left\{ Z_\ell (Z_m), \mathcal{S}_\ell^* \right\}. \qquad (2.10)$$

Consider now a BTS, assume it to be simple, i.e. substitute
$Z \to Z_\tau^{(\alpha)}$ and choose as extensive quantity the total mass
$M_\tau^{(\alpha)} = \sum_k M_{\tau k}^{(\alpha)}$ of the system. Neglecting exchange processes of heat
and work, i.e. considering only mass exchange, we get from (2.10)

$$\dot{M}_\tau^{(\alpha)} = \sum_{k=1}^{N} \left(\mathcal{F}_{\tau k}^{(\alpha)} \{\div\} + \pi_{\tau k}^{(\alpha)} \{\div\} \right), \qquad (2.11)$$

$$\{\div\} = \{ \mu_{\tau\ell}^{(\alpha)} \left(M_{\tau m}^{(\alpha)} \right), \mu_m^{*(\alpha)} \}.$$

Here $\mu_{\tau\ell}^{(\alpha)}$, $1=1...N$ indicate the chemical potentials of the various components. Equation (2.11) is a generalized biological evolution equation, various forms of which have been discussed in the literature [7-9] .

We now would like to emphasize that Eqs.(2.8,2.9) combined with Eqs.(2.4, 2.5) form a complete set of (non linear integro-differential) equations from which on principle the values of the extensive and the intensive quantities Z_k, z_k of Σ can be calculated at any time t, given initial conditions.

The formalism given in this section likewise can be formulated for the total system Σ°, any BTS $\Sigma_\tau^{(\alpha)}$ of type α or the environmental system $\Sigma^{*\alpha}$ as introduced in Sect.1.

3. An integral equation for the number of biothermodynamic systems within a population

Consider the time of Eqs.(1.6). It reads

$$\sum_k {}_{z_k}^{*\alpha} \Big| \dot{Z}_k^\circ(t) = Z_{tk}^{(\alpha)}(t)\, \nu^{(\alpha)}(t) + \int_{-\infty}^{t} \partial_t Z_{\tau k}^{(\alpha)}(t)\, \nu_{(\tau)}^{(\alpha)} d\tau + \dot{Z}_k^{*\alpha} . \quad (3.1)$$

Choosing as extensive quantity Z=S, for the entropy of the respective systems we get

$$\dot{S}^{(o)}(t) = S_t^{(\alpha)}(t)\, \nu^{(\alpha)}(t) + \int_{-\infty}^{t} \dot{S}_\tau^{(\alpha)}(t)\, \nu^{(\alpha)}(\tau)\, d\tau + \dot{S}^{*(\alpha)}(t). \quad (3.2)$$

Let us denote Gibbs' equations for the various systems

$$\Sigma^\circ(t) : \dot{S}^{(o)} = \sum_k z_k^{(o)} \dot{Z}_k^{(o)} , \qquad (3.3)$$

$$\Sigma_\tau^{(\alpha)}(t) : \dot{S}_\tau^{(\alpha)} = \sum_k z_{\tau k}^{(\alpha)} \dot{Z}_{\tau k}^{(\alpha)} , \qquad (3.4)$$

$$\Sigma^{*\alpha}(t) : \dot{S}^{*\alpha} = \sum_k z_k^{*(\alpha)} \dot{Z}_k^{*(\alpha)} , \qquad (3.5)$$

144

and besides the Gibbs-Duham relation

$$\sum_{t}^{(\alpha)}(t): \quad S_t^{(\alpha)} = \sum_k z_{tk}^{(\alpha)} Z_{tk}^{(\alpha)}. \tag{3.6}$$

Inserting (3.3 - 6) in (3.2), we combine the resulting equation
with Eq.(3.1) which itself has been multiplied by $z_k^{+(\alpha)}$ and added
up over k=1...N+2. We thus get

$$\Lambda^{(\alpha)}(t)\, r^{(\alpha)}(t) = L^{(\alpha)}(t) + \int_{-\infty}^{t} K^{(\alpha)}(\tau,t)\, r^{(\alpha)}(\tau)\, d\tau, \quad \alpha = 1...A, \tag{3.7}$$

with

$$\Lambda^{(\alpha)}(t) = \sum_k \left(z_k^{*(\alpha)}(t) - z_{tk}^{(\alpha)}(t) \right) Z_{tk}^{(\alpha)}(t), \tag{3.8}$$

$$L^{(\alpha)}(t) = \sum_k \left(z_k^{+(\alpha)}(t) - z_k^{(0)}(t) \right) \dot{Z}_k^{(0)}(t), \tag{3.9}$$

$$K^{(\alpha)}(\tau,t) = \sum_k \left(z_{\tau k}^{(\alpha)}(t) - z_k^{*(\alpha)}(t) \right) \dot{Z}_{\tau k}^{(\alpha)}(t). \tag{3.10}$$

The relation (3.7) also can be written as balance equation for
the total number $n^{(\alpha)}(t)$ of BTS of type α present in a population at time t. Indeed, using (1.3) we get

$$\dot{n}^{(\alpha)}(t) = \frac{L^{(\alpha)}(t)}{\Lambda^{(\alpha)}(t)} + \int_{-\infty}^{t} \frac{K^{(\alpha)}(\tau,t)}{\Lambda^{(\alpha)}(t)}\, \dot{n}^{(\alpha)}(\tau)\, d\tau. \tag{3.11}$$

The quantity $\Lambda^{(\alpha)}$ characterizes the initial or birth state of the
BTS, $L^{(\alpha)}(t)$ describes external influences on the overall system
Σ°, whereas the kernel function $K^{(\alpha)}(\tau,t)$ includes the evolution
process of the BTS, i.e. their exchange process of heat, work and
mass with their surrounding system $\Sigma^{*\alpha}$. Hence the first term on
the r.h.s. of (3.11) is a flux term, whereas the integral indicates
a production term of BTS.

If the fundamental equation (2.2) and the constitutive equations
(2.8, 2.9) of the BTS $\Sigma_{\tau}^{(\alpha)}$ are known and if the external influences on the overall system Σ°, i.e the heat and mass supplies,
are known, the auxiliary quantities $\Lambda^{(\alpha)}$, $L^{(\alpha)}$, $K^{(\alpha)}$ are also
known and Eqs.(3.7, 3.11) become integral equations of the
Volterra-type from which the growth rate $\dot{n}^{(\alpha)}$ of the total number
$n^{(\alpha)}$ of BTSs within a population can be calculated on principle.

Using a very simple example, the procedure will be demonstrated
in the next section.

4. Example

Let us consider the evolution process of a set of very simple BTSs
of the same type which however at any time t may be in different
states of their individual evolution process. A member system Σ_τ,
starting its evolution at time τ , consists of a certain amount
$M_{\tau 1}(t)$ of mass "1". The environmental system Σ^* may contain two
different components "0" and "1", the latter being exchanged with
Σ_τ by isothermal absorption-desorption processes. We consider only
this process, i.e. neglect the exchange of work and chemical
reactions. We then have for Σ_τ the fundamental equation (2.2)

$$ S_\tau = S_\tau \left(U_\tau, V_\tau, M_{\tau 0}=0, M_{\tau 1} \right), \tag{4.1} $$

the thermostatic equation of state

$$ -\frac{\mu_{\tau 1}}{T} = \left(\frac{\partial S_\tau}{\partial M_{\tau 1}} \right)_{U_\tau, V_\tau} , \tag{4.2} $$

the conservation equation (2.5) for

mass: $\qquad \dot{M}_{\tau 1} = J_{\tau 1} , \tag{4.3}$

energy: $\qquad \dot{U}_\tau = J_{\tau u} = \mu_{\tau 1}^* J_{\tau 1} , \tag{4.4}$

and the constitutive equation (2.8)

$$ J_{\tau 1} = \mathcal{F}_{\tau 1} \left\{ \frac{\mu_1^* - \mu_{\tau 1}}{T} \right\} , \tag{4.5} $$

indicating a certain non-linear functional of its argument.
For the sake of simplicity we assume that due to appropriate mass

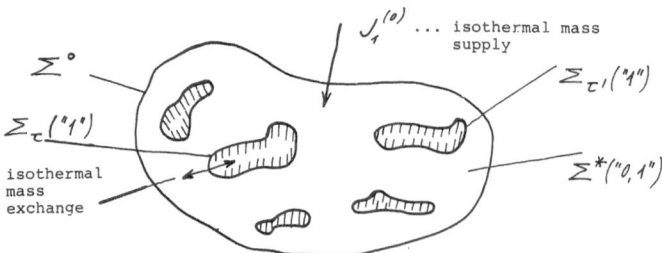

Fig.2. Population of simple BTS growing by isothermal mass exchange

supply to the overall system Σ^0, the chemical potential of component "1" in the environmental system, i.e. outside of Σ_τ, is constant

$$\mu_1^* = const.$$ (4.6)

Besides, we assume the mass density in Σ_τ to be constant, from which

$$\mu_{\tau 1} = g_\tau = const$$ (4.7)

can be inferred.

We rewrite the constitutive equation (4.5)

$$\begin{aligned} J_{\tau 1} &= \overset{\circ}{\sigma}_\tau \, \mathcal{P}_{\tau 1} \left\{ \frac{\mu_1^* - \mu_{\tau 1}}{T} \right\} \\ &= k_\tau \, M_{\tau 1}^{2/3} \, \mathcal{P}_{\tau 1} \left\{ \frac{\mu_1^* - \mu_{\tau 1}}{T} \right\} \\ &= \alpha_1 \, M_{\tau 1}^{2/3} . \end{aligned}$$ (4.8)

Here $\overset{\circ}{\sigma}_\tau$ indicates the active surface of Σ_τ, exchanging mass "1" with its surroundings, $\mathcal{P}_{\tau 1}$ is the mass flux density, k_τ indicates a shape-dependent coefficient whose value is assumed to be constant. Thus $\alpha_1 = k_\tau \, \mathcal{P}_{\tau 1}$, due to (4.6, 4.7), is also a constant irrespective of the structure of the functional $\mathcal{P}_{\tau 1}$!
Equations (4.3, 4.8) deliver the growth rate equation

$$\dot{M}_{\tau 1} = \alpha_1 \, M_{\tau 1}^{2/3} .$$ (4.9)

It has the solution

$$M_{\tau 1}(t) = M_{\tau 10} \left(1 + \alpha \, (t - \tau) \right)^3 \dots \quad t \geqslant \tau ,$$ (4.10)

with $M_{\tau 10} = M_{\tau 1}(\tau)$ being the initial mass of the BTS Σ_τ and

$$\alpha = \frac{\alpha_1}{3 M_{\tau 10}} = k_\tau \frac{\mathcal{P}_{\tau 1}}{3 \, M_{\tau 10}} = const.$$ (4.11)

Of course, the result (4.10) can be valid only during a certain juvenile period of the system Σ_τ, after which assumptions (4.6, 4.7) and k=const. have to be replaced by more realistic ones, avoiding the divergence of $M_{\tau 1}(t)$ for $t \to \infty$.

147

The auxiliary quantities Λ, L, K are (cp.(3.8-10))

$$\Lambda = -\frac{1}{T}\left(\mu_1^* - \mu_{\tau 1}\right) M_{\tau 10} = const, \qquad (4.12)$$

$$L(t) = -\frac{1}{T}\left(\mu_1^* - \mu_{\tau 1}\right) J_1^{(0)}(t), \qquad (4.13)$$

$$K(\tau, t) = \frac{1}{T}\left(\mu_1^* - \mu_{\tau 1}\right) \dot{M}_{\tau 1}(t), \qquad (4.14)$$

leading to the evolution equation

$$\dot{n}(t) = j_1^{(0)}(t) - 3\alpha \int_0^t \left[1 + \alpha(t-\tau)\right] \dot{n}(\tau) d\tau. \qquad (4.15)$$

Here

$$j_1^0(t) = J_1^{(0)}(t) / M_{\tau 10},$$

where $J_1^{(0)}$ is the supply of mass "1" to the overall system Σ^0, which itself is assumed to be in thermodynamic equilibrium prior to t=0. The Laplace-transform solution of (4.15) reads

$$\dot{n}(t) = j_1^{(0)}(t) - \int_0^t \left(\sum_{i=1}^3 A_i(\alpha) e^{P_i(\alpha)(t-\tau)}\right) j_1^{(0)}(\tau) d\tau. \qquad (4.16)$$

The coefficient A_i, p_i are functions of the parameter α defined by

$$F(\alpha, p) = \frac{3\alpha p^2 + 6\alpha^2 p + 6\alpha^3}{p^3 + 3\alpha p^2 + 6\alpha^2 p + 6\alpha^3} = \sum_{i=1}^3 \frac{A_i(\alpha)}{p - P_i(\alpha)}. \qquad (4.17)$$

Due to the second law (2.7) we have $\alpha > 0$ and hence $Rep_i < 0$!
For constant mass supply, i.e. $j_1^0 = const.$,we get from (4.16) and $F(\alpha,0) = -\sum_i (A_i/p_i) = 1$,

$$\dot{n}(t) = -j_1^{(0)} \sum_i^3 \frac{A_i}{P_i} e^{P_i t} \geqslant 0 \qquad (4.18)$$

and

$$n(t) = n_o - j_1^{(0)} \sum_i^3 \frac{A_i}{P_i^2} \left(e^{P_i t} - 1\right), \qquad (4.19)$$

leading to an asymptotic number of systems

$$n_\infty = n_o + j_1^{(0)} \sum_i \frac{A_i}{P_i^2} \geqslant n_o.$$

These results are sketched in Fig.3.

148

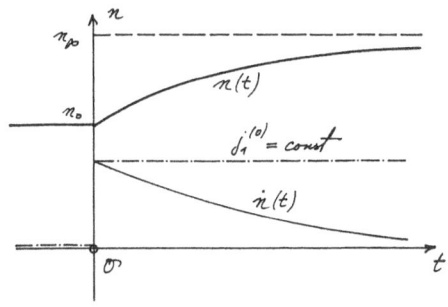

Fig.3. Time behaviour of the total number n of simple BTS growing only by isothermal mass exchange

References

1 W. Schottky, H. Ulrich, C. Wagner:
   Thermodynamics (Springer, Berlin 1973)

2 J.U. Keller:"Thermodynamics of Processes and the Evolution of
   Simple Biothermodynamic Populations", in Thermodynamics of
   Biological Systems (NAUKA, Moscow 1983 and de Gruyter,
   Berlin 1984)

3 J.U. Keller: Thermodynamik der irreversiblen Prozesse, mit
   Aufgaben, Rechenweg und Lösungen, Teil 1, Thermostatik und
   Grundbegriffe (W. de Gruyter, Berlin-New York 1977)

4 H.B. Callen: Thermodynamics (Wiley, New York 1960)

5 J. Keller:"Ober den zweiten Hauptsatz in der Thermodynamik
   Irreversibler Prozesse" (de Gruyter, Berlin-New York 1975)

6 J.U. Keller:"A general antisymmetry property of the constitutive equations of non-equilibrium thermodynamics", Int.J.
   Eng.Science, 17(1979), 715.

7 L.V. Bertalanffy: Biophysik des Fließgleichgewichtes (Vieweg,
   Braunschweig 1977)

8 H.-J. Scharf:"Wachstum", Verh.Anat.Ges., 71(1977), 29-58.

9 R. Walter, I. Lamprecht:"Modern Theories Concerning the
   Growth Equation", in Thermodynamics of Biological Processes,
   I. Lamprecht, A.I. Zotin (Eds.) (W. de Gruyter, Berlin 1978)

Self-Organising Dynamic Models of Human Systems

P.M. Allen, G. Engelen, and M. Sanglier
Chimie-Physisque II, C.P. 231, Université Libre de Bruxelles
B-Bruxelles 1050, Belgium

Abstract

The conceptual difficulties of modelling human systems are briefly
discussed, contrasting the new methods emerging from the theory of
self-organizing systems with the descriptive methods usually used.
A generic model of an urban system is described and its behaviour
explored.

1. Introduction

Over recent years new concepts have emerged governing the evolution of complex systems. From the natural sciences a whole new
range of phenomena have been discovered which, ultimately, correspond to a much deeper understanding of the mechanisms underlying
evolutionary change. From chemistry and physics (1)(2)(3)(4), it
has been shown how non-linear systems, operating far from thermodynamic equilibrium, can undergo successive structural instabilities leading to a progressive complexification of their organization and functioning. Such phenomena involve a dialogue between
stochastic and deterministic aspects of the system, as the interactions lead to a 'self-organization' process of successive
periods of stability and instability, where new qualities, traits
and characteristics emerge over time. It offers us a new scientific paradigm, involving real non-conservation, of particular interest for the human sciences. In this article we shall briefly
review some applications of these ideas in the domain of the human
sciences, where modelling has, on the whole, been somewhat unsuc-
cessful in planning and policy decisions in the real world of
human affairs.

The reasons for this lack of success, we believe, are deeply roo-
ted in the scientific method itself when applied to human systems.
For the scientific method is that of induction and intuition. We
define a class of situations which are analogous (by intuition)
and having observed certain behaviour in certain circumstances, we

suppose that in a seemingly similar case that behaviour will again be observed. If it is, then we can keep our 'rule'. If not, then we must try something else. We learn that what we thought to be similar systems in fact are not. We learn of essential differences.

While this idea may seem possible in the case of systems made up of simple, structurless particles under various conditions, clearly, it becomes much more questionable if the elements of the systems themselves have 'memory'. In that case, each system may be unique, with its own past and therefore its own future, and science, reduced to a classification by ones, may be rendered impotent.

The question is one concerning 'description' as opposed to 'understanding', where this latter term implies some 'generic' or 'transferable' law.

Let us examine closer the problem of modelling dynamically a complex system. An equation is always about 'accounting', about a conservation between left and right hand sides, and in the case of a system changing over time, a differential equation relies on a conservation of the 'changes' in each side. The variable x changes at precisely the rate that the processes of growth, decline, aggregation and dilution impose.

$$\frac{dx}{dt} = \text{rates of(creation - destruction + arrival - exit)}$$

These mechanisms written explicitly, contain 'parameters' that must be 'calibrated' in any particular application. But there are two sources of change in this equation. The first is the disequilibrium that may exist at time t ($dx/dt \neq 0$), which will lead to a change in x over time, even when the parameters remain constant. The second is the evoution of the parameters which, even if dx/dt is initially zero, will generate a change in x. We have in one limit the evolution of a differential equation under fixed boundary conditions, in the other that of a 'solution' to the differential equation under changing boundary conditions. In general, a real system will exhibit a subtle mixture of both, but in fact most 'models' in the human sciences retain only this second mechanism of change. Systems are calibrated assuming that $dx/dt = 0$ (that meaningful simultaneous relations exist between the variables) corresponding, it is supposed, either to a maximised utility or entropy in the circumstances. This assumes that the time scale of the dynamic equation is much more rapid than that which is of interest for the study. Typically, therefore, the dynamics of an urban model results from the exogenous change of population, of income or of economic structure, which is distributed over space by means of the model according to the structure corresponding to the 'solution' present initially. It is for this reason that these models are essentially descriptive and of fixed structure, a structure which is extrapolated into the future in a quasi-equilibrium manner.

The question of course is, firstly, whether the initial structure corresponds to a 'solution', and secondly whether this 'solution' will remain stable over time. As is now well known complex systems involving nonlinearities and feedback may possess multiple solutions, and the bifurcation of a particular solution is a common phenomenon. In order to be able to comment on the stability or instability of the structure which is present initially, it is necessary to use the dynamic equation itself to explore system behaviour around the existing solution.

In this way the system can probe the 'validity' of its own state of organization, and either retain it nevertheless (stability) or move away to some other branch of solution, some other state of organization and perhaps of complexity.

In the models which we have developed we propose to retain both sources of change in the differential equations, in order to be able to discuss the stability of our system over time, and to explore the future in a manner which includes the possibility of structural change.

A key point that arises, once we admit that a complex system may evolve both through the progressive displacement of a particular branch of solution under exogenous changes and also by jumping to a new branch, is that branches can differ qualitatively in their nature, possessing different characteristic pleasures and problems. We are faced with a fundamental non-conservation of emergent properties involved in such an evolution.

This is well illustrated by the amusing example of 'origami' which is the subject of figure (1), where an initially uninteresting

152

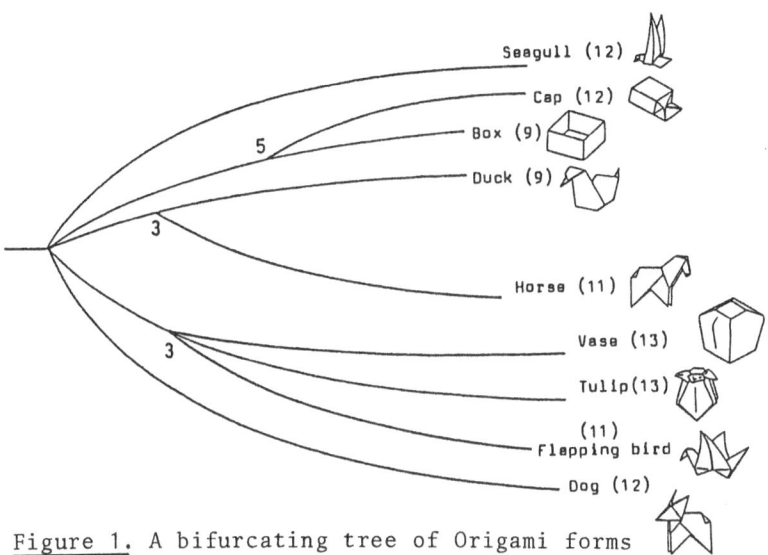

Seagull (12)

Cap (12)

Box (9)

Duck (9)

Horse (11)

Vase (13)

Tulip(13)

(11)
Flapping bird

Dog (12)

Figure 1. A bifurcating tree of Origami forms

piece of paper can be folded into many striking forms according to
a 'tree' of evolution, where characteristic traits emerge over
time. We have discussed this analogy in detail elsewhere (5), and
we shall not therefore discuss it further here.

Returning to our intial reflections concerning the sometimes dis-
sappointing performance of past models in real applications, we
perhaps can now see good reasons for this lack of success, in that
they are essentially statistically fitted descriptions
(calibration) which are extrapolated into the future under various
scenarious for exogenous events.

The dynamic models which we propose pose the invariance necessary
to any attempt at modelization, not at the level of urban structures, but instead at that of actors preferences which are fashio-
ned by their professional and private roles, and by the beliefs
and perceptions which condition them. It is through successive
spatial instabilties that complex behaviour evolves in a system of
interacting actors, each of which may have very simple preferences
and criteria. It is in the occurrence of these instabilities, in
their probabiltiy frequency and type,that unpredictable events and
aspects of the world play a vital role as we shall discuss. Also
we shall show how the simple 'generic' preferences of actors can
give rise to quite different urban forms, where the actors exhibit
different behaviour. In this way our approach attempts to go beyond the 'particular', which must inevitably be only descriptive,
and to attain some degree of general applicability.

2. Evolutionary Human Systems

In this section , a brief sketch will be given of some recent
applications of these ideas. They represent just a scratch on the
surface of a whole new domain of understanding, a domain in which
science and art truly meet, in which both logic and intuition find
a natural place.

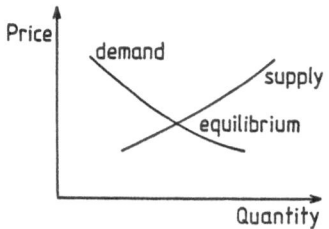

Figure 2. The basic diagram of economics,
showing the market equilibrium where demand
and supply are equal

In the first example we describe briefly a simple model of an economic market in which different products are in competition. Tra-
ditionally, the basic diagram expressing the market interaction is
that shown in figure (2), where price and quantity form the ortho-
gonal axes, and where 'supply' and 'demand' curves intersect at a
point called 'market equilibrium'. This equilibrium state however,
results really from the dynamics generated by the underlying individual, microscopic level of consumer choice and entrepreneurial
strategy. Thus the number and type of products available in a market are the attributes of the particular branch may become unstable and the dynamics of consumer choice may lead to some other
stationary state, characterized by other numbers and types of pro-
duct.Our model consists in a set of differential equations which
express consumer behaviour when faced with a particular range of
possible acquisitions, involving different prices and qualities.
Here we can restrict our study to that of only 2 axes, price and
quality, but clearly, this can be generalized to higher dimensions. Each consumer, we suppose, would most like to have the
highest quality imaginable at no cost. In reality, however, the
'supply' side will only offer products at some non-zero price.
Customers are attracted to the various products to different
degrees, depending on the relative importance they accord to price
and quality. We have supposed for simplicity that everybody would
put equal weight on their desire for quality, but different individuals assign a different importance to the price they are wil-
ling to pay. Roughly speaking poor people must on average assign a

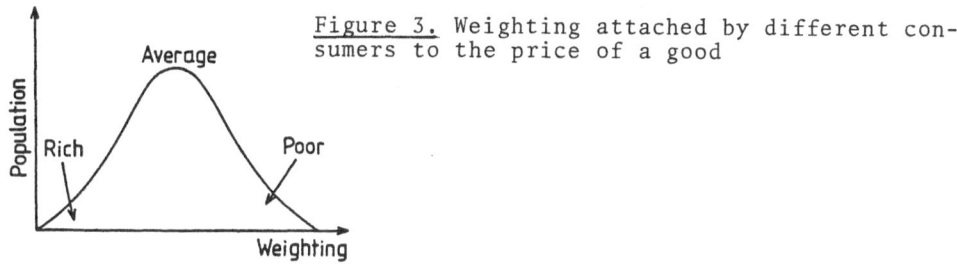

Figure 3. Weighting attached by different con-
sumers to the price of a good

higher weighting to the problem of price than the rich. The
assumption we have made corresponds approximately to assuming a
Gaussian distribution of revenue for the consumer population.

Details of this work have been given elsewhere (5), and here we
shall show just a few of the market structures which can emerge.
If there were no market thresholds, economies of scale, multiplier
effects or 'fashions', then it is true that an almost infinite
number of products could exist, one for each consumer ! However,
in the real world all these factors exist, and different market
structures can emerge depending on the history of the system, the
products launched, the profit strategies, market size etc. The
same equations of interaction can give rise to monoploy, duopoly
or oligopoly, and indeed to several realizations of each.

In figure (5) some of these possible market equilibria are shown,
which could be attained by our simple model when up to 3 particular products can interact. Of course it must be born in mind that
in a real market other products could appear of higher, lower or
intermediate quality and that one 'dimension' of a firm's strategy
could be to move its product 'up or down market' to compete with
or complement his rivals.

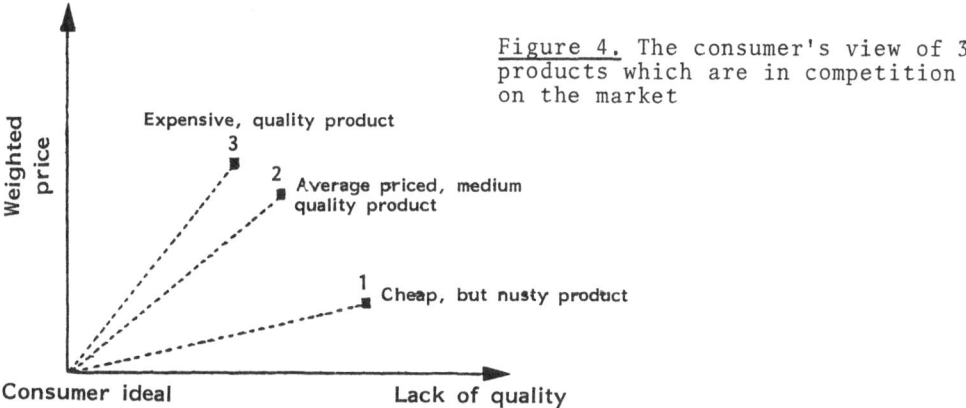

Figure 4. The consumer's view of 3
products which are in competition
on the market

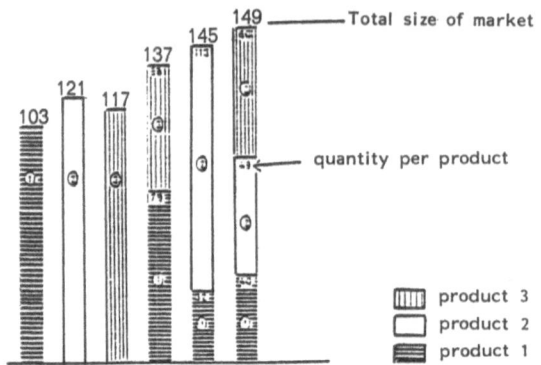

Figure 5. Several possible market equilibria for a system with up to 3 products in competition

quantity per product

Total size of market

▥ product 3
☐ product 2
▤ product 1

The study reveals that the later a firm arrives in a market the greater the initial investment required to become established. The 'future' of a market therefore depends not only on the strategies of those firms competing in the market, but also on the size of potential investors that have so far remained outside the market. Clearly, a whole host of market phenomena can be studied using such a model. In particular, we can explore the different types of structure which are most probable in markets of very different economies of scale, of different elasticities, of luxury goods where fashion plays an important role, of a market where government intervention is considered necessary etc. Also, any particular market could be specifically modelled, and the market strategy of any particular firm studied in this dynamic context. In figure (6), as an example, we show the various possible consequences of a decision on the part of one firm to halve its profits in order to capture a larger market share. If other firms do not retaliate in time, then it is beneficial to the firm, but if they do, then the stationary state to which the system moves is worse for all the firms including the one initiating the process. This seems a particularly realistic result in the light of the recent history of transatlantic air fares, of automobile prices and promises, of steel and of oil prices. The problem for an entrepreneur is to estimate the likely speed and type of response that may come from the otherfirms on the market, and this gives of course a probabilistic dimension to the future evolution of the market.

What should be clear from these brief comments is that even such a simple system is enormously rich, and apart from simple strategies of price and quality there are all the dimensions of collusion, merger and hierarchy which the model can show would be a natural consequence of such a system.

156

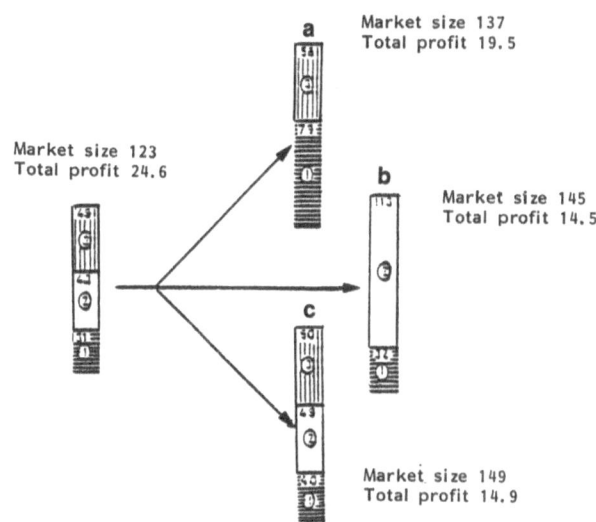

Market size 137
Total profit 19.5

Market size 123
Total profit 24.6

a

b

Market size 145
Total profit 14.5

c

Market size 149
Total profit 14.9

Figure 6. Three of the possible outcomes of the action of 1 to increase
market share

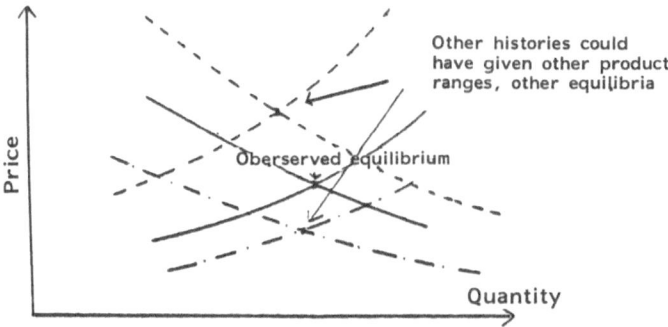

Price

Other histories could
have given other product
ranges, other equilibria

Oberserved equilibrium

Quantity

Figure 7. The observed equilibrium could have been different

The important point of principle demonstrated by these simple
simulations is that for the same population, having the same
'value system', for the same technology and the same potential
products, the flows of goods observed in a market can be radically
different as a result of the 'history' of the system. Thus the
fundamental diagram of 'supply' and 'demand' is misleading because
it can only be constructed in retrospect. It refers to a particular outcome, and the intersection could have been elsewhere. Figure (7).

These quantitative differences in total market volume hide in reality much richer structural differences between outcomes. In one
case we may have a market dominated by small firms offering consi-

derable choice, while in another, there may be only a few very
large firms offering standardized products at low costs. Each
situation, once established is self-consistent and logical, as
long as it remains stable.

The dynamic model referred to above also shows other interesting
effects that may be observed in an evolving market For example,
under slowly changing conditions such as market size, economies of
scale, or interest rates a relatively sudden reorganization can
occur as the branch of solution becomes unstable and the system
moves to some new branch, and market structure. In such changes
very small differences, possibly of random origin can prove decisive in selecting the new structure that emerges, and also if such
a change can be anticipated then it is a moment when small, weak
firms can penetrate the market and become established. The image
of a market system is therefore that of a 'dynamic game' with a
varying number of players, where periods of adaptive jockeying are
separated by successive crises during which major reorganization
occurs.

3. An Urban Model

Over recent years several examples of these ideas have been
studied in fields as varied as ecology (6), regional geography
(7), oil exploration (8), and the case we shall develop here,
intra-urban evolution (9). This has already been described in some
detail elsewhere and so we shall simply give a brief summary here.
It is based on the dynamic interaction of the different urban
actors, in their constant cooperative and competitive efforts to
survive and to functions successfully, which lead to behaviour
which expresses not only the 'needs' (utility) of each actor, but
also his limited means of action and the subjective nature of his
perceptions. Each variable of the model represents the density at
a particular location of a certain type of actor, each type having
its own locational criteria reflecting the 'role' the actor has
been assigned (or believes he has been assigned) by society.The
model expresses the changes induced in these densities by the
mutual interaction of the actors through their conflicting or complementary criteria. In this model we have considered 7 variables
consisting of 5 types of employment and 2 types of resident. Other
disaggregations are of course possible, and may be found to be
more suitable for different circumstances, the choice depends on

the study requirements, and more fundamentally on the existence of
distinctive locational criteria which may characterize an actor.

    The 5 types of employment are:
            - heavy industry (Steel, Automobile,Paper...)
            - light industry (electronics,   )
            - exporting tertiary(business, finance .....)
            - infrequent specialized tertiary
            - elementary tertiary

    The 2 types of population are:
                - blue collar
                - white collar

The criteria of location of these different actors are those which
are usually supposed, taking into account their particular needs.
In the scheme of figure (8), we see how the external demand produces jobs, the jobs attract residents, and the residents concentr--
te further economic demand generating more jobs, which if the tertiary centre becomes sufficiently important becomes an attractor
of regional economic demand, complementing the original economic
base of the city.

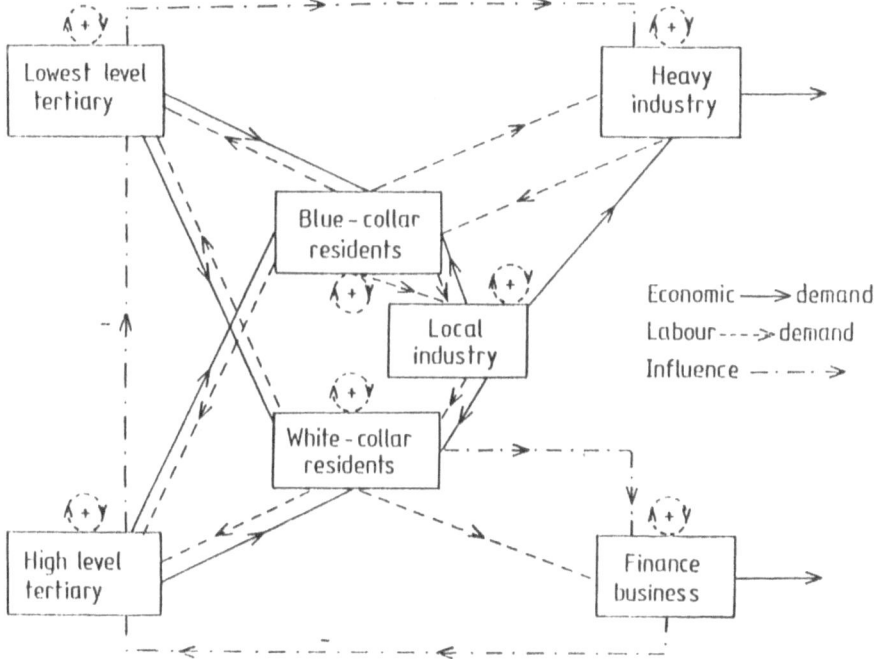

Figure 8. The interaction scheme for our intra-urban model

Our model, a basic set of 'urban mechanisms', is repressented by a
set of non-linear differential equations each of which describes
the time evolution of the number of jobs or residents af a particular type at a given point. In a homogeneous space one possible
'solution' of such equations would be to have an equal distribution of all variables on all points. Such a 'non-city',although
tneoretically possible, corresponds to an <u>unstable</u> solution, and
any fluctuations by actors around this state will result in a
'nigher pay-off', and this will drive the system to some structured distribution of actors, with varying amounts of aggregation
and decentralization.

However, in reality, there are two reasons for the structure of
the system: the first is due to the non-linear interaction mechanisms which give rise to instabilities as mentioned above; the
second is due to the spatial inhomogeneities of the terrain and of
the transportation network.

In the model simulations which we shall present here, we have supposed two transportation networks on which interactions and accessibilities are calculated. One is a road network for private
transportation taking into account 3 different qualities of road,
the other is a set of 4 public transport networks, describing the
possibilities of travel by train, bus, metro and tram. We have
tnerefore, a dynamic land use/transportation model which permits
the multiple repercussions involved in the various decisions concerning land use or transportation to be explored as the effects
are propagated, damped or amplified around our interactive scheme
of figure (8).

The simulations described here have been based broadly on the evolution of a city ressembling Brussels. The global characteristics
of employment and population are shown in table (1), and the spa-

Transportation networks

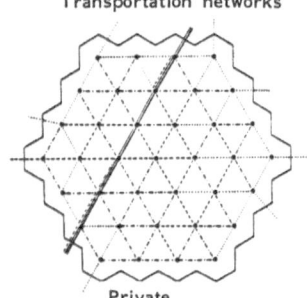

Private

<u>Figure 9.</u> The road network of out city, based
on that of Brussels

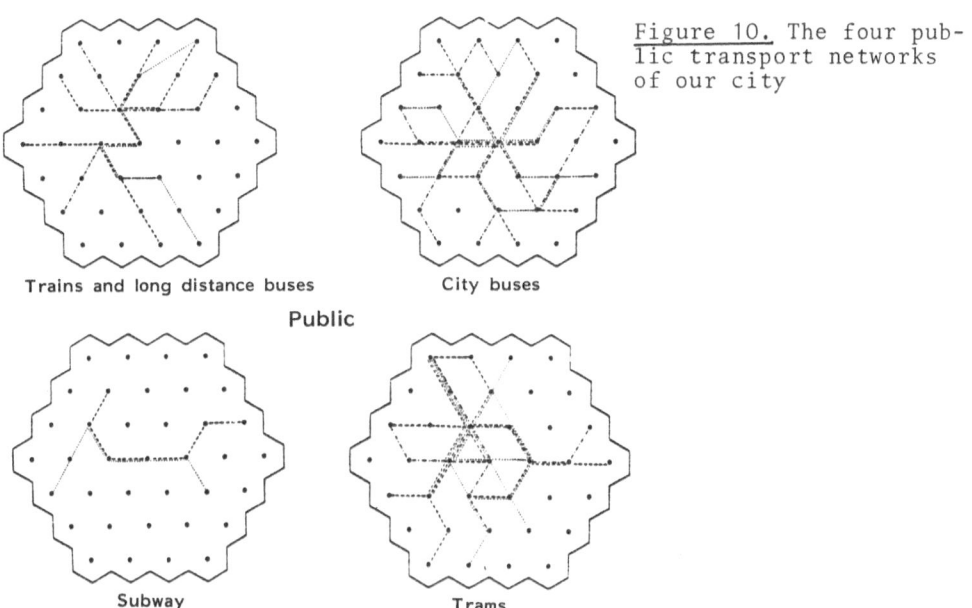

Figure 10. The four pub-
lic transport networks
of our city

Trains and long distance buses   City buses

Public

Subway   Trams

Each link of each network is characterized by a "time and money"
cost. Travel behaviour depends on the relative sensitivity of an
actor to these.

Table 1. Global figures generated by our model simulation,
in order to explore the evolution of a city which
ressembles Brussels.

VARIABLE	T = 10	T = 20	T = 30
Total Employment	729,600	669,500	674,300
Total number of Active Residents	462,670	411,560	414,200
Coefficient of Employment	1.58	1.63	1.63
Employment Structure:			
- Industry	25%	22%	22%
- Tertiary	75%	78%	78%
Structure of commuter flows from outside the urban centre			
- Blue Collar	40%	33%	33%
- White Collar	33%	44%	44%

tial evolution of urban structure is shown in figures (11),(12),
(13) and (14) for successive times of the simulation. The simulation times of 0, 10, 20 and 30 are of course somewhat unreal but
they are supposed to describe changes of urban structure which
could occur over some 40 or 50 years. The initial condition of our
simulation, which of course affects the structure which evolves is

161

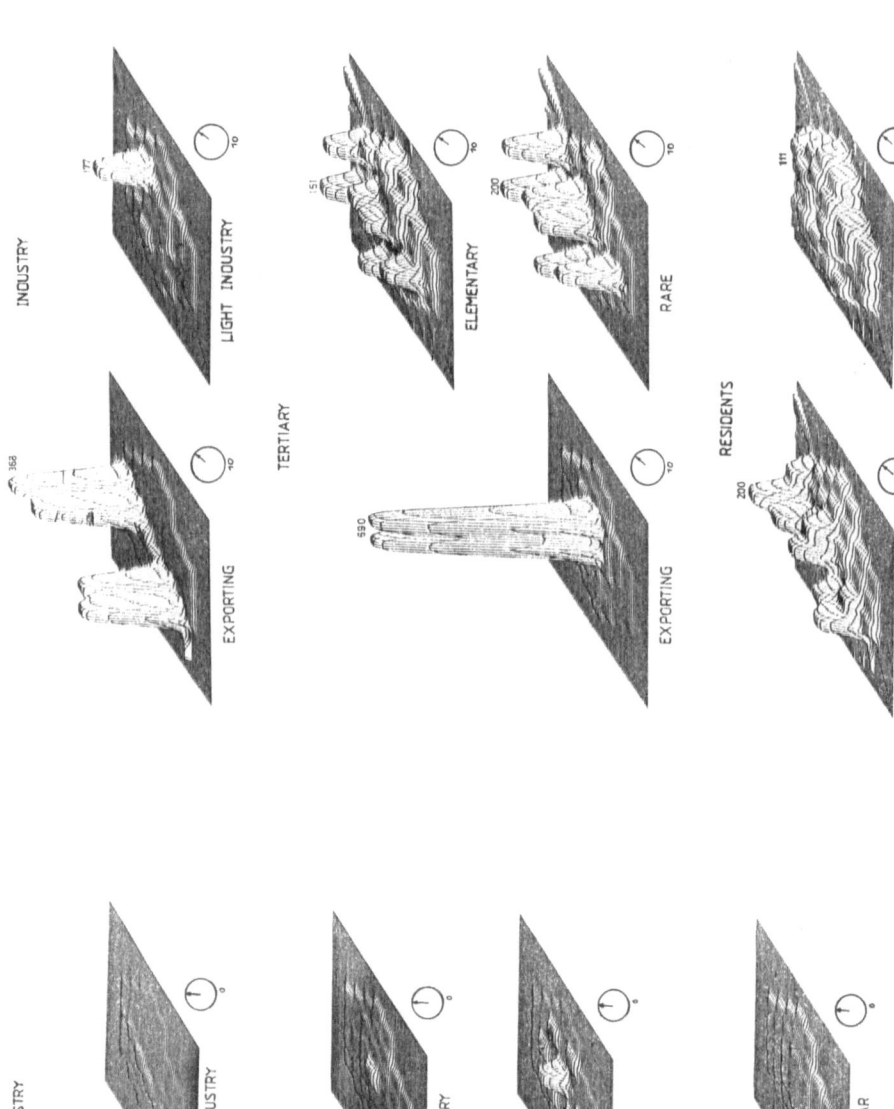

Figure 12. The distribution of employment and residence at t = 10

Figure 11. The initial distribution of employment and residences in our urban space

162

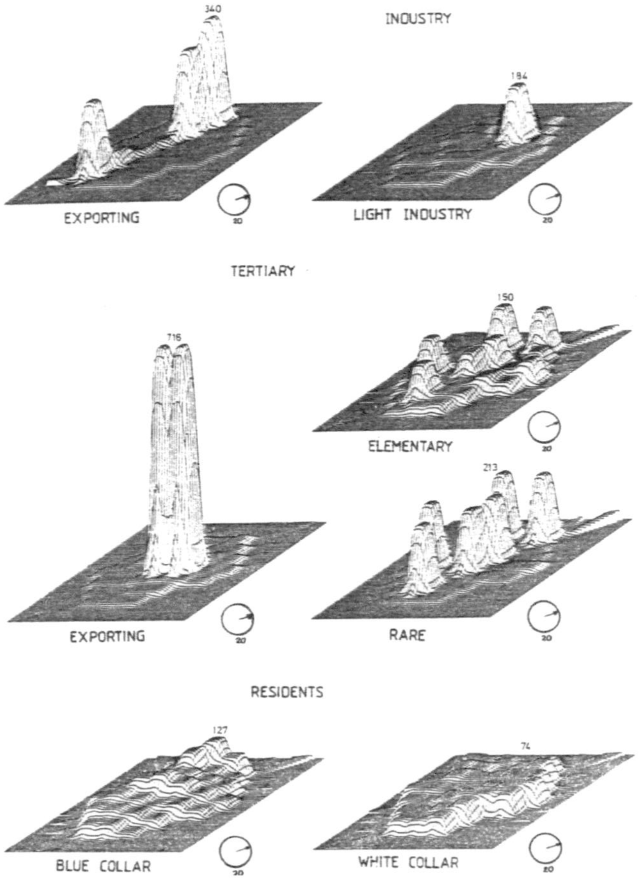

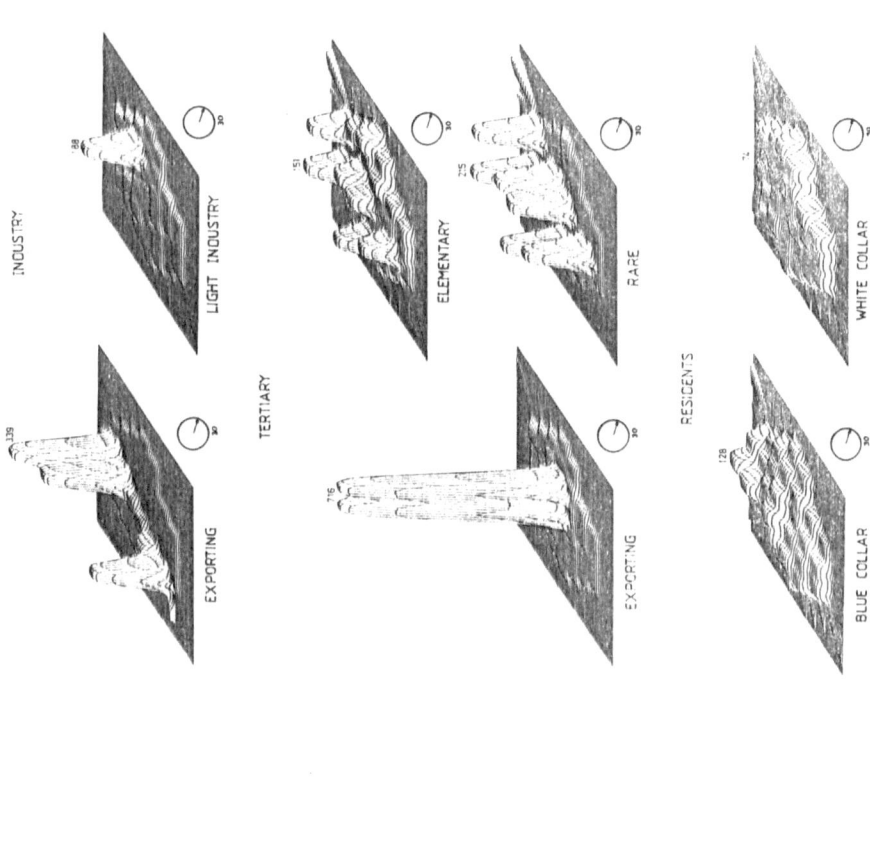

Figure 14. The distribution of employment and residences at t = 30

Figure 13. The distribution of employment and residences at t = 20

163

in fact taken from the result of an earlier simultion made without
the transportation network. We see that our urban system evolves
to a complex interlocked structure of mutually dependent concentrations. We have two poles of heavy industry, and a distribution
of blue collar residents reflecting this. Light industry, after
remaining diffuse for some time concentrates in one pole in the
north east. Financial and business employment in the city centre,
at around t = 7 , begins to spread through the urban space. Then,
it exceed a threshold on the point adjacent to the centre and
grows dramatically there, causing the decentralized locations
thoughout the city to decrease. The white collar and blue collar
residents spread out (many live outside the system) according to
the accessibilities of the networks, and a spatial hierarchy of
shopping centres appears, serving the suburban population and
encouraging further urban sprawl.

This completes our description of the evolution of our system
according to the deterministic equations of our model, and starting from the particular initial conditions that we have used. In
the next section we shall discuss why such a simulation would be
insufficient as a basis for decision making and city planning. In
reality such a system will not run its course undisturbed by
external and internal disturbances. This is the point that we
shall examine in the next section.

4. Discussion

The ideas sketched out in our first section tell us that the
deterministic equations governing the average behaviour of the
elements of a complex system are in fact insufficient to determine
precisely the state of the system and even its qualitative character. This is because of the existence of many possible branches of
possible solution. Only the effects of factors and events not
included in the differential equations break this ambiguity and
'decide' which branch precisely the system will really be on ! In
this way an event of historical significance is one which is not
contained in the average behaviour of the elements. Just as the
events which are important in origami are not the mechanical properties of the paper, but rather the intervention of human hands
which decide, at times and places when it is possible, where and
when to fold the paper.

This tells us that 'choice' really exists and that 'planning',
'policy' and 'intervention' need not be empty words used by perhaps well-meaning people attempting to stem inevitable tides of

change. However, in order that these words should be meaningful, it is necessary to know the consequences of a plan, a policy or an intervention.

We must always compare the evolution following an action with that following other actions or indeed inaction, and we must make the comparison in as many dimensions as possible, identifying the effects as they will be experienced subjectively by the different actors of the system.

Such comparisons are possible using our approach, but of course the actual decision concerning which action or policy should be pursued is a value judgement which must be made by political decision makers on behalf of the collectivity. The weighting accorded to differentsocial groups, to the long or the short term, and to the degree of disparity between groups that is reasonable, are matters of political judgement. However, in the absense of a model this judgement can be exercised on entirely fictitious future perspectives. Developers may depict the desperate need for some installation, with future demand soaring, job creation, local economic revival and increasing local property values, and all this with apparently no harm, indeed possitive good, for the environment. Objectors, however, will paint an image of the same installation in terms of the destruction of natural beauty or of an area of historical and architect-

ural interest, of threatened ecological collapse, of future overcapacity in the domain offering therefore only slight short term economic benefits which certainly would not off-set the serious reduction in property values that must be expected.

Since, in a democracy, political success seems to depend on not offending the side commanding the greatest number of voters, then there is a tendency for unpleasant facilities necessary for dense urban centres to be either thrust on sparse rural communities, or simply not to be installed. Of course, this is sometimes off-set by the possible financial power of other interests,who can be very persuasive in the political game. Whatever the precise details of such mechanisms however, this is clearly a very poor way for society to plan and make policy. Without for a moment suggesting that disagreement and discord could ever be entirely banished from the domain of human affairs (of course a ridiculous and unpleasant prospect), the existence of better, more successful and firmer based models could at least change the nature of the debate. Instead of presenting different perhaps equally fictitious, views of the future, the discussion could at least centre on the relative merits of the different possible actions.

Our models, hopefully, may provide a step towards such a situation, enabling the consequences of policy to be explored, not just in its narrow context, but also in its wider 'systemic' one, in which the action may set off a chain of events and repercussions throughout the system. Most disagreements concerning decisions are not about the immediate short term effects and the narrow context of construction costs, floor area, Kilowatts required, immediate traffic changes etc. but instead concern the long term and wider implications of the decision. Our model is aimed at exploring these kind of 'knock on' effects in the wider system.

Here we shall briefly illustrate different types of urban decisions which can be explored in an evolutionary context of possibly changing spatial organization and travel patterns.

In the first example, we show in figure (15) several possible outcomes of the implantation of a new shopping centre in our theore-

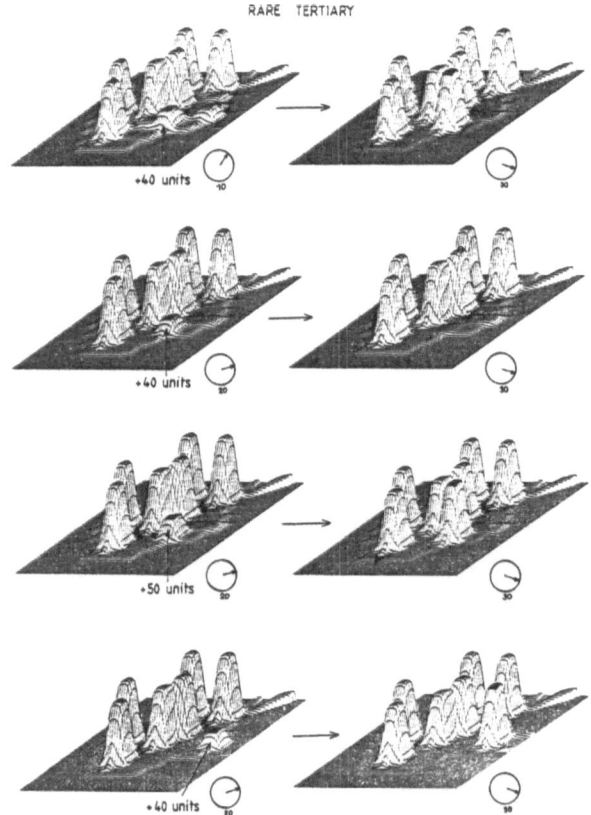

RARE TERTIARY

+40 units

+40 units

+50 units

+40 units

Figure 15. A series of simulations showing the importance of the size, the moment and the location for a successful investment in the retail structure

tical city (Brussaville). Our model shows us that if we launch a new centre at the location indicated, of size 40 units (implying a total involvment of some 4000 jobs) at time t = 10, then it will grow and stabilize in the retail structure, preventint similar initiatives nearby from succeeding afterwards. We have supposed here that prices practised are the same as those elsewhere, but clearly if the developer were prepared to accept lower profits during the starting up period, then the centre could be launched with a smaller initial size. This question can of course be explored using our model. However, we see from the second part of figure (15) that if the same investment of 40 units is made at the same place but at a later time t = 20 in the urban development then it does not succeed, regressing to zero as a total loss. Clearly, it could be maintained at some level by lowering prices and profits, but intrinsically we see that the time t=20 is less propitious than t=10.

In the event, at t= 20, an investment of 50 units does in fact succeed at this point, but our model shows us that if only 40 units were available, and normal profits necessary, then it could succeed if the location of the proposed shopping centre were shifted to the point shown in the fourth part of figure (15).

In the second example, figure (16), we show the long term impact of adding a new metro line across our city. Blue collar workers tend to increase in the neighbourhoods at each end of the line, and white collar residents return to the central core - some gentrification occurs. Clearly, our model could be used to make a cost/benefit analysis of different possible routes, frequencies and speeds, in order to weigh the possible long term censequences and arrive at a decision before embarking on such a major upheaval of the urban tissue that such an implantation implies.

The next example shows that if the increasing use of computer/telecommunication systems leads to a decrease in the need for ousiness, finance and administration employment to aggregate and form the C.B.D., then our city can undergo a major structural revolution. In figure (17) we see that at a certain critical value the C.B.D. disappears and office jobs are dispersed through and outside the city. Clearly, traffic flow patterns, residences and retail distribution will be vitally affected by this and we see some possible outcomes.

Another type of effect which it is interesting to observe is that of 'happenstance' - natural intervention.Our model could be used to explore the effect of various types of 'disaster' on the func-

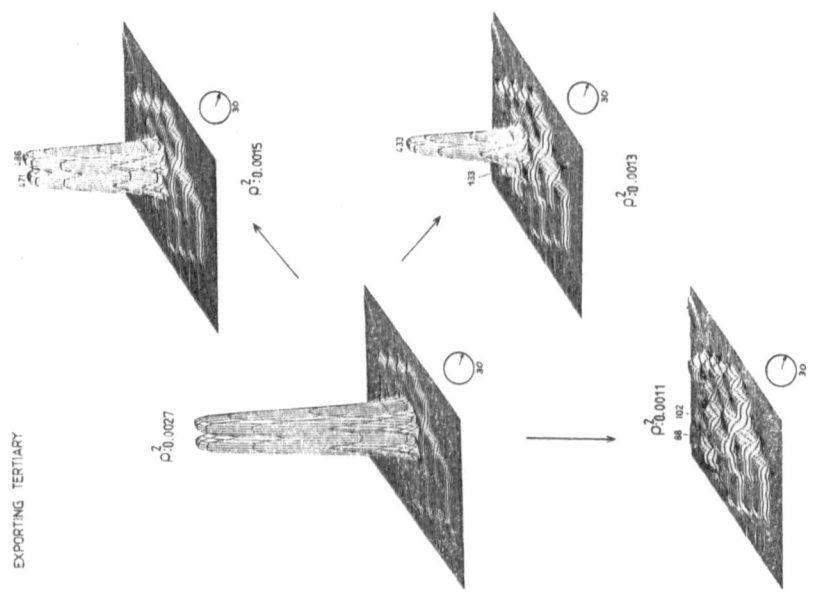

Figure 17. Possible effects on employment in the central business district of the increasing use of telephone linked computers for information exchange

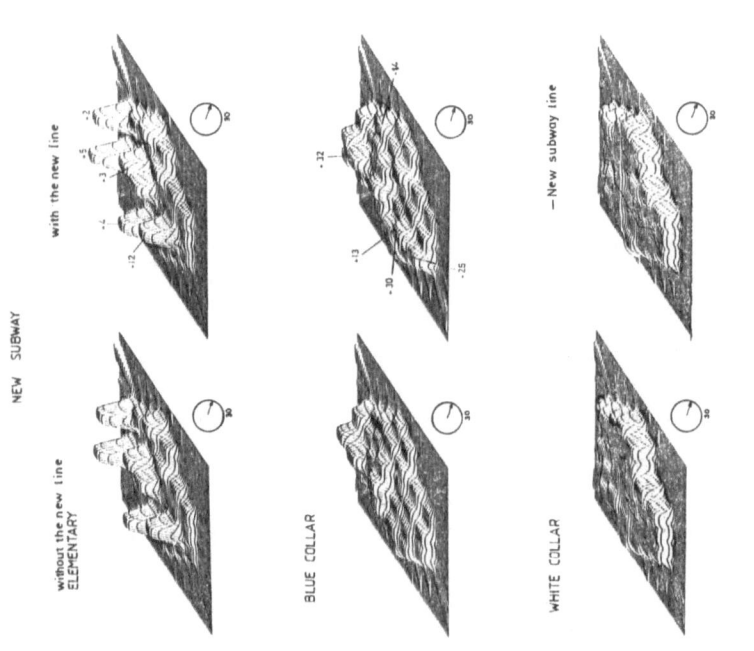

Figure 16. Some of the effects resulting from the implantation of a new subway line between the north and the south

tioning of the city, of pieces removed from the transportation
system, of neighbourhoods devastated by earthquake or floods, and
also of course the effects of some man made catastrophe such as
the closure of some large industry on which the city depends, or
of war and bombardment.

One particular example of less dramatic 'happenstance' is shown
in figure (18) where an urban centre grows from exactly the same
initial conditions as before, with identical locational criteria
of the actors and parameter values, except that instead of the
city developing with a canal/river/railway crossing it as before,
we have in its place a line of hills. The only effect of this on
the model is that the accessibility of these points, for the functioning of heavy industry, instead of being priviledged is reduced.

We see that the city that evolves, figure (18), is totally different from that of our refernce simulation shown in figures (11) to
(14). Industry is dispersed throughout the urban tissue, and with

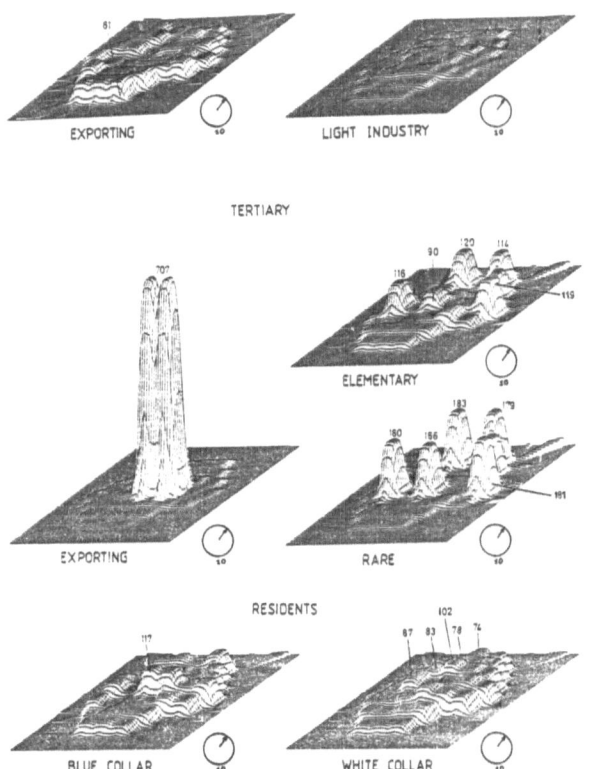

Figure 18. The distribution
of employment and residences
in another city, where instead of the river/canal we
have supposed a line of
hills

169

it the blue collar residents, and local industry. White collar residents, instead of locating in the south-east, aggregate along the line of hills which have replaced our canal. The distances travelled to work are not the same as before, nor are the 'costs' of shopping trips or the distribution of retail centres. Furthermore, not only are the spatial distributions of the variables different, but also the global quantities of industrial and tertiary activity, and of white and blue collar residents are modified. In fact this has been true for all our exploratory simulations, but here it is perhaps more striking because none of the parameters are different, only the terrain has changed. In this case the absense of an axis of good accessibility for industry lowers the 'attractivity' of the whole city for this type of investment and our model takes into account this change in its global performance.

This underlines the fact that global quantities are not constraints on an evolution, but on the contrary are observables which are generated by the local events in the system. Our approach is 'generic' in the sense that it should be contrasted with one based on 'observed behaviour' of a particular system. For example, a model based on this city of figure (18) would suppose as part of the 'utility' function of white collar residents that they 'wish' to live on the hills. This is not necessarily the case as we see from our simulation.

Instead of viewing the system as 'closed' and capable of predicting the future course of events in a mechanical fashion, we are led to another view. Our model provides a set of dynamic relationships which can beused to explore the many possible futures which could occur depending on the perturbations and shocks which it may experience.

Instead of 'stochastic' effects being simply minor irritations for the modeller, producing a deviation from the 'mean' of his 'prediction', we see that in fact they drive the system from one state of organization to another - they are the vital force of evolutionary change.

Such a view stresses the dangers of short term, narrow, optimization procedures often used to make decisions in the economic, social and political spheres. Such methods threaten society either with fossilization, or, in the long term with collapse. It is the adaptive possibilities of societies and markets that allow them to survive in the long term, and this requires stochasticity, irrationality,and redundance in order to produce originality.

170

Perhaps, in stochasticity, we see the source of evolutionary change, and ultimately man's creative intelligence is its fruit.

References

1) P.Glansdorff and I.Prigogine,(1971) 'Structure, Stability and fluctuations', Wiley Interscience,   London.
2) G.Nicolis and I.Prigogine, (1977) 'Self-Organization in Non-Equilibrium Systems', Wiley Interscience, New York.
3) H.Haken,(1977), 'Synergetics' Springer-Verlag.
4) I.Prigogine, P.M.Allen and R.Herman, (1977) 'Complexity and the laws of Nature', in 'Goals for a Global Community', Eds.Lazslo and Bierman.Pergamon.
5) P.M.Allen, (1982) 'Evolution, Modelling and Design in a Complex World', Environment and Planning, B Vol.9, p95-111.
6) P.M.Allen, (1980) 'The evolutionary paradigm of dissipative structures', AAAS Selected Symposium,Ed.E.Jantsch Westview Press, Boulder.
7) P.M.Allen and M.Sanglier,(1981) 'Urban Evolution, Self-Organization and Decision-making',Environment and Planning,A, Volume 13.p167-183.
8) B.Braunschweig, (1983) Proceedings of the 7th Int. Systems Dynamics conference, Brussels, 1982, Large Scale Systems, North Holland.
9) P.M.Allen, M.Sanglier, G.Engelen and F.Boon, (1983) 'Evolutionary Spatial Models of Urban and Regional Systems', Systemi Urbani.To appear.

Part IV

Social Sciences

Some Applications of Basic Ideas and Models of Synergetics to Sociology

Arne Wunderlin and Hermann Haken

Institut für Theoretische Physik, Universität Stuttgart, Pfaffenwaldring 57/IV
D-7000 Stuttgart 80, Fed. Rep. of Germany

Abstract

We shall propose a general procedure how to apply synergetics to social
processes, i.e. in which way one may translate social behaviour into
mathematical structures. This is achieved by merely using macroscopic
quantities like order parameters. This may at least partly remove the
difficulties connected with the problem of measurability in social
systems. Our ideas are exemplified by using models on the formation
of public opinion. Furthermore we shall construct a minimal model for
social processes which summarizes all ingredients of a synergetic
system.

1. Introduction

Various links and relations exist between social sciences like history,
economy, psychology, and sociology [1,2]. The common subject of all
these disciplines is the human being, that of an individual, of a group,
a society, eventually even mankind. They try to analyze a wealth of
complex interactions. Examples are provided by the analysis of the
situation of an individual within a group, the competition and/or
cooperation of different groups, the coexistence of distinct social
forms, etc. Quite often pronounced accentuation of special cords of
this complicated and interlaced net of connections caused different
systems of instruction. Up to the end of the nineteenth century the
main efforts were devoted to state universal principles. Scientists
believed in the possibility to derive conditions, forms, and developments of human beings from these principles in a systematic fashion.

Undoubtedly these hypotheses have contributed valuable knowledge to our
understanding of, for example, social processes. On the other hand it
remains fair to say that those theories have failed which try to provide us with universal principles and generally valid methods. Instead
of discussing possible reasons for that failure we take this fact as
admonition. We shall renounce developing a global theory of social
processes and rather study the following aspect: the basis of our considerations is given by the fundamental principles of synergetics [3,4],
an interdisciplinary science which may be viewed as a theory of mathematical structures. Consequently we shall ask whether there exist social
processes which have its equivalent in these mathematical structures.
It is not our aim to answer this question in its whole generality.
Rather we shall try to derive answers from concrete examples on the
formation of public opinion. We furthermore assume that we can do this
without normative pretension and valuation.

In natural sciences the application of synergetic methods systematically
starts with a mathematical description of the underlying subsystems.

The basis of such a formulation is formed by fundamental principles
like conservation laws, etc. A similar attempt can be found in traditional "mathematical sociology" or "quantitative sociology" [5,6].
There scientists try to give reasonable translation of physical notions and conceptions of the "microlevel", i.e. the level of the individual subsystems, onto the subsystems of a society, e.g. the single
human.

For example we may attribute to each single human an attitude vector
$\underline{i}$ in an aspect space (or property space) which is spanned by different
aspects like education, religion, citizenship, etc. The "phase space"
of a society may then be obtained as a direct sum of the aspect spaces
of the individuals. Slight variations of this concept which take into
account possible group structures lead to other spaces like the "socio-
configuration space", etc.

In modern literature it is often overlooked that the concept of aspect
space is a very old one [1]. It was introduced in the seventeenth
century as social or moral space, when people tried to develop social
mechanics. Indeed, the analogy between phase space in physics and
aspect space in sociology seems compelling. However, there are serious
deficiencies [9] in this correspondence which immediately become obvious if one asks questions like how to define distances in that space,
or more generally a topology, what about completeness, etc. Therefore
it is fair to say that not very much is known about this space and
indeed, it is only of formal or a esthetic value as a starting point
of a mathematical formulation of the developments, forms, and movements in a society. We use this fact as a motivation to seek for
different attempts. To provide a basis for our following considerations
we shall first give a brief review of some basic ideas and principles
of synergetics.

2. Some Basic Concepts of Synergetics

In order to keep the article self-contained we note that synergetics
deals with open systems which are composed of many subsystems. On a
mezoscopic level such a system can be completely described by a state
vector $\underline{q}$ in a high-dimensional configuration space. Typically the state
vector evolves in time and space due to the prescription

$$\underline{\dot{q}} = \underline{N}(\underline{q}, \nabla, \{\sigma\}) \qquad \text{+ fluctuations,} \qquad (1)$$

where $\underline{\dot{q}}$ denotes the time derivative of $\underline{q}$. $\underline{N}$ is a non-linear function
of $\underline{q}$ and the Nabla symbol refers to spatial variations. The set $\{\sigma\}$
pays attention to external or control parameters. We note that these
parameters are considered as macroscopic quantities which vary on
time scales much larger than $\underline{q}$, i.e. they are treated as quasistatic.

The main interest of synergetics concentrates on the evolution of a
given state $\underline{q}_0$, say a stationary state, through sequences of instabilities into new states which evolve by spontaneous self-organization
on a macroscopic scale. Indications for such a possibility can be
gained from linear stability analysis. Writing

$$\underline{q} = \underline{q}_0 + \delta\underline{q}, \qquad (2)$$

we immediately obtain

$$\delta\underline{\dot{q}} = L(\underline{q}_0, \nabla, \{\sigma\})\delta\underline{q} + O(\|\delta\underline{q}\|^2). \qquad (3)$$

The linear matrix L certainly depends on q_o, also on spatial variations, as well as on the macroscopic control parameters. L determines
in connection with the boundary conditions the collective modes of the
system which we shall denote by $O_j(x)$ and the index j distinguishes
between the different collective modes. Furthermore the linear matrix
yields a set of corresponding eigenvalues λ_j which are obtained from
the condition

$$\text{Det } [L - \lambda I] = 0 \tag{4}$$

(I: unit matrix). Obviously

$$\lambda_j = \lambda_j (\{\sigma\}). \tag{5}$$

The reference state q_o is locally stable as long as the real part of
all of the eigenvalues fulfills

$$\text{Re } \lambda_j < 0. \tag{6}$$

Because the λ_j depend on $\{\sigma\}$ we expect that by a suitable variation of
say one of the σ's the real part of only few of the eigenvalues
crosses zero and becomes positive. This situation clearly indicates
an instability of q_o along the corresponding directions. Using the
hypothesis

$$\delta q = \sum_j \xi_j (x,t) O_j(x) \tag{7}$$

we may split the amplitudes

$$\{\xi\} \xrightarrow{\qquad} \begin{array}{c} u \\ \underline{s} \end{array} \tag{8}$$

into a small set of unstable ones (u) and a large set of amplitudes
referring to the stable directions (s). As a consequence of the slaving
principle we may express the values of the amplitudes of the stable
modes as a function of the unstable ones. A vivid description says
that the slow modes which become unstable dominate the motion of the
whole system, i.e. become the order parameters of the system.

Mathematically this has the consequence that the complex behaviour
of the system is solely described by appropriate equations for the
order parameters which may be given the form

$$\dot{u} = N(u, \nabla, \{\sigma\}) \qquad + \text{ fluctuations.} \tag{9}$$

Formally the order parameter equations look like the original equations
(1). But what has been achieved is that the number of variables has
been reduced drastically. We therefore may discuss a large class of
instabilities by means of a simple geometric picture. Furthermore eqs.
(9) describe processes of self-organization on a macroscopic level. It
is this fact which becomes important for our following considerations.

3. On the Formation of Public Opinion

The problem of quantification of social processes and behaviour is
intimately connected with the problem of measurement. On these
grounds it appears rather difficult to give a reasonable description
of the single subsystems, i.e. the single humans. This difficulty may
at least partly be removed if we confine ourselves to collective behaviour on macroscopic scales.

Our main interest is devoted to processes of spontaneous self-organization which spread out over the whole society. The corresponding macroscopic quantities which yield an appropriate description of such collective phenomena are the order parameters which then dominate the evolution and behaviour of a society. Our initial point is therefore to single out macroscopic quantities, the control parameters and the order parameters. We note that this way of reasoning is in some sense inverse to the systematic method used, when synergetics is applied to systems of natural sciences. Instead of starting from the single subsystems we propose to identify directly, e.g., the order parameters phenomenologically.

The justification of this method rests on a fundamental result which has been found from the application of synergetics to well-defined systems in natural sciences [3,4]. The investigation of such systems near a critical point, beyond which self-organized behaviour occurs, shows that the critical behaviour is not affected by the detailed interactions and structure of the subsystems. Rather there exist few macroscopic quantities which characterize an instability exhaustively. Our discussion therefore is confined to the behaviour of a society near such an instability point.

To exemplify this we shall discuss a simple model for the formation of public opinion. The important rôle of public opinion in the development of a society was especially emphasized by Walter Lippmann in 1922 [1]. Furthermore we note that public opinion can be measured by votes, polls, etc. The model we shall discuss to some extent has been proposed by W. Weidlich in 1971 [7]. Our presentation includes a slight reinterpretation and extension of its original formulation.

We consider a society which may be decomposed into a rather complicated gathering of groups, classes, etc. We imagine that the society under consideration is confronted with an important problem. If we go to the extreme, the members of the society may believe that the survival of the society as a whole depends on the solution of that problem. Modern examples are provided by the peace movements, green movements, etc.

Experience now tells us that for the members, in concentrating to the solution of the given problem, the detailed group structure, citizenship, etc., loses its significance. Indeed, opposite opinions may appear within the same group. The behaviour of the churches in their settlement with the peace movement may serve as one example. Following Weidlich we assume that only two solutions are available, + or - . The decision of the society may then be measured by counting the numbers of +, n_+, and -, n_-, votes. The problem may be further simplified by the assumption

$$n_+ + n_- = n ,$$ (10)

where n, the total number of members, is considered as constant. Then just one macroscopic variable, namely

$$\xi = \tfrac{1}{2} \left(n_+ - n_- \right)$$ (11)

serves as an order parameter. To keep our problem simple we take n very large. This allows us to treat ξ approximately as a continuous variable. We now have to seek for the equation of motion of the order parameter. For example we may construct a probability density $p(\xi,t)$ which is assumed to fulfill a master equation. Using as elementary processes only that one where just one member of the society changes his opinion with a certain transition probability per unit time, the

problem becomes isomorphic with the Ising model of a ferromagnet.
Details can be found in [6].

We shall take a different point of view by stressing again the rôle
of the control parameters as macroscopic quantities and shall discuss
their measurability. In the construction of the transition probabilities there appears for example an important parameter, the so-called
adaption parameter κ [6]. κ measures the influence of people on
their neighbours. We notice that it is by no means obvious how to
attribute κ in a given state of a society a precise value. Instead of
giving a quantitative picture we therefore prefer to give a qualitative
discussion of the possible states of the society. It is well known
that there exists a whole class of equations which qualitatively yield
the same situation. Mathematically this class can be precisely defined through the notion of topological equivalence. We are led to
the conclusion that we may use the most simple representative of the
corresponding class, and this naturally yields the concept of normal
forms [8]. Loosely speaking, giving equal value to both solutions, +
and -, the corresponding equation (the symmetry preserving unfolding
of the normal form) reads

$$\dot{\xi} = (\kappa - \kappa_c)\xi - \xi^3 \qquad (\text{ + fluctuations}). \qquad (12)$$

This equation has been discussed in detail in [3]. One observes a
symmetry-breaking transition from a state of indifference to the state
of polarization in a society. Connected with this transition are the well-
known phenomena such as critical slowing down, critical fluctuations,
etc. We note that in the case of polarization ($\kappa > \kappa_c$) it is not predictable which state eventually will be occupied by the society. The
final state is determined solely by the fluctuations. This shows the
special importance of the probabilistic viewpoint. However, it should
be realized that in describing a definite state of a society by
probabilistic methods, we usually have only one sample to test our
theory.

We have emphasized the special importance of a qualitative discussion
of social problems. This can be continued by looking for possible
dangerous disturbances to the transition which do not preserve symmetry. Then one state, + or -, may become prefered against the other,
the transition may become smooth or show bistable behaviour with a
hysteresis.

4. Outline of a General Approach

We have argued that macroscopic variables are of special importance
if one undertakes a translation of social processes into mathematical
structures. In our case at first the control parameters have to be
identified. These control parameters must clearly be separated from
the order parameters by time scale arguments. If we treat the control
parameters as quasistatic quantities, their characteristic time
scale yields an estimation of the local domain where the model can
be applied.

Variations of the control parameters now cause certain instabilities
of the present status of a society. What may happen typically in
many cases can be discussed in terms of linear stability analysis.

Indeed by writing (4) explicitly in the form

$$\lambda^n + a_1 \lambda^{n-1} + \ldots + a_{n-1} \lambda + a_n = 0, \qquad (13)$$

we observe that the coefficients a_i are functions of the control parameters. An instability as discussed in Weidlich's model is observed if one real eigenvalue becomes positive, i.e. at the critical point

$$a_n = 0. \tag{14}$$

The corresponding critical surface in the parameter space obviously is of codimension 1. Therefore if we wish to cross the critical surface it is (typically) sufficient to change one control parameter. Just the other way round, changing one control parameter we would expect as one typical situation which may arise, one order parameter which dominates the behaviour of the system as described in the model above. A similar reasoning holds, e.g. for the Hopf bifurcation. This method can be systematically extended to critical surfaces of higher codimension which clearly yield more complicated instabilities.

The order parameters must then be identified phenomenologically. Their equations of motion may then be modelled in the form (compare (9))

$$\dot{\underset{\sim}{u}} = A\underset{\sim}{u} + \underset{\sim}{f}(\underset{\sim}{u}) = \underset{\sim}{X}(\underset{\sim}{u}) ; \quad \underset{\sim}{X}(0) = 0. \tag{15}$$

We then may choose the most simple representative of the corresponding system whose existence is guaranteed whenever $\underset{\sim}{X}(\underset{\sim}{u})$ has a polynomial form by the normal form theorem [8]. There exists a local diffeomorphism Φ

$$\Phi \circ \underset{\sim}{X} = \hat{\underset{\sim}{X}} \tag{16}$$

such that

$$\dot{\underset{\sim}{\xi}} = A\underset{\sim}{\xi} + \hat{\underset{\sim}{f}}(\underset{\sim}{\xi}) \tag{17}$$

is 'simpler'. This method yields a qualitative picture of possibilities which may arise if one or several control parameters are varied. We emphasize again that we do not give quantitative results but confine ourselves to gain insight into the qualitative behaviour. The crucial point, however, follows from the fact that the specific rôle of non-linearities in social processes has found an adequate description. Predictions can be given only as possible 'scenarios' which may be caused by the non-linearities.

5. Construction of a Minimal Model

There may arise situations in a society where synergetics can go far beyond the description given above. As a first step we shall analyze a typical situation which serves as a predecessor to the order parameter equation (12). This equation usually results from a coupling between unstable and stable collective modes of the form

$$\dot{\xi} = (k - k_c)\xi - \xi \sum_i g_i s_i , \tag{18}$$

$$\dot{s}_i = -\gamma_i s_i + \xi^2 \tag{19}$$

(g_i are coupling constants), where we have neglected fluctuations for simplicity. Indeed, if $k \approx k_c$ we have

$$\gamma_i \gg |k - k_c| \tag{20}$$

and may put

$$\dot{s}_i = 0 \qquad (21)$$

which expresses the most simple form of the slaving principle. We then
have from (19)

$$s_i = \frac{1}{\gamma_i} \xi^2 \qquad (22)$$

and arrive with (18) at

$$\dot{\xi} = (\kappa - \kappa_c)\xi - \xi^3 \sum_i g_i/\gamma_i \qquad (23)$$

which is, up to a trivial scaling, identical with (12). Because the s_i
still represent collective modes they may also be considered as macroscopic quantities. Therefore we conclude that if it is phenomenologically known which stable or slaved modes couple to the order
parameters we may give assertions how they behave. In the situation
under consideration we observe that in a conflict situation as
described by (23), it is quite unimportant for the slaved modes
whether the order parameter finally comes to the + or the - state.
One possible interpretation for social systems would be the following.
If we identify + and - with a totalitarian regime of the left and the
right, respectively, and furthermore consider the sum of communication
channels like newspapers, television, etc. as a collective slaved
mode, our result seems reasonable.

These considerations naturally lead us to the problem constructing
a minimal model [9] of a social process which contains all ingredients
of a synergetic system. The model is meant   to provide an idea how
complicated the detailed processes may be which finally enter the
order parameter equations.

Different interpretations of the following model are available. We
shall take it as a possible mechanism how ecological ideas may come
up and finally dominate a society through a single group. The society
is assumed to be distributed over a country which is divided into
approximately equal parts. These different regions are enumerated by
an index μ . By s_μ we denote the local activities of people and $s_\mu > 0$
means that the local activities are higher than usual ($s_\mu = 0$). We
furthermore introduce a variable n_λ which measures ecological activities of a central group λ . The activities of these central groups
are not local but spread out and influence the whole society. These
central groups are forced by the sum of the activities of all local
groups and we may write

$$\dot{n}_\lambda = -\kappa_\lambda n_\lambda + \sum_\mu g_{\mu\lambda}^{(1)} s_\mu . \qquad (24)$$

To simplify the interpretation we add that $n_\lambda > 0$ indicates group λ
is positively engaged in ecological problems. The first term on the
rhs of (24) shows that all activities decay if there is no forcing
from the local groups. Their effect is expressed through the second
term, where g denotes a coupling constant. n_λ should act back onto
the activities of the local groups. We assume that this happens via
the local newspapers. If their activity is measured by p_μ this
may be modelled through

$$\dot{s}_\mu = -\beta_\mu s_\mu + \sum_\lambda g_{\mu\lambda}^{(2)} p_\mu \cdot n_\lambda + F_\mu , \qquad (25)$$

where we have additionally added a force F_μ in order to take into
account special local problems. The plus sign in the coupling shows
that the local activities are reduced if the local newspaper and
the central group are acting against each other. The local newspapers on their own are activated by their knowledge about the average
pollution measured through p_o. In addition they are driven from the
local groups via the activity of the central groups:

$$\dot{p}_\mu = \gamma_\mu (p_o - p_\mu) - \sum_\lambda g_{\mu\lambda}^{(3)} n_\lambda \cdot s_\mu . \tag{26}$$

The sign of the non-linear term is chosen in a way that the interest
of the local newspapers is reduced if local activities and central
or global activities point in the same direction and enhanced in
the other case.

This model summarizes indeed all ingredients of a synergetic system:
the local subsystems, collective modes might be introduced, e.g. via

$$s_\lambda = \sum_\mu g_{\mu\lambda}^{(4)} s_\mu . \tag{27}$$

If, for example, $p_\mu = p$ (independent of μ) one may observe selection
of one central group which dominates the behaviour of ecological
problems and symmetry breaking may be observed again. Furthermore
more complicated solutions including chaotic behaviour exist for this
model. We conclude that models of this kind may give an idea how to
trace back qualitatively from the macroscopic motion to the motion
of the slaved modes and possibly to the single subsystems which in
our example are local groups of people.

6. Conclusion

We have proposed along brief lines in which way synergetics may be
applied to social problems. Our treatment might be generalized in
several respects. Indeed there may occur situations in which after
a transition several control parameters are effectively influenced
by the order parameters, the time scale of their motion might be reduced considerably and they also have to be considered as time dependent [10]. Furthermore there exist many effects known in synergetics for systems from natural sciences which might have their
counterpart in social dynamics. One possible example is that there
must be a minimal number of subsystems in order to allow for collective
behaviour [4]. This might explain that there is no observation of self-
organized behaviour in small villages but only organized movement.

Our discussion started from macroscopic quantities. Predictions have
only qualitative character which seems to be the only possible way
which is consistent with the limited measurability of social developments. These predictions are by no means deterministic but take -
apart from the probabilistic point of view - only the character of
possible 'scenarios'.

References

1. P.A. Sorokin: Contempory Sociological Theories (Harper and
   Row, New York 1928)
2. M. Horkheimer: Kritische Theorie (S.Fischer 1968)
3. H. Haken: Synergetics. An Introduction (Springer, Berlin-
   Heidelberg-New York 1983)
4. H. Haken: Advanced Synergetics (Springer, Berlin-Heidelberg-
   New York 1983)

5. J.S. Coleman: Introduction to Mathematical Sociology (The Free Press, New York 1964)
6. W. Weidlich, G. Haag: Quantitative Sociology (Springer, Berlin-Heidelberg-New York 1983)
7. W. Weidlich: Collective Phenomena 1, 51 (1972)
8. P.J. Holmes: Physica 2D, 449 (1981)
9. A. Wunderlin, H. Haken: Materialien XXXII, Univ. Bielefeld Schwerpunkt Mathematisierung p 93 (1981)
1o. Karmeshu, A. Wunderlin: unpubl. manuscript, Stuttgart 1978

A Synergetic Approach to Energy-Oriented Models of Socio-Economic-Technological Problems

Ali Bülent Çambel

The George Washington University, 801 22nd Street N.W.,
Washington, DC 20052, USA

> "...mathematics is the art of giving
> the same name to different things."
>
> Jules Henri Poincaré

1. Introduction

The introduction of microquantities into the analysis and synthesis of macro-systems is one of the most crucial issues that challenges the scholarly community regardless of discipline. The assertion applies equally well in the physical, biological, behavioral and social sciences, as well as in modern technology and policy planning.

To some extent, this "linkage" of microquantities with macrosystems is relatively a new realization. The word linkage is placed within quotation marks because I do not wish to imply that either the study of microquantities or the synthesis and use of macrosystems by themselves alone are new revelations. For example, the statistical analyses of Ludwig Boltzmann (1844-1906) cannot be considered recent. Nor would one consider Sir Charles Algernon Parson's (1854-1931) invention of the modern steam turbine (1884) a recent event. However, in the development of modern electric power, which has certainly changed the course of society, Parson did not have steam property tables and Mollier charts at his disposal. Indeed, one might conjecture if scientific research must always precede technical innovation.

The need for introducing micro- or statistical information into the analysis and synthesis of macro- or large systems is not only an intellectual challenge, but it is forced upon us for a variety of reasons. These include the complex interactions within present day society and the dwindling of resources making it necessary to understand, devise, and/or perfect the sophisticated devices and processes that are part and parcel of the post-industrial society or what has been called the Creativity Revolution.

There are a number of contributions to the literature wherein microdata are used in the study of macrophenomena. The most systematic efforts along these lines are those of H. HAKEN best summarized in his seminal book [1] and followed by his school of synergetics.

The present paper is modest in scope, breadth and depth. Specifically, I wish to highlight some studies undertaken by the author and his students. The objective of these is to develop mathematical computer models to identify optimal avenues of social and economic development by evaluating proper energy mixes. Cardinal to this program is the assertion that socio-economic development (macro) must be based on the quality of life (Q.O.L.) of the individuals involved which clearly involve microconsiderations.

In formulating our Gedankenspiel we remind ourselves that in neo-classical economics one generally excludes the so-called externalities and deals with economic equilibrium. In an analogous manner, in classical fluid mechanics one excludes the so-called external body forces while in classical thermodynamics one stipulates reversible processes among equilibrium states. While in each of the above cases the fcrmalism is elegant and powerful, it is far from reality where one should consider irreversibilities, nonequilibrium and external effects. Here the twin subjects of plasma physics and magnetohydrodynamics appear revealing. Thus, for example, charged particles experience attraction and repulsion, and when exposed to electromagnetic effects undergo cyclotron motion, etc. Space does not permit a detailed explanation here and these may be found in [2] and [3].

Nor is it possible to describe here the lengthy computer programs which we write either in FORTRAN or BASIC depending on the needs of the country being studied [4,5]. Instead I shall present the basic algorithm and describe how the input data are evaluated.

However, before embarking on this discussion, I present in Fig. 1 actual data from HENDERSON [6] because actual data confirming socio-economic model validity are difficult to obtain to say the least. It is hoped that Henderson's figure reproduced here will provide some confidence, some belief in the suggestion that equations of the physical science may at times be applicable to phenomena involving the behavior of individuals.

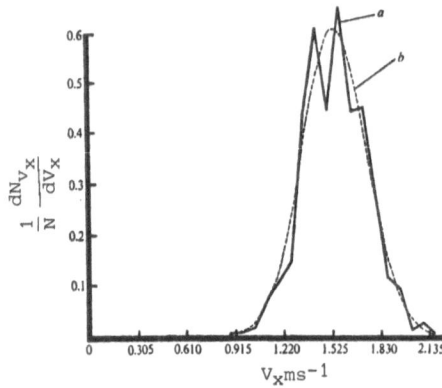

Fig. 1. Henderson's statistics of crowd fluids [6]

Fig. 1 Probability density function of the velocity component V_x distribution of 693 students in walk mode on a footpath outside Fisher Library, University of Sydney. Curve a, measured distribution; curve b, Maxwell-Bolzmann distribution. $V_x = 1.53$ m s^{-1}; $v_{r \cdot m \cdot s.} = 0.201$ ms^{-1}

2. The Basic Algorithm

The basic model is both a demand and supply model for the energy sector. Thus

$$\text{Demand} = \text{Supply} . \tag{1}$$

However, we break down each side into pertinent components. Thus, the l.h.s. is defined as follows:

$$\text{Demand} = \text{Absolute Human Needs} + \text{Quality of Life} + \text{Industrial Mix} . \tag{2}$$

In turn, the r.h.s. is defined as follows:

$$\text{Supply} = \text{Nonrenewable Energy} + \text{Renewable Energy} - [\text{Efficient Use of Energy}] . \tag{3}$$

Equation (3) is used because technology makes it possible to reduce the actual supply needed to meet the demand. In (4) the term in brackets, i.e., the efficient use of energy, can be broken down still further. How we do this depends on whether we want to use an economic approach or a technical approach, although, of course, the distinction is not clear-cut.

Consider the economic view first:

Efficient Use of Energy = Conservation + Cogeneration + Substitution + Storage + Technological Innovation . (4)

However, from an engineering or resource view, thermal energy storage, so success-ful in Germany and in the U.K.,must comply with the first law of thermodynamics, namely energy balance. Thus, thermal storage (or any form of storage) is really a deferral of energy use to a time when it can be purchased more cheaply and/or with-out accruing new capital and/or operational costs. Thus, we can write for the en-gineering version of (4)

Efficient Use of Energy = Conservation + Cogeneration + Substitution + Technological Innovation . (5)

3. Interpretation of the Terms

I shall now attempt to suggest means of determining the various microscopic quanti-ties that go into the terms of (2). It should be emphasized that in actuality these cannot be perfectly separated and there is coupling among them. This occurs even in the physical sciences. For example, a plasma which contains negatively charged electrons and positively charged ions exhibits different microproperties depending on whether or not an external magnetic field is imposed on the plasma. In the latter case, the properties assume tensorial forms.

In my discussion, I shall emphasize the terms on the l.h.s. of (2), namely those that contribute to the demand, because it is these where the coupling of micro-and macroquantities is particularly important.

Absolute Human Needs: These are those that are mandatory for survival in the most primitive sense of living, and it is basically food to sustain life.

Per Capita Requisite Food = Basal + Work Level + Environment + Disease . (6)

The basal need is the number of kilocalories needed to keep the body functioning at rest. In general, this depends primarily on body size. The energy consumed in do-ing work depends on the type of work, i.e.,manual or sedentary. Of course, manual work in a developing country can be expected to be higher than in the case of an industrialized society that has the benefit of machinery and computerized automa-tion including robotics. Environmental energy needs are due to the prevailing cli-mate. Finally, in many developing countries the inhabitants suffer from disease and parasites. One well accepted medical figure will be given as an example: 7% more energy per $1°[F]$ fever above the normal temperature. Thus, a fever of $104°[F]$ ($40°[C]$) would require about 38% additional food energy [7,8]. Assuming a basal requirement of about 2200 kcal, this would be 836 kcal per sick individual. In highly populated developing countries where vaccines and other modes of health care are unavailable, this can amount to a huge amount. However, not all people are sick all the time. Thus, we need to concern ourselves with two issues: (a) popu-lation demography; and (b) the quality of life. We shall consider (a) first in this subsection and (b) in the next subsection.

Of course, the population increases differently in different countries. We use two approaches: (a) concerning population changes we deal primarily with people already living because that is a reliable guide, although sufficiently long-term to make our preliminary findings reasonably acceptable. See Fig. 2 from COATES [9]; (b) further, we adapt the species continuity equation from reacting flow dynamics where instead of the particle number, we use the population number P [2].

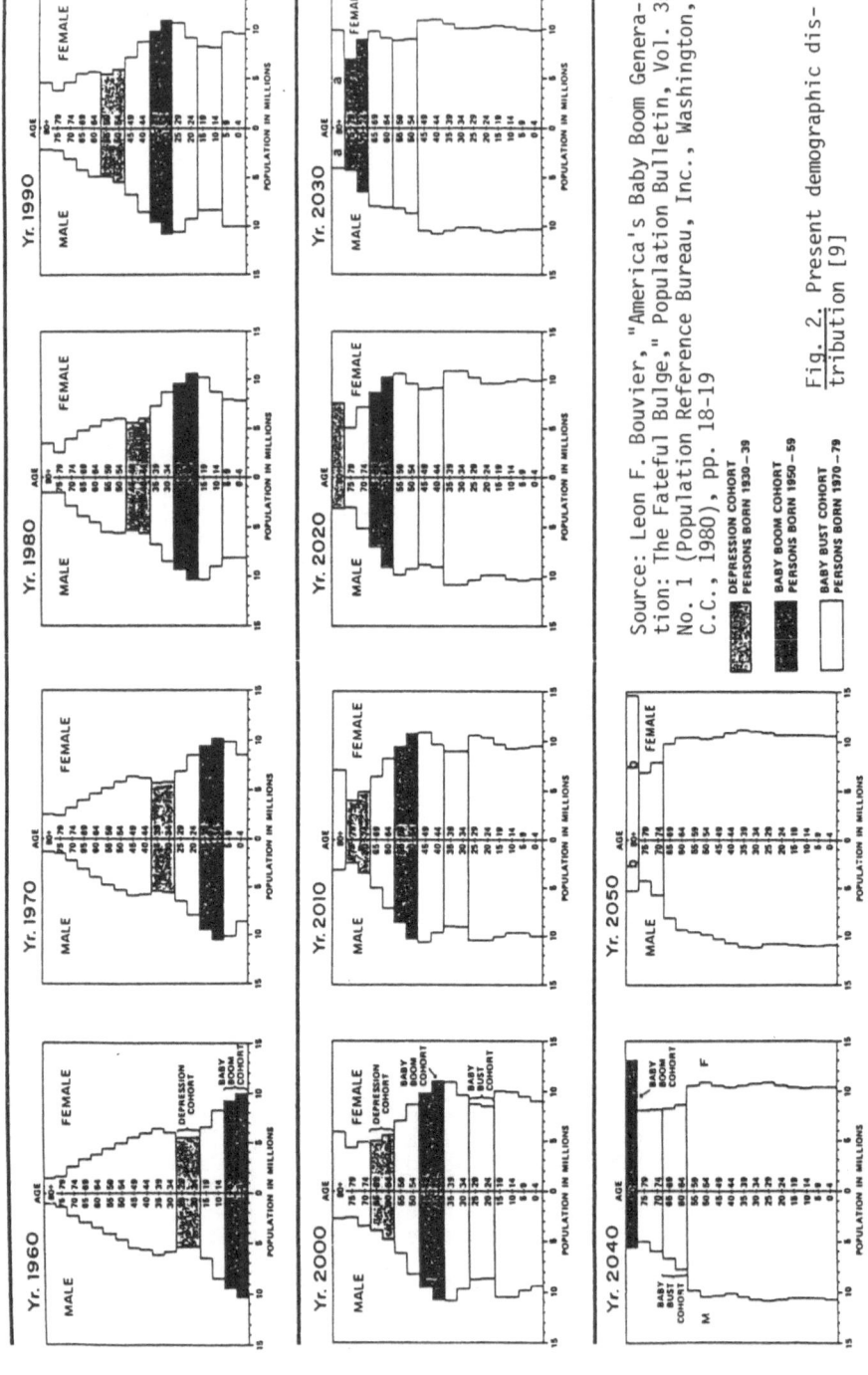

Progress of Depression Cohort, Baby Boom Cohort, and Baby Bust Cohort Through U.S. Population Age-Sex Pyramid: 1960-2050

Source: Leon F. Bouvier, "America's Baby Boom Generation: The Fateful Bulge," Population Bulletin, Vol. 35, No. 1 (Population Reference Bureau, Inc., Washington, C.C., 1980), pp. 18-19

Fig. 2. Present demographic distribution [9]

186

Thus

$$\frac{\partial P_i}{\partial t} + \vec{v} \cdot (P_i \vec{v}) = \dot{w}_i \tag{7}$$

where P_i are the various types of population and $i=1,2...n$; and where $\dot{w}$ is the net population production due to birth rates and death rates. As a rule, these numbers can be obtained from the host country, the United Nations or the World Bank. One can also use in (7) population fluctuations due to immigration/emigration or due to the "Gastarbeiter" prevalent in some Western industrialized countries or those in the Persian Gulf.

In (7) we can differentiate between male and female, for example. However, we do not generally do so, not just because we agree with the anthropologist Margaret Mead that, "Men and women are equal, but not similar"; but rather because of the latest advice of physician colleagues [7]; the differentiation as far as caloric intake is concerned is determined by (6). Thus, two persons, one male and one female, weighing the same amount and having no disease or parasites would have differing caloric intake levels, depending only on the type of work (manual or sedentary) that each does. This observation is important in developing countries where women may be relegated to certain types of work, but also is meaningful in developed countries where a marked shift is occurring from the heavy manufacturing industries to the high technology and service industries. It is also important to take account of age because this is related to the energy demand components in (6).

Many developing countries are adopting industrial agriculture which carries with it a high amount of embodied energy (cultural energy) [10,11]. In analyzing this we know, for example, that to grow an equal amount of protein, pork, fish and poultry requires far less cultural energy than does beef. Yet, we do not advocate that Islamic countries try to eat pork which would be alien to their religion and customs. However, we may suggest that they consider shifting to chicken and fish thus reducing beef and lamb imports.

The Quality of Life: This is difficult to quantify. We in the U.S. often have the penchant to compare our lifestyle with that in other countries. Are the Soviets worse off for a low coffee consumption, because they have a preference for tea from their samovars? Are France, Germany or Italy less well off for their low scotch whiskey consumption because they prefer beer or wine?

A number of measures of the quality of life have been proposed. These metrics are useful, but their choice is not immediately apparent. Accordingly, we use as a first approximation, Abraham Maslow's five levels of "needs hierarchy", which are shown in summary form in Table 1, and we assume a Darwinian philosophy of evolution that the different segments of society will want to better themselves to a higher level through an evolutionary process. (Even a revolution, which constitutes a step function, is followed by an evolutionary period of reconciliation and getting on with the goals of the revolution.) At this time, our work is too immature to use the more elaborate categories of C. Graves.

It must be noted that some social indicators are easier (e.g., those that have a tangible worth such as a monetary or energy value) to quantify than others. Others such as personal safety or health are more difficult to quantify. Thus, we use their inverse values. For example, the number of reported doctors' visits or hospitalizations is generally available. Of course, some crimes are more serious than others, as are certain illnesses. Here we substantially use standard statistical techniques, incorporating various importance levels of the statistical data. Certainly, infant mortality is more serious than some other sicknesses.

My analogy of the Maslowian levels in the physical world is a simple one. Thus at very low temperatures a gas has only three translational degrees of freedom. As the energy level is raised, it acquires rotational and vibrational degrees of free-

Table 1. The five Maslowian levels

Level	Characteristics
1 -- Survival	Here the dominant driving force is simply the struggle to sustain life, to secure food, drink, sleep, warmth. This state involves scarcity and extreme poverty. Primitive man, social outcasts, and severe defectives fall into this category as do prisoners in POW camps -- to these persons survival is all important. Little else counts.
2 -- Security	Persons living within this socioeconomic environment are motivated by a desire for protection -- from beasts, from people, from natural or economic catastrophies. Individuals are afraid of chaos and seek the familiar, security in numbers -- the outlook is basically status quo. Minorities, the poor, marginal farmers, and small businessmen fall within this category.
3 -- Belongingness	People at this level of social activity strive to be a part of something bigger, conform to group norms ("organization men"), depend upon the opinion of others. The mass middle class falls into this category.
4 -- Esteem	People in this category are goaded by achievement that is "visible," ostentatious; they want to stand out and have others think well of them; materialism, and "keeping up with the Joneses" pervades this social strata. Within this category are business executives, politicians, and nouveaux riches.
5 -- Growth or Self-Actualization	Living up to one's full potential is the primary concern at this level of development. Individuals within this social strata are tempered by idealism, motivated by ends not means, willing to follow -- or to lead. They have an abiding conviction that the world can be better, their outlook is dominated by social awareness, and with a future-oriented and global perspective.

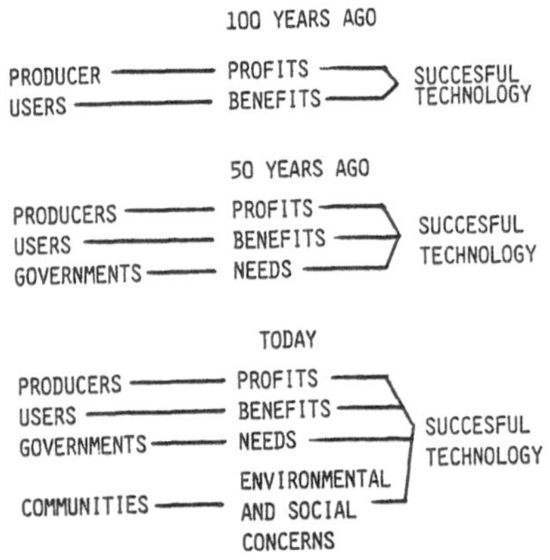

100 YEARS AGO

PRODUCER ——— PROFITS ——→ SUCCESFUL
USERS ——— BENEFITS ——→ TECHNOLOGY

50 YEARS AGO

PRODUCERS ——— PROFITS ———
USERS ——— BENEFITS ——— SUCCESFUL
GOVERNMENTS ——— NEEDS ——— TECHNOLOGY

TODAY

PRODUCERS ——— PROFITS ———
USERS ——— BENEFITS ———
GOVERNMENTS ——— NEEDS ——— SUCCESFUL
ENVIRONMENTAL TECHNOLOGY
COMMUNITIES ——— AND SOCIAL ———
CONCERNS

Fig. 3. Pluralization of socio-economic-technical actors [12]

dom, eventually dissociates, is ionized and then becomes a meson gas. The computations of properties at the various energy levels become more complicated. So it is with a social system. As it advances, it becomes more pluralistic. This is shown dramatically in Fig. 3, for which I am indebted to Fred KOOMANOFF [12].

Having decided (more or less) at what level a certain country is, we ask what it will take in energy requirements to go on to the next level? As noted, socioeconomic progress is measured to some extent by social indicators, i.e.,health, education, etc. Let us denote generically any indicator by I and rewrite (7) generally as follows:

188

$$\frac{\partial I}{\partial t} + \vec{\nabla} \cdot (I\vec{v}) = \dot{w}_I \ . \tag{8}$$

There have been made attempts, particularly in the U.S., to quantify the contributions to the so-called quality of life. However, the criteria seem to change as may be noted from Table 2. Clearly, greater emphasis is being placed on education, science, and technology, etc. The matter has been studied by GROSS and SPRINGER in considerable detail [13]. This suggests that people's aspirations cannot be ignored.

Table 2[a].

	Number of Goal Areas[b] in 1933 Report	Number of Goal Areas[c] in 1960 Report
1. The Individual	2	6
2. Equality	1	3
3. Democratic Process	4	11
4. Education	0	5
5. Science	1	8
6. Democratic Economy	6	9
7. Economic Growth	5	9
8. Technological Change	1	5
9. Agriculture	1	5
10. Living Conditions	6	10
11. Health and Welfare	10	10
	37	81

[a]Social Indicators and Goals by Albert D. Biderman p.68 in Social Indicators. Edited by Raymond A. Bauer
[b]President's Research Committee on Social Trends, Recent Social Trends in the United States. McGraw-Hill Book Co., Inc. New York, 1933.
[c]President's Commission on National Goals, Goals for Americans. Prentice-Hall, Inc., 1960. Englewood Cliffs, New Jersey.

Clearly, the variation in social dynamics depends on time. But it also depends on techno-economic locus, more than on geographic locus, as is evident from the composition of the attendees of various summit meetings. For a more systematic approach, see Spreng's triangle elsewhere in this volume. Therefore, we write the so-called total or convected derivative. Thus,

$$\frac{DI}{Dt} = \frac{\partial I}{\partial t} + \vec{\nabla} \cdot (I\vec{v}) = 0 \ . \tag{9}$$

It should be noted that in (9) one cannot always use the chronological time t, as measured by clock, watch or calendar. With a shrinking world due to high speed transportation, satellite communications and international trade we therefore consider the socio-economic time or as MURPHY [14] calls it, entropy time as shown in Fig. 4 which shows the time constraints when there is international trade among nations whose Sabbaths are respectively Sunday, Friday and Saturday. Similarly, the weekly load curves of electric utilities are very sensitive to the type of time [15].

Other units or metrics are important. These include the standard engineering units of mass, length, temperature and current. For economic reasons a monetary value must be added. While we do use currency such as Dollars, we find this at times somewhat rheologic and dependent on world circumstances. Thus we use both monetary and energy values, the latter which is fixed, to define economic dynamics. With H.G. Wells: "Ultimately the government...fixed a certain number of units of energy as the value of a gold sovereign..."

From (9) we can to some extent predict the change of any indicator of the quality of life, I, as follows:

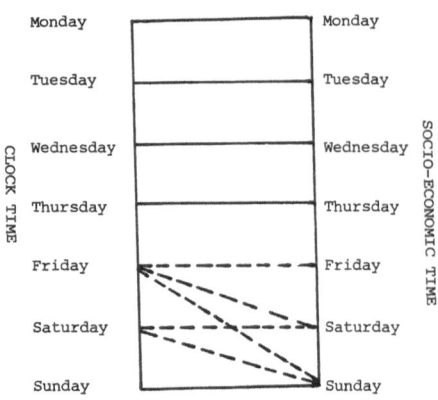

Fig. 4. Chronological vs. socio-economic time

$$\frac{dI}{dt}\bigg|_{\underline{future}} = \left(\vec{v} \cdot \vec{v}\right) I + \frac{\partial}{\partial t}\left[I(t) + I(\tau+t) + \exp\left(-\frac{E}{CA}\right)\right] + I\left[I^1(t)\right] . \qquad (10)$$

It should be noted that substantially this equation has been described with greater erudition by Haken elsewhere in this volume.

4. Entropy Management

For progress -- which is dynamic in nature -- there must also exist a certain type of socio-political-economic equilibrium, while the process of progress is not in perfect equilibrium. Both are necessary, neither is sufficient. For example, J. JACOBS has compared Birmingham and Manchester [16] while this author has studied urban growth and decay [17]. To explore this issue let us recall Shannon's formulation of the Second Law:

Information = Signal - Noise . (11)

Let us further recall the classical definition of available energy:

Energy Available for Useful Work = Total Energy Supplied - Energy Rejected
- Losses . (12)

Let us now apply these concepts to foreign aid (the North-South dilemma), aid to cities in the U.S.A., or the "bail out" of corporations.

In the most generic sense, let information be indicative of a progress indicator I; let signal be indicative of external aid E.A.; and let noise be indicative of internal mismanagement M. Then,

Progress = External Aid - Mismanagement . (13)

There may then be defined an

$$\text{Efficiency of Progress} = \frac{\text{Progress}}{\text{E.A.}} \qquad (14)$$

$$= 1 - \frac{M}{\text{E.A.}} . \qquad (15)$$

If the indicator I is properly chosen, it is necessary that

$$\frac{\partial I}{\partial t} > 0$$

190

$$\frac{\partial I}{\partial t} = \frac{-M \frac{\partial (E.A.)}{\partial t} - (E.A.) \frac{\partial M}{\partial t}}{(E.A.)^2} \qquad (16)$$

$(E.A.)^2 > 0$

$M \frac{\partial (E.A.)}{\partial t} > (E.A.) \frac{\partial M}{\partial t}$.

Case 1: M = constant, $\frac{\partial M}{\partial t} = 0$, $M \frac{\partial (E.A.)}{\partial t} > 0$.

Case 2: E.A. = constant, $\frac{\partial (E.A.)}{\partial t} = 0$, $(E.A.) \frac{\partial M}{\partial t} < 0$.

The early repayment of loans by successful corporations indicates the importance of superior management.

Further, Mismanagement $\propto$ Entropy .

The Boltzmann expression of entropy is written

$S = k \ln N$

where k is the Boltzmann constant and N denotes the number of particles or other sorts of subensembles. For a socio-economic system the basic equation applies, but in a more complex form:

$M = \text{History} + C(t) \ln P + C(t) \ln (\text{Vested Interest Groups})$. $\qquad (17)$

Any evaluation of the "quality of life" must take into consideration the social system in conjunction with its surrounding environment (e.g., national governments or international relations).

In the words of Schrödinger: "...a living organism continually increases its entropy...and thus tends to approach the dangerous state of maximum entropy, which is death. It can only keep aloof from it, i.e., alive by continually drawing from its negative entropy." However, in a politico-economic sense then, foreign aid, state aid to cities, and government aid to corporations are all manifestations of sharing negative entropy.

In view of the aforementioned, we can depict in Fig. 5 the situation as follows.

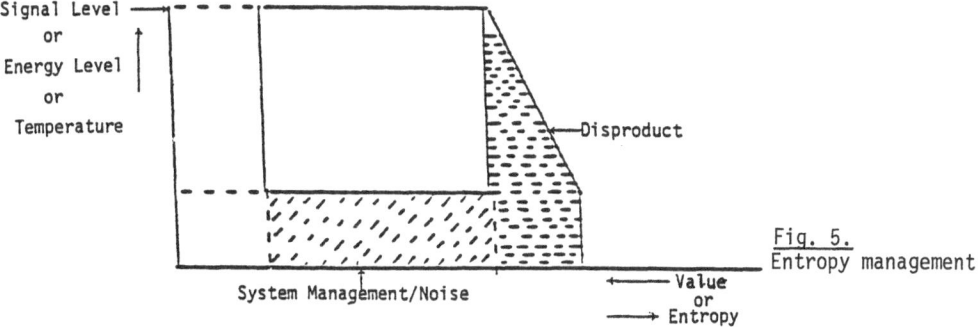

Fig. 5.
Entropy management

Finally, we define a "socio-economic" Gross National Value as follows:

Gross National Value = Gross National Product - Gross National Disproduct . (18)

It would be interesting to define a second law of thermodynamics type of efficiency or effectiveness. Superficially, it is not difficult to do so. Thus, the following is suggested:

$$\text{Social Second Law Effectiveness} = \frac{\text{Actual Social Benefits}}{\text{Maximum Social Good}} . \qquad (19)$$

It is immediately obvious that neither the numerator nor the denominator in (19) is easily quantifiable. This is due to two reasons: (a) peoples' aspirations change; and (b) more importantly, peoples' value concept even for the same quantifiable social indicator is different. Accordingly, at this time we use a Social First Law Efficiency which to some degree can be evaluated:

$$\text{Social First Law Efficiency} = \frac{\text{Gross National Value}}{\text{Gross National Product}} . \qquad (20)$$

For people to experience an improvement in their quality of life, it is necessary that there exists a reasonably stable, but nevertheless dynamic regime. The very term "dynamic" implies that society is not stagnant. It is slightly away from the equilibrium state. Therefore, we can write two simultaneous, linear, differential equations of the Lancaster type as follows:

$$\frac{dE}{dt} = C_1 R - C_3 E + C_5 + f(E,R) \qquad (21)$$

$$\frac{dR}{dt} = C_2 E - C_4 R + C_6 + f(E,R) \quad \text{where} \qquad (22)$$

E - potential of government to bring about reform and progress, e.g., economic ability, progressive policies, etc.

R - potential of reformers to bring about pressure, e.g., voting strength, economic pressure, etc.

f(E.R.) - human factors

C_1 - "progressivity" coefficient of government

C_2 - "influence" coefficient of reformers

C_3 and C_4 - "conservation" coefficients (government and reformers respectively).

If C_3 is high, reformers are slowed down; if C_4 is high, government is slowed down.

C_5 and C_6 - "confidence" coefficients. If the signs are the same, confrontation is less likely. If the signs are opposite and $|C_5|$ and $|C_6|$ are high, polarization and confrontation are likely to occur.

If $\left|\frac{dE}{dt}\right| > \left|\frac{dR}{dt}\right|$ Stable government   Inequality (A)

If $\left|\frac{dE}{dt}\right| < \left|\frac{dR}{dt}\right|$ Unstable government .   Inequality (B)

It should be admitted at once that various coefficients C cannot be quantified in absolute terms, (we don't even have a metric for them) but they can be compared on some arbitrary scale 1 through 5. Therefore, one might be able to estimate the relative magnitudes of inequalities (A) and (B).

Another approach that is useful is to go back to Schrödinger's desire for negative entropy. While Schrödinger also mused whether or not classical thermodynamics is sufficiently adequate for living systems and suggested the need for a more general "Theory of Life", I shall take liberties by suggesting a different interpretation.

The well-known argument of Clausius that entropy must either increase or remain constant applies only to isolated systems, those surrounded by a Caratheodory type adiabatic wall. But socio-economic systems are not isolated, as any glance at the daily headlines will demonstrate. The impacts of political forces, satellite communications and the transfer of technological innovation cause constant transfers between the system and its surroundings oblivious to the demarkations on a political map. Thus,

$$ds \gtreqless 0 . \tag{23}$$

For beneficial progress we want

$$ds < 0 . \tag{Inequality (C)}$$

The total change in entropy for socio-economic systems may then be expressed as follows:

$$\Delta s = S_2 - S_1 = \int_1^2 ds + Ck\ln N_i + \sum s_{m_i} + \sum s_{e_i} \tag{24}$$

where: C = quality of life coefficient. $C = f (I_1, I_2, \ldots I_n)$; N_i = Jane Jacobs contributions*; s_{m_i} = material entropy crossing boundaries; s_{e_i} = energy (or information) entropy crossing boundaries.

For progress we demand that $\Delta S = s_2 - s_1 < 0$. Hence,

$$\left[Ck\ln N_i + \sum s_{m_i} + \sum s_{e_i} \right] > \int_1^2 ds . \tag{Inequality (D)}$$

Further, the bracket term should be negative. We call this a "rejuvenating" term, and demand inequality C already mentioned.

Industrial Mix: It has been shown [18, 19, 20] that in the industrialized countries there is a shift from the heavy manufacturing industries to the high technology and service industries. This shift is due primarily to: (a) ascendency of groups on the socio-economic stratification grid; and (b) the impact of modern technology. Again, these two are not completely separable.

For people to achieve their rising expectations, one must then ask if there are any limitations. Such limitations are due to "resource availability" be they natural resources (e.g., energy, minerals, etc.) or human resources (e.g., health, education, employment, etc.). History shows that there ultimately occurs a gap between resources and needs or between achievements and expectations. This situation is sketched qualitatively in Fig. 6 which is self-explanatory and for which I am indebted to STAGNER [21].

Next we wish to devise a tool to measure the situation mathematically. Thus, we write a dimensionless level of well-being coefficient π, as follows:

$$\pi = \frac{\text{Resource} - \text{Need}}{\text{Need}} \tag{25}$$

$$= \frac{R}{N} - 1, \text{ where } \frac{R}{N} \text{ should increase.}$$

---

*Without her permission, we named these after the urbanologist Jane Jacobs for her incisive analysis of the decay and viability of cities.

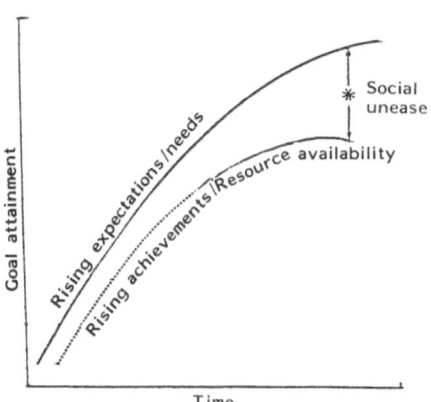

Fig. 6. The widening gap [21]

Then the rate of change towards progress can be obtained by differentiating (25).

$$\frac{\partial \pi}{\partial t} = \frac{N\frac{\partial R}{\partial t} - R\frac{\partial N}{\partial t}}{N^2}$$

(26)

where $0 < N$ and $N \gtreqless 0$.

It follows that:

N will increase with constant technology and increasing population or standard of living; while N will decrease with technological substitution or technological innovation. In turn, R will decrease through consumption when the technology is constant; but R will increase by technological innovation. While SAMUELSON basically adheres to economic equilibrium theory, his comparison between thrift and innovation is most pertinent here [22].

While new technologies that are "add ons" will, of course, result in added energy requirements, technological innovation that serves as a substitution for an older one, or that causes structural shifts in the industrial economy may slow down the rate of increase of energy demand. Although historical precedents may be helpful, true innovation implies qualitative change. Thus, historical antecedents are not generally applicable quantitatively. This is particularly so, because over the years populations have tended to increase and there also have been improvements in Maslowian levels. Also, in a changing world, the common rules of "scaling" do not apply [23]. Indeed, they have not worked even in the case of nuclear power plants which are about as scientific and as well organized as may be found. Further, progress whether it be social or technological implies some sort of flux $\vec{J}$, some driving potential $\vec{P}$, and some resistance $\vec{R}$ to the change, whether or not this is progress or regression. Thus, one can write in accordance with general transport theory:

$$\vec{J} \propto L \frac{\vec{P}}{\vec{R}}$$

(27)

where L is some appropriate conductivity or diffusivity coefficient. Such coefficients are microscopic in nature. However, there are many progresses and these are generally coupled to one another, just as the Peltier and Seabeck effects in thermoelectricity. Thus, after Onsager one might write a set of linear, simultaneous equations as follows:

$$J_1 = L_{11}X_1 + L_{12}X_2 + \ldots L_{1n}X_n$$

$$J_2 = L_{21}X_1 + L_{22}X_2 + \ldots L_{2n}X_n \quad \text{etc.}$$

(28)

Of course, Onsager's approach requires that the coefficients L_{ij}, adhere to the reciprocity condition, i.e., $L_{ij} = L_{ji}$; and furthermore that they vary linearly. It is not clear that this is so in socio-economic situations. While we are embarking on understanding the phenomena better, we are also considering Prigogine's more sophisticated theories. Because our explorations are too recent, I shall not comment further at this time.

5. Dimensionless Reaction Times

In analyzing high speed fluid flows with accompanying physico-chemical reactions, it is customary to write the well-known Damköhler ratio:

$$\text{Damköhler's Parameter} = \frac{\text{Time of Translation of Fluid Element}}{\text{Physico-Chemical Relaxation Time}} \gtrless 1 . \qquad (29)$$

If Dam >> 1, the time spent in travel is sufficiently long for the nonequilibrium gas to return to its equilibrium state. Therefore, we speak of equilibrium flow. When the converse situation exists, we speak of frozen flow [2]. It appears that this concept might be applied to socio-economic problems. Let us write the following [24].

$$\text{Technical Parameter} = \frac{\text{Time to Commercialize Technical Innovation}}{\text{Time for Device to Become Obsolete}} . \qquad (30)$$

At this time, the view regarding the numerator of (30) remains unsettled; some authorities believe that it is becoming shorter while others claim the opposite. Examples can be cited to prove either contention.

$$\text{Education Parameter} = \frac{\text{Time to Acquire Knowledge (Education)}}{\text{Time for Education to Become Obsolete}} . \qquad (31)$$

While education does not necessarily become obsolete, the limited amount of knowledge acquired during the formal school years surely becomes insufficient in the high technology areas. Thus, a love for learning must be inculcated in university students and, hence, the need for faculty research or employer professional development programs.

Finally, it is conceivable to consider a leadership parameter as follows:

$$\text{Leadership Parameter} = \frac{\text{Time in Office}}{\text{Time for Progress to Take Place}} . \qquad (32)$$

Here again the jury is out concerning the optimum time in office. Until conclusive data become available, I suggest that any decision be made on the basis of (30) and (31) which are indicative of entrepreneurmanship and intellectual resilience; really the ability to adapt to the rapidly changing circumstances. As Carlos Romulo stated: "Time is ruthless with living beings who cannot change and will not adapt to change. It is equally cruel with institutions that freeze in the mold of the past."

6. Observations

It appears that modelling socio-economic systems from carefully chosen examples in the physical sciences may offer some useful semi-quantitative insights. The analysis and evaluation of socio-economic systems requires synergizing of micro-variables and macro-phenomena.

Mathematical analysis tempered by humanistic interpretation is necessary before resorting to the brute force of computers.

Any study of socio-economic systems requires an appreciation of synergetics and the harmonious cooperation among discipline trained persons who must work in an interdisciplinary mode and environment.

7. Acknowledgements

I am indebted to numerous colleagues in academia, government, private industry and the independent sector for stimulating discussions. Unfortunately, space does not permit me to mention them all, but a few must be cited.

I owe the late Theodore von Kàrmàn the intellectual fortitude to confront problems in at times an unconventional manner as well as the difference between microscopic and macroscopic effects. Walter Heller and Jerome B. Wiesner made me develop an appreciation for differentiating between scientific potentialities and economic realities, while Eugene Webb showed me the way to the behavioral sciences outside the traditional textbooks. Maxwell D. Taylor was always receptive to mathematical representation of crucial issues. Bertram M. Gross seduced me intellectually to try the mathematical analysis of the quality of life indicators, while Bruno Fritsch impressed on me the possibilities of thermodynamic analysis for understanding socio-economic issues. Finally, I must thank my graduate students at previous institutions as well as presently at George Washington University for their willingness to do their research with me. Of course, I take full responsibility for the statements I made in this essay.

8. References

1. H. Haken: Synergetics (Springer-Verlag, New York 1981).
2. A.B. Cambel: Plasma Physics and Magnetofluidmechanics (McGraw-Hill, New York 1963).
3. Y.L. Klimontovich (trans. A. Dobroslavsky): The Kinetic Theory of Electromagnetic Processes (Springer-Verlag, New York 1983).
4. L.S. Alvarez: "Allocating Energy Resources: Computer Program for an Energy Model for Developing Countries (Mexico)", M.S. Thesis (The George Washington University, Washington, DC 1983).
5. J. Benefield: "Energy Development Modelling", M.S. Thesis (The George Washington University, Washington, DC (in preparation)).
6. L.F. Henderson: Nature 229, 381 (1971).
7. H. Pollack: Personal communication.
8. J. Chino: "Energy Mix Model for the Peoples' Republic of China", D.Sc. Dissertation (The George Washington University, Washington, DC (in preparation)).
9. J.F. Coates: Personal communication.
10. B. Fritsch: Wir werden überleben (Günter Olzog Verlag, München-Wien 1981).
11. B. Fritsch: The Energy Demand of Industrialized and Developing Countries Until 1990 -- Toward a World-Wide Burden Sharing (Institut Für Wirtschaftsforschung E.T.H., Zurich 1982).
12. F.A. Koomanoff: Personal communication.
13. B.A. Gross: Social Intelligence for America's Future (Allyn and Bacon, Inc., Boston 1969).
14. R.E. Murphy, Jr.: Adaptive Processes in Economic Systems (Academic, New York-London 1965).
15. A.B. Cambel, et al.: "The Techno-Socio-Economics of Distributed Thermal Energy Storage as a Part of Load Management" (The George Washington University, Washington, DC 1982).
16. J. Jacobs: The Economy of Cities (Random House, New York 1969).
17. A.B. Cambel: "An Introduction to the Deterministic Formulation of Synergistic Social Change", Proc. of the United Nations Interregional Seminar on the Use of Modern Management Techniques in the Public Administration of Developing Countries, ESA/PA/MMTS/31 (Washington, DC 1970).
18. R. Reich: The Next American Frontier (Times Books, New York 1983).
19. B. Bluestone and B. Harrison: The Deindustrialization of America (Basic Books Inc., New York 1982).
20. J. Naisbitt: Megatrends (Warner Books, New York 1982).
21. R. Stagner: Personal communication.
22. P.A. Samuelson: Economics (McGraw-Hill, New York 1982).
23. E. Benvenuti: Personal communication.
24. A.B. Cambel: Journal of Engineering Education 53, 2, p. 113 (1962).

New Synergisms in Socio-Economic Systems:
More Questions than Answers

Bruno Fritsch

Swiss Federal Institute of Technology, Center for Economic Research
CH-8092 Zürich, Switzerland

Discussions about the future course of world economics and world politics became
diverted from a balanced analytical assessment into a predominantly pessimistic view
mainly because of two events: one occurred in the field of academia, the other in
the field of power politics. The academic event was Dennis Meadows' et al. book on
"The Limits to Growth" [1] submitted to the Club of Rome; the political event was
the first oil crisis brought by the OPEC countries upon the oil-consuming nations
including less developed economies as well as industrial countries in West and East
in November 1973. Since that time, concepts related to the notion of "limits" have
become fashionable. The world was conceived as a closed system (spaceship earth),
and the resources available to mankind (or as the Americans prefer to say: human-
kind) were declared to be finite and hence depletable. Pollution of the environment
was assumed to be proportional to industrial growth, and since the ability of the
environment to absorb pollutants must also be considered to be finite, the message
was clear: either we stop growing or we will destroy the life support system which
serves as the base for all known civilization. The stage for doomsday philosophies,
ranging from Heilbronner's Inquiry into the Human Prospect [2] to Amory B.Lovin's
numerous writings on energy issues was set. Ten years later the world's resources
were not depleted as predicted by Dennis Meadows; on the contrary: the prices of
many raw materials plunged to a low level due to lack of demand, causing severe
problems to many less developed countries who were depending upon the export of
these commodities. However, the growth rates of most industrial countries dropped
to almost zero, recovering only slowly after the second oil shock of 1979. But this
drop was not the result of any political decision following Meadows' recommendation,
it was, in fact, the consequence of the quintupling of the real price of oil within
a comparatively short period of eight years; an event which was in no way causally
related to this study.

The limits to growth theories were based on the false assumption that the earth
is a closed system, and on the underestimation of technological advances related
to man's innovative capacities. Although these wrong assumptions were essentially
responsible for the wrong conclusions, forecasts and recommendations derived from

these theories, it would be unwise to dismiss them altogether. Paradoxically, the discussions initiated by the limits to growth paradigm broadened the horizon of social scientists, and particularly that of economists, engineers and many other highly specialized scholars, and also evoked their awareness towards certain factors which they, for the sake of purity of theoretical argumentation, often "assumed away". Today there seems to be little disagreement about the variables which have to be included in the analysis when approaching problems of growth, environmental pollution, resource availability and the long-term prospects of mankind. These variables, including their mutual relationships and interactions are: energy, matter, information, knowledge, and the learning capabilities of political systems.

The following is a modest attempt to consider some of the interrelationships between these strategic factors. It seems to be appropriate to distinguish three levels or system areas: the *physical level, the societal level,* including economic and political strata, and the *socio-ecological level.* On the physical level the following relationships hold:

$E = m.c^2$ which is the famous and well-known Einstein formula.

The relationship between energy and information may be derived from thermodynamics and Shannon's information theory according to which 1 bit of information equals k.ln2; k being the Boltzmann constant $(1,38 \times 10^{-23}$ per $^\circ K)$. Hence 1 bit $= 10^{-23}$ joule (1 J = 1 W/s). For derivation see M.Tribus et al. [3].

The *open* system earth receives an equivalent of approximately 175000 TW a/a through the radiation of the sun, or an amount corresponding to 10^{38} bit/s.

The first conclusion from this is, as evident as it is fundamental, albeit as yet not fully realized by the public, that there exists an abundance of energy in physical terms. When speaking of the "energy crisis", we must realize that shortages of availability of primary and/or secondary energy are the result of the lack of energy generating systems due to insufficient investments rather than to a shortage of energy *per se.* Hence, any shortfall of supply must not be related to the shortage of physical availability of energy to this planet but rather to the capital generating capability of a given economy. Thus, the problem has to be discussed on the next level mentioned above, i.e. the economic level.

In this context we have to keep in mind the fact that any energy generating system uses energy in two ways: first, energy is directly used for the transformation of primary into secondary energy. Secondly, energy is needed by the energy generating system itself for the maintenance and development of the system components. For the above reasons, the energy which will be delivered to the non-energy part of the economy by the energy-generating system is a function of the following variables and parameters:

198

Y_0 = initial income at time t_0

Y_t = income at time t

r = growth rate of GDP

r^T = rate of technical progress

C_t = consumption at time t

I_1 = investment in the non-energy sectors of the economy

I_2 = investment in the energy system

I = total investment $I=I_1+I_2$

E_t = energy production in time t

L = lead time of investment

α = fraction of income consumed

λ = fraction of income available for I_2

σ = productivity of investments in the energy system.

The ratio between E_t and C_t can be easily obtained from the following equations:

$$C_t = \alpha Y_0 e^{rt} \dots\dots\dots\dots\dots\dots (1)$$

$$C_{t-K} = \alpha Y_0 r^{r(t-L)} \dots\dots\dots\dots\dots (2)$$

$$I_2 = \lambda(1-\alpha)Y_0 e^{r(t-L)} \dots\dots\dots\dots (3)$$

$$E_t = \alpha_t I_2 \dots\dots\dots\dots\dots\dots (4)$$

$$\sigma_t = \sigma_0 e^{r^T t} \dots\dots\dots\dots\dots\dots (5)$$

$$E_t/C_t = \frac{\sigma_0 e^{r^T t} \lambda(1-\alpha)Y_0 e^{r(t-L)}}{\alpha Y_0 e^{rt}} = \frac{\sigma_0 e^{r^T t}\lambda(1-\alpha)}{\alpha e^{rL}} .$$

These equations *define* the relationship between the consumption of the non-energy part, and the production of the energy system of the economy. This ratio is deter-mined by I_2 the fraction of GDP going into the energy system, by the productivity of investments in the energy system σ, the rate of growth of the economy r (which in turn depends *inter alia* on the capital/output ratio, i.e. the productivity if I), and on the leadtime of investments L. Since λ is a function of α and σ depends on the available technology r^T, the *strategic factors* governing the E_t/C_t ratio are: (1) the *saving ratio* α of the economy and (2) the *technical progress function* r^T. From the above consideration it becomes evident that at any given set of parameters, the energy system and the non-energy system of the economy are mutually interdepen-dent and therefore one sector cannot outgrow the other.

Since productivity σ depends on the technical progress function r^T, and since this function is determined by the transformation of general knowledge K into tech-nical advances, whereby K in turn depends partly on E, we have $\sigma e^{r^T} = f[K(E)]$. If on the other hand the growth of the economy (Y) depends *inter alia* on the saving rate σ and the productivity of investments, the total energy availability $E = f[\sigma e^{rT}\{K(E)\}, \alpha(Y)]$.

A part of the derived conclusion is that, at any given time, the energy sector cannot outgrow the rest of the economy, and *vice versa,* it is important to observe that this function does not say anything about the *kind* of energy source the economy makes use of. For historical reasons, more than 90 per cent of the primary energy used in the world is still derived from fossil sources - either coal and/or natural gas. Given the present technology, a 1000 MW coal power station releases annually approximately 10 million tons CO_2 and 7 thousand tons SO_2 into the atmosphere, apart from 18 thousand tons of nitric oxides, 1.5 thousand tons of dust (aerosols) and 11 mrem of radioactivity which is about ten times more than is released by a nuclear power plant. It is because of the effects of these pollutants, particularly the CO_2 and the SO_2 components, that the use of this particular energy source is limited: the SO_2 released into the atmosphere causes, together with other pollutants, the acid rain which in turn is responsible for the widespread destruction of forests and the decrease in the pH values of lakes which reduces drastically the number of species living in these lakes. The emission of CO_2 causes the well-known "greenhouse effect". Currently about 740 billion tons p.a. of C are released worldwide into the atmosphere thus increasing the CO_2 concentration from 315 ppm by vol. in 1958 to approximately 340 ppm by vol. in 1982.

The fact that the use of one particular source of energy is limited for reasons of environmental protection does not mean that the use of energy, as such, has to be limited. The CO_2-caused greenhouse effect is often mixed up with the waste heat which is the ultimate result of the use of *any* energy irrespective of its source. However, the increase in the average temperature caused by this physical law of dissipation of high grade energy into waste heat differs in orders of magnitude from the increase in temperature which is caused by the release of CO_2. In order to increase the temperature on earth by $1^\circ C$ by waste heat, we would have to use energy to the amount of approximately 10^{23}J, i.e. 10^{11}TJ. This corresponds to a level of energy consumption of 3000 TW a/a as compared with the present world energy consumption of 10 TW a/a. (For computation see P.Staub in B.Fritsch 1981).

At this point we approach the cumbersome problem of entropy, negentropy, information and knowledge. Without attempting to present an exhaustive analysis of the very complex interrelationships between these notions and the process of economic growth - an attempt which was first tried by Georgescu-Roegen [4] - one should observe two basic facts. First, "All order states of the earth - be they natural or technical - result from the openness of the earth's system" [5]. By implying the wrong notion of a closed "spaceship earth" and by ignoring the cognitive capacity

of man to generate suitable states of order for his life support by the use of capital, the depletion hypothesis considers as a loss any use of resources which produces waste and thus increases the state of disorder, i.e. the entropy of the system. Although this hypothesis seems, at first glance, plausible, because it implicitly states an analogy to the second law of thermodynamics, i.e. the ultimate and
irreversible dissipation of high grade energy into waste heat, it is nevertheless
wrong. As is rightly observed in the IIASA Energy Study: "Since resources are not
consumed but are, in fact simply degraded by use to materials with less order, it
is only order states that can determine the usefulness of these resources." (ibid.)

In its most general form, the problem can be formulated in the following way:
each system requires for its functioning the input of a certain number of "ordered
states" which can be represented by a vector with elements set in a specific concentration pattern. These inputs are "used up" by the system according to its internal needs which are governed by the laws of its metabolism. What is released as
output can be considered as a vector with elements representing another concentration pattern, usually - but not necessarily - with a somewhat smaller concentration.
By applying energy, capital and knowledge, this "waste vector" can be, in principle,
transformed into the original concentration pattern.[1] This would correspond to
total recycling. Parts of this transformation can be performed by auxiliary systems
which use the "waste vector" as their inputs. What is waste for one system may be

---

1) In a somewhat more formalized way the question can be stated in terms of matrix
algebra. Multiplying a vector of inputs V^i with a distribution of its elements
γ, i.e. V^i_γ with the matrix of the system S the elements of which being $a_{ij} - a_{in}$
an a_{im} respectively (whereby $S = f(E_c, N)$ with $E_c \in N$, i.e. S depends on the level
of economic activities E_c and the constraints given by nature N), we obtain an
output vector V^o with a distribution δ, i.e. V^o_δ such that $\gamma \neq \delta$. The question then
is whether there is a transformation T subject to inputs of capital C, energy E,
and knowledge K, derived from system S, which would restate δ into γ. Or, in an
even more general way: when can a linear mapping $S: \mathbb{R}^m \rightarrow \mathbb{R}^n$ be reversed by $T:$
$\mathbb{R}^n \rightarrow \mathbb{R}^m$, i.e. $T(\vec{Sx}) = \vec{Sx}, \forall \vec{x} \in \mathbb{R}^m$? If $S: \mathbb{R}^m \rightarrow \mathbb{R}^n$ is injective, i.e.
$\vec{x} \neq \vec{y} \Rightarrow S\vec{x} \neq S\vec{y}$ there exists an inverse mapping $T: \mathbb{R}^n \rightarrow \mathbb{R}^m$ whereby $m \leq n$.
Only if $m = n$, the usual transformation $T = S^{-1}$ always holds. If the mapping
$S: \mathbb{R}^m \rightarrow \mathbb{R}^n$ is not injective then an inverse mapping $T: \mathbb{R}^n \rightarrow \mathbb{R}^m$ does not exist.
However an $T: \mathbb{R}^n \rightarrow \mathbb{R}^m$ mapping may exist for $\vec{x}_o = T(S\vec{x})$ if $\vec{x}_o$ has similarities
with the vector $\vec{x}$ in such a way that both vectors could be mapped under S
upon one and the same vector ($S\vec{x} = S\vec{x}_o$). Further investigation of these relationships is required and will be pursued.

a valuable input for the metabolism of another system. Thus, many agricultural
wastes are valuable inputs for pigs which in turn results in meat used as one form
of energy input by the "system man". The following diagram shows a simplified
version of these interactions.

However, there may be a "rest" of disorders such as the dissipation of lead into the énvironment which cannot be "recovered" with the available amounts of energy and technical knowledge. It is then again a question of the formation of knowledge which may offset the adverse synergetic effects on the environment caused by
this dissipation as well as of compensation for the "loss" of this particular
source by seeking other materials with similar properties. If the occurrence of
these dissipations per unit of time exceeds the accumulation of knowledge to deal
with this problem, then we do have a limit of the *speed* with which such dissipations are allowed to occur. Hence, deacceleration of certain activities (material
transformations) may be required: again it is a question of lack of knowledge
rather than a question of the exhaustion of available resources.

From these observations strong evidence can be derived for the hypothesis that
energy and knowledge are the two strategic factors governing the processes of an
evolving and yet sustainable societal system. Such a system comprises the elements
shown in Fig. 2.

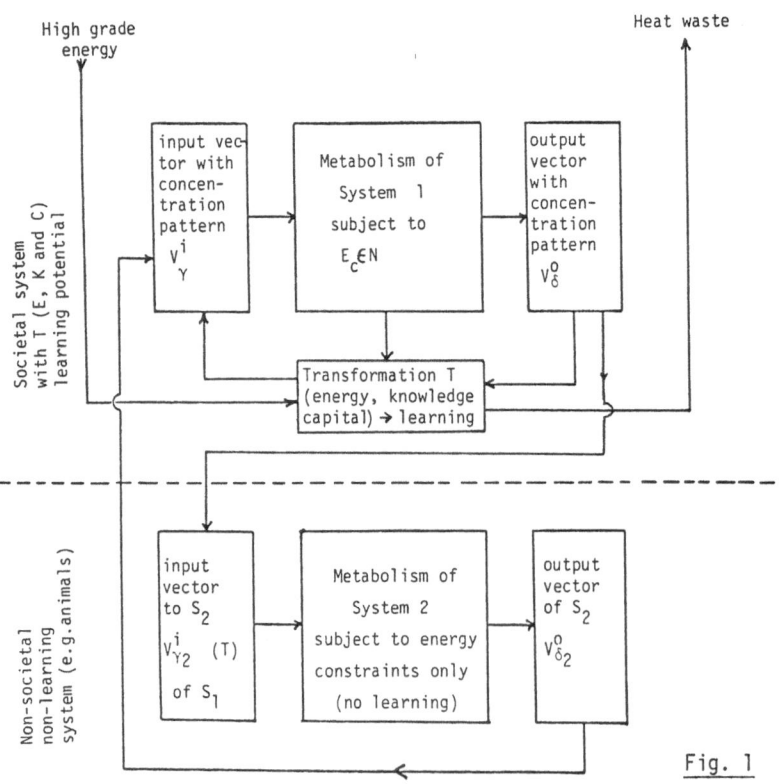

Fig. 1

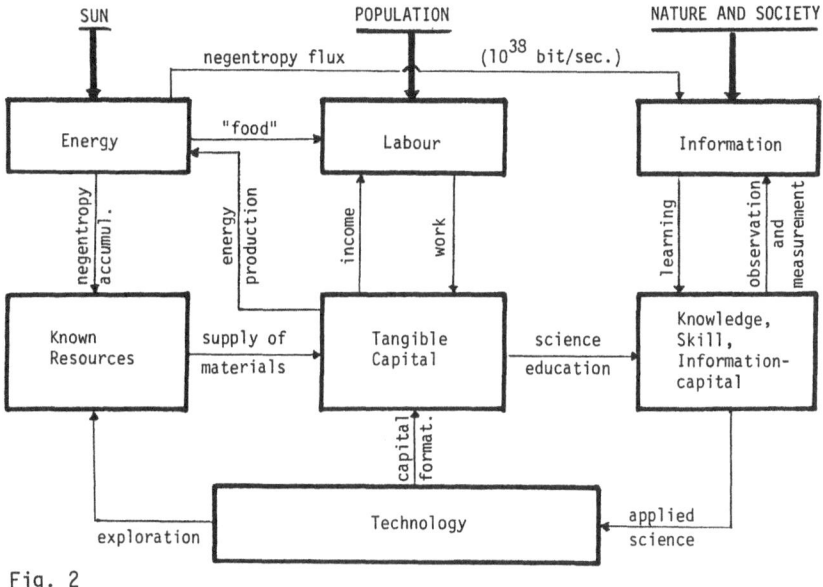

Fig. 2

Scholars who advocate low level soft energy paths refer to the falling energy elasticities prevailing in modern "post industrial" societies. Frequently it is even maintained that the upcoming "information society" will be able to operate with less and less energy, simply because the storage, transformation and retrieval of information requires only very little energy. Although the last part of this statement is true, it is nevertheless erroneous to believe that the future information society will need less energy - be it in absolute or in relative terms. The reason for this is quite simple: while the storing, processing and transmitting of *information* requires little energy indeed, the generation and acquisition of *knowledge* does need more and more energy. Space exploration is one example, high energy physics another. Therefore, if we accept the acquisition of knowledge as an open system, more and more energy will be needed. For instance, the experimental verification or falsification of theoretical concepts such as, e.g., the concept of supersymmetry in physics, may require a new generation of accelerators which are many times more powerful than even the largest accelerators in use today. If, in contrast to this concept of openness of knowledge, the opposite concept of "speculative knowledge" detached from empirical evidence is accepted, then certainly less energy will be needed to keep a society alive. Such a society will be based on belief rather than on facts. We would then find ourselves in the vicinity of the right hand bottom corner of Dr.Spreng's triangle, i.e. in the "starving philosopher" corner.

There is no reason to assume that this decision will ever be subject to our free will. But if we ever should embark on this "metaphysical trajectory" it will surely not be imposed upon us because of any physical scarcity of energy, but by other reasons emerging from the many yet unknown synergisms generating a set of societal evidences asserted less by empirical evidence but more likely by conclusive behaviour of the members of the society (Zeitgeist). It seems that these phenomena, i.e. the mutual interference between empirical (physical) and societal (value oriented) evidence, create a set of new synergisms not yet fully understood. It is therefore worthwhile to pursue them further by also applying the concepts and tools developed in the realm of synergetics.

References

1 Meadows, D. et al. The Limits to Growth, New York 1972.

2 Heilbronner, R. An Inquiry into the Human Prospect. London 1975.

3 Tribus, M. et al. Energy and Information, in: Scientific American, Vol. 225, No.5, September 1971.

4 Georgescu-Roegen, N. The Entropy Law and the Economic Process. Harvard University Press, Cambridge Mass. 1971.

5 IIASA, Energy in a Finite World. Report by the Energy Systems Group of the International Institute for Applied Systems Analysis, Laxenburg. Ballinger Publishing Company, Cambridge Mass. 1981 (p. 697).

Suggestions for further reading

1. Adams, R.N., Energy and Structure. A Theory of Social Power. Univ. of Texas Press, Austin and London 1975.
2. Afifi, A.A., Sagan, L.A. Energy Consumption as an Indicator of Longevity. International Institute for Applied Systems Analysis, Professional Paper. PP-78-6, Laxenburg Austria, August 1978.
3. Afifi, A.A., Sagan, L.A., Energy and Literacy: An Index of Health Development. International Institute for Applied Systems Analysis, Professional Paper. PP-78-7, Laxenburg, Austria August 1978.
4. Alchian, Armen A., Uncertainty, Evolution, and Economic Theory. The Journal of Political Economy, Vol. LVIII, Feb.-Dec. 1950, pp. 211-221.
5. Barber, T.Y. (ed.). Biofeedback and Self-Control. Aldine, Atherton, Chicago 1970, 1971.
6. Bateson, G. Steps to an Ecology of Mind. Ballantine Books. New York 1978.
7. Bauer, B., Wert und Bewertung in der Wirtschaftsgesellschaft. Ein Beitrag aus der Sicht der Thermodynamik. Dissertation, FB Wirtschafts- und Sozialwissenschaften der Universität Hohenheim, Juni 1978.
8. Bazykin, A.D., Structural and Dynamic Stability of Model Predator-Prey Systems. Internat. Institute for Applied Systems Analysis, Research Memorandum RM-76-8, Laxenburg, Austria, March 1976.
9. Bertalanffy, L. v. General System Theory. Foundations, Development, Applications. George Braziller, New York 1968.
10. Blauberg, I.V., Sadovsky, V.N. and Yudin, E.G., Systems Theory. Progress Publishers, Moscow, 1977.
11. Boulding, K.E. The Image. Knowledge in Life and Society. Ann Arbor Paperbacks. The Univ. of Michigan Press 1956.
12. Boulding, K.E. The Meaning of the 20th Century. The Great Transition. Harper Torchbooks, New York 1964.
13. Chedzey, C.S., Holmes, D.S., Soysal, M., System Entropies of Markov Chains. General Systems Vol. XXI, 1976, pp. 73-85.

14. Chedzey, C.S., Holmes, D.S. System Entropy and the Monotonic Approach to Equi-
    librium. General Systems Vol. XXII, 1977, pp. 139-142.
15. Clapham, W.B.Jr., Pestel, R.F. A Common Framework for Integrating the Economic
    and Ecologic Dimensions of Human Ecosystems. International Institute for Applied
    Systems Analysis, Research Memorandum RM-78-31, Laxenburg, Austria, June 1978.
16. Cleveland,H. Information as a Resource. The Futurist, December 1982, pp. 34-39.
17. Cornacchio, J.v. Maximum-Entropy Complexity Measures. Int. J. General Systems,
    Vol. 3, 1977, pp. 215-225.
18. Darlington, C.D., The Evolution of Man and Society. George Allen and Unwin Ltd.
    London 1969.
19. Dubois, D., Prade, H. Fuzzy Sets and Systems: Theory and Applications. Mathemat-
    ics in Science and Engineering. Vol. 144, Academic Press, New York 1980.
20. Eigen, M., Winkler, R. Das Spiel. Naturgesetze steuern den Zufall. Piper, Mün-
    chen/Zürich, 1978.
21. Georgescu-Roegen, N. The Entropy Law and the Economic Process. Harvard Univer-
    sity Press, Cambridge, Mass. 1971.
22. Grümm, H.R. (ed.), Analysis and Computation of Equilibria and Regions of Stabil-
    ity. A Record of a Workshop July 21 - August 1, 1975. International Institute
    for Applied Systems Analysis, Conference Proceedings. Laxenburg 1975.
23. Grümm, H.-R., Thermal Radiation and Entropy. International Institute for Applied
    Systems Analysis, Laxenburg, Austria, Research Memorandum RM-78-2, Januar 1978.
24. Haynal, Miklos, Révész, György, Learning Aspects of Interactive Control Systems.
    in: Hamza, M.H., Tzafestas, S.G. (eds.) Advances in Measurement and Control,
    Vol. 3, Proceedings of the Internat. Symposium MECO 78, Acta Press, Anaheim,
    Calgari, Zürich 1979.
25. Heal, Geoffrey. On a Relation Between Economic and Thermodynamic Optima. In:
    Resources and Energy 1 (1978), pp. 125-137, North-Holland Publishing Co.
26. Iberall, A.S. On a Thermodynamic Theory of History. In: General Systems Vol.
    XIX, 1974, pp. 201-207.
27. Jantsch, E. Anwendungen der Theorie dissipativer Strukturen. Neue Züricher
    Zeitung, Nr. 281, 3. Dezember 1975, pp. 45-46.
28. Jantsch, E., Design for Evolution. George Braziller, New York, 1975.
29. Köhler, A., Grundlagen der Umwelttheorien - eine kritische Beurteilung.
    Institut der Deutschen Wirtschaft, Köln 1980.
30. Laszlo, E., Goals in a Global Community. Pergamon Press, New York, 1977.
31. Lotka, A.J., Elements of Mathematical Biology, Dover Publications, New York,
    1977.
32. Maruyama, M. The Second Cybernetics: Deviation-Amplifying Mutal Causal Processes.
    In: American Scientist Vol. 51, 1963, pp. 164-179.
33. Nicolis, K., Prigogine, J. Self-Organization in Non-Equilibrium Systems: From
    Dissipative Structures to Order through Fluctuation. Wiley-Interscience, New
    York, 1977.
34. Oldershaw, R.L. Empirical and Theoretical Support for Self-Similarity Between
    Atomic and Stellar Systems. In: Int. J. General Systems, Vol. 8, 1982, pp. 1-5.
35. Pacault, A., Vidal, D. Synergetics. Far from Equilibrium. Springer, Berlin-
    Heidelberg-New York, 1978.
36. Papentin, F. On Order and Complexity. I. General Consideration. In: J.Theor.
    Biol. 87 (1980), pp. 421-456.
37. Piaget, J., Behaviour and Evolution. Pantheon Books, New York 1978.
38. Prigogine, I., From Being to Becoming. Time and Complexity in the Physical
    Sciences. W.H. Freeman & Co. San Francisco, 1980.
39. Prigogine, I., Stengers, I., Dialog mit der Natur. Piper-Verlag, München 1981
40. Radnitzky, G. Knowing and Guessing. In: Zeitschrift für allgemeine Wissen-
    schaftstheorie XIII/1, 1982, pp. 110-121.
41. Radnitzky, G., Die Evolution der Erkenntnisfähigkeit, des Wissens und der
    Institutionen. Aus: Riedl, R., Kreuzer, F. (Hrg.) Evolution und Menschenbild.
    Hoffmann & Campe, Hamburg, 1983. pp. 82-120.
42. Reyer, H.-U. Soziale Strategien und ihre Evolution. In: Naturwissenschaftliche
    Rundschau, Vol. 35, Januar 1982, pp. 6-17.
43. Riedl, R. Biologie der Erkenntnis. Die stammesgeschichtlichen Grundlagen der
    Vernunft. Parey, Berlin und Hamburg 1979.
44. Schwefel, H.P., Evolution und Optimierung. Arbeitspapier der Gruppe Grundlagen-

forschung-Energietechnologie, Kernforschungsanstalt Jülich GmbH (keine weiteren Angaben).

45. Slesser, M. Energy Analysis: Its Utility and Limits. Research Memorandum International Institute for Applied Systems Analysis, RM-78-46, Laxenburg, Austria, September 1978.

46. Small, M.G., Rifkin's Entropy Paradigm: A Critique. In: World Future Society Bulletin, Vol. XVI, No. 3, May-June 1982, pp. 29-36.

47. Smuts, J., Holism and Evolution. Republished, Viking, New York 1961.

48. Stumm, W., Störungen der globalen Kreisläufe als Folge der Energiedissipation. Referat und Diskussion der 5. Sitzung der Studiengruppe Energieperspektiven, Dokumentation Nr. 5, Eidg. Institut für Reaktorforschung, Würlingen 1982.

49. Taschdjian, E., The Third Cybernetics. In: Cyberneticy Vol. XIX., No. 2, 1976, pp. 91-104

50. Thom, R., Structural Stability and Morphogenesis. Addison Wesley, Reading, Mass. 1976.

51. Unger, R.M., Knowledge and Politics. The Free Press, New York, 1975.

52. Varela, F.G. Maturana, H.R., Uribe, R., Autopoiesis: The Organization of Living Systems, its Characterization and a Model. In: Biosystems, Vol. 5, No. 4, May 1974, pp. 187-196.

53. Voigt, H., Evaluation of Energy Processes through Entropy and Exergy. International Institute for Applied Systems Analysis, Laxenburg, Austria, Research Memorandum RM-78-60, November 1978.

54. Watanabe, S., Knowing and Guessing. A Quantitative Study of Inference and Information. John Wiley & Sons Inc. New York, 1969.

55. Weidlich, W., Dynamics of Interactions of Several Groups. In: Co-operative Phenomena (H. Haken, ed.), North-Holland, Amsterdam 1974.

56. Weizsäcker, E.v. (Hrsg.), Offene Systeme I: Beiträge zur Zeitstruktur von Information, Entropie und Evolution. Klett, Stuttgart 1974.

On the Entropy of Economic Systems

Daniel T. Spreng

Scientific Consultant on Energy and the Environment and

Research Associate, Power Systems Group, Swiss Federal Institute of Technology
CH-8092 Zürich, Switzerland

This is a qualitative discussion on how the concept of entropy relates to human affairs,in particular to economic processes and systems.
The starting point of the discussion is Georgescu-Roegen's book "The
entropy law and the economic process" [1]. Although published a year
earlier, Georgescu-Roegen's book has survived "The Limits to Growth"
[2] as a neo-Malthusian argument in the great debate over our future:
growth vs. no-growth, high technology vs. appropriate or intermediate
technology, centralized vs. decentralized society.

For Malthus 185 years ago it was food that would run out unless the
growth (of population) would be curbed drastically, for Meadows it was
mostly the scarcity of resources that put an end to growth of wasteful
technology and for Georgescu-Roegen it is his very general notion of
entropy increase that limits technology-based affluence. "The Limits
to Growth" has lost some of its punch due to the fact that the resource figures chosen in that publication have proven to be much too
small, so that most of the limits which we should encounter in the
near future have drifted further away now than they were 10 years ago.
Georgescu's argument, on the other hand, has been taken up by people
like E.F. Schumacher, Barry Commoner and some of the ideologists of
the green movement.

Georgescu-Roegen's main point is that the second law of thermodynamics can be applied to a very general notion of entropy, relevant to
our economy. For Georgescu-Roegen entropy embraces all forms of dis-
organisation relevant to the economy: dissipation of resources, pollution, depletion of nutritive elements in the soil, disruption of
social structures, etc. Affluent living by many people, based on a
non-agrarian economy, will sooner or later result in the dissipation
of all "low entropy" (Georgescu-Roegen's expression for negative entropy) and result in a kind of heat death. The entropy problem will not
only be felt on earth when the sun stops shining, but the dissipation
of concentrated materials, i.e. the transformation of resources to
effluents, is according to Georgescu-Roegen an imminent danger for a
society that depends on a fast growing, resource and technology-based
economy.

Although Georgescu-Roegen's book is thoughtful and interesting, Jeremy
Rifkin's popularized version of the message, "Entropy, a New World
View" [3], does not match the original. On the cover of the paper-back
version of Rifkin's book, two sentences from the text illustrate its
over-simplicity: "Entropy is the supreme law of nature and governs
everything we do" and "Entropy tells us why our existing world view is
crumbling and what will replace it". The fact that Rifkin's book has
been so widely read makes it to be of some interest all the same.

Physicists on the whole (with a few exceptions, see for instance [4])
have not paid any attention to Georgescu's argument. They simply point

out that this kind of general definition of entropy has little to do with physics and the second law of thermodynamics holds only in closed systems; the economy especially its subsystem agriculture, due to the sunshine it depends on, is certainly not a closed system.

The present paper is an attempt to examine Georgescu-Roegen's thesis without any blinkers, risking the danger not only of leaving the realm of physics, but also of being not always as precise as physicists have learned to be. In doing so, one is, however, in good company. One only has to recall the formulation of the second law of thermodynamics by the great physicist Clausius before a lay public in Zürich in 1865: "The entropy of the universe tends towards a maximum." This was a bold extension of the entropy concept beyond its original field of application.

Thermodynamic entropy

In 1824, when Carnot published his classic "Reflections", the genuine scientific advance it contained remained almost unnoticed for some time. Only well after 1840, when the first law of thermodynamics was formulated, was Carnot's insight discovered by the theoreticians and named the second law of thermodynamics. Carnot's interest was in applied research. He discovered that the most advantageous steam engines were those which made the best use of the "fall of caloric" (la faculté de rendre utile une grande chute du calorifique). Carnot, at this time, still accepted the view that the phenomena of heat were the effects of a material fluid, caloric, whose natural elasticity accounted for the elasticity of gases. He introduced the notion of an ideal heat engine with an efficiency that could not be surpassed by any actual engine (and that this ideal efficiency was equal to the temperature difference between the boiler and the condensor, divided by the temperature of the boiler).

The ideal Carnot engine produces the maximum of work possible, when the heat in the boiler is cooled down to the temperature of the condensor. The engine Carnot constructed worked without any loss; the work produced could be used again to heat the boiler to the starting temperature. That there are no engines more efficient than the Carnot engine in producing mechanical work is the original formulation of the second law of thermodynamics. Thus, actual engines have some irreversible losses; what is lost is not energy but capacity to do work. Entropy is a measure for that irreversible loss. Negative entropy, negentropy for short, is a measure for the capacity to do work. All we need to know here about thermodynamic entropy is that it is closely connected to the idea of constructing a useful, economic machine and that it measures irretrievable losses.

The exact thermodynamic definition of entropy was first given by theoretical phycisists only for closed systems in or near states of equilibrium. Engineers, however, have always successfully used the entropy concept in situations where its definition was on less firm grounds.

By analogy it is possible for open systems to split the total entropy change, dS, into two parts: d_iS, the internal entropy change inside the system and d_eS, the entropy exchange of the system with its surroundings. The second law of thermodynamics then simply is: $d_iS \geqslant 0$, i.e. the internal entropy can only grow, any entropy change that occurs inside the system, by itself, is an irreversable loss of negentropy.

The global entropy balance

Let us first examine the earth as an open thermodynamic system. The system has to be regarded as an open system due to the solar radiation it receives and the heat radiation it emits. The energy balance is very nearly zero, but the entropy balance certainly not; the incoming radiation has the temperature of the surface of the sun ($\simeq 5000$ K) and the outgoing radiation the surface temperature of the earth ($\simeq 300$ K). What happens to the negative flux of entropy that enters this system or the positive flux that leaves it?

The solar negentropy flux is mostly used to drive the climate machine: the winds, the water cycle, the ocean currents, etc. However, all the negentropy input that goes into the climate machine is ultimately dissipated by mechanical friction. It looks like the total entropy of the earth remains, in the very first approximation, about constant since

$-d_e S$ (from the sun) $\simeq d_i S$ (due to friction of the climate machine).

However, about 10 % of the solar negentropy flux falls on green leaves, photosynthesizes carbohydrates (at an efficiency of ~1 %), and leads to many highly organized biological structures. It is the principle of self-organization that makes it possible to store the incoming entropy in the biosphere. (In case of the climate the situation is of course similar: the climate machine itself is a somewhat ordered structure that could build itself up due to the incoming negentropy flux.) The biosphere does however not grow and accumulate for ever; there is not only photosynthesis, but also respiration and decomposition. The total mass of the biosphere may be about constant, temporal decreases (for instance at the start of an ice age) are likely to be followed by temporal increases. Therefore, we may write in a first approximation also for the sunshine that reaches green leaves:

$-d_e S$ (from photosynthesis) $\simeq d_i S$ (due to respiration and decomposition).

It thus seems that the incoming negentropy flux from the sun is dissipated on earth by natural processes. How does man with all his technological activities fit into this picture? Let us examine this by looking at a typical example of man's production machine.

A process chain

Figure 1 shows the waste heat production associated with the production, use and disposal of an unspecified aluminium product [5]. Does this figure resemble a plot of the corresponding entropy production? Can we assume that the waste heat production, ΔW, is about equal to the entropy production times the temperature of the environment, $\Delta S.T_0$? One may argue this to be the case in a qualitative way, at least for the first three process steps. The energy dissipation is likely to be the largest contribution to the entropy production. Contributions to the entropy change from mixing and concentrating the materials involved are smaller and partially cancel each other out: in each step some waste is dissipated into the environment, in the reduction step some oxygen as well, and aluminium is concentrated in the course of these steps from a state of relative dispersion (25 % in the bauxite ore, compared to 8 % average concentration in the earth crust) to a state of concentration (usually about 99.8 % purity).

Georgescu-Roegen probably would say that Fig. 1 resembles a plot for the entropy production in the process chain, if it includes the entropy

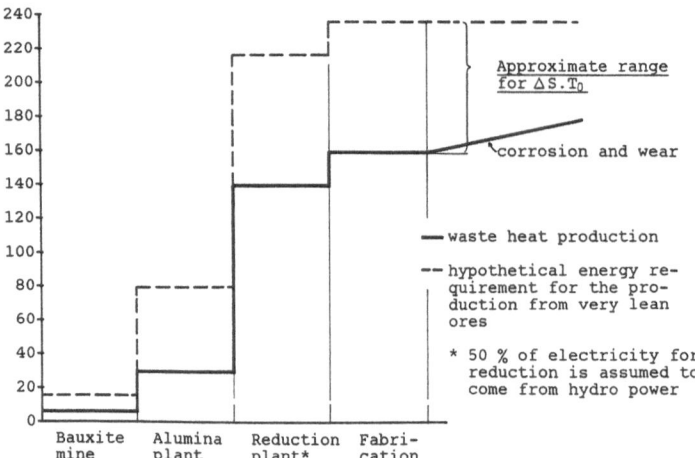

Energy requirement/waste heat production
[MJ/kg]

240 ┐
220 ┤
200 ┤
180 ┤   Approximate range
160 ┤   for $\Delta S \cdot T_0$
140 ┤
120 ┤   corrosion and wear
100 ┤
80 ┤
60 ┤   ──── waste heat production
40 ┤
20 ┤   ─── hypothetical energy re-
0 ┘   quirement for the pro-
                                             duction from very lean
     Bauxite Alumina Reduction Fabri-   ores
     mine   plant   plant*   cation
                                          * 50 % of electricity for
                                            reduction is assumed to
                                            come from hydro power

Fig. 1: Entropy production in a process chain (an object made out of aluminium)

production due to wear (dispersion of aluminium) and corrosion (dispersion plus more waste heat production) during and after the use of the particular product. This effect should however not be overestimated. As long as the worn off particles are macroscopically identifiable, so that they can be collected with fine tweezers, the entropy production is zero. Should the dispersion be more complete, the dispersed atoms will be lost in the average earth crust concentration. If we would not use bauxite, but "ore" of average earth crust concentration, the energy requirement of the first two process steps might double or perhaps even triple but the overall energy requirement would then only be 50 % higher. Goeller and Weinberg [6] discussed this question more realistically for most minerals that are in use. Their finding is that for iron and aluminium there are leaner but not very lean ores that will last for many thousand years and that the use of other metals can be substituted by these two metals.

The energy requirement for the reduction step is larger than the waste heat production, because some of the energy is put into the metal as chemical energy. When corrosion takes place this energy is given off to the environment. The dotted line in Fig. 1 is therefore an upper limit for the entropy production times the temperature of the environment, $\Delta S \cdot T_0$.

But to end the investigation at this point is like adding up all the costs of a process chain and concluding it would have been cheaper not to do anything at all. We have to examine also the gains of negentropy resulting from the activities of the process chain.

Thoma [7] attempted to do this. He estimated the negentropy increase that resulted from machining a piece of equipment, within given tolerances, to have an exactly specified shape. The entropy concept on which such a calculation is based is very different from thermodynamic entropy. Before proceeding with the discussion of our process chain, this other idea of entropy will briefly be discussed.

Entropy as hidden information

Two quantities are often used to explain the abstract term entropy: disorder and lack of information.

Consider a gas with known macro state, i.e. with known volume, pressure and temperature. Boltzmann showed, over one hundred years ago, that the entropy of that gas is equal to the logarithm of the large number of all microstates (spacial arrangements and velocity of all atoms) that are possible within this macrostate. If, for instance, all atoms are compressed in half the volume, there are fewer possible places the atoms can be and the entropy is therefore smaller.

The idea of the entropy being lower when the atoms in that particular "Gedankenexperiment" are all to be found in half the volume, the other half being empty, led to the notion of entropy being generally lower in ordered states. This notion, to be found in most textbooks, is not always correct. Even in the case of the gas we only have to introduce some attracting forces between the atoms and clusters may be formed - presumably this would correspond to an increase of order - while the Boltzmann entropy is increasing [8]. It is interesting to note that to measure structure or order, v.Weizsäcker regards it a necessity to introduce, besides the micro- and the macro-levels, at least a third level of description against which the micro- and macro-levels can be measured.

The idea of entropy having something to do with information, on the other hand, is well founded. Not only has the formula Shannon developed for the expectation value of information a striking resemblance to the Boltzmann formula for entropy. The relationship is much closer than that. We can regard the Boltzmann entropy to be a special case of information: it is the information contained in the microstate that is hidden to the macroobserver. If one macrostate, or macroappearance, has many possible microstates the entropy, or the hidden information, is large.

In information theory, information relates to a sender and a receiver. In the case of the Boltzmann entropy, which is firmly and clearly related to the thermodynamic entropy, the sender is at the atomic level on which all movement is known and the receiver at the macrolevel where man's senses and instruments detect only a very muddled image of that movement. The muddled image is the density, temperature and perhaps some other macrovariables like an electromagnetic field. The information lost between the microlevel sender and the macrolevel receiver is equal to the thermodynamic entropy.

For the following reasons thermodynamic entropy is very special information:

- The two levels of description are clearly and naturally defined and they are distinct from each other (we can only start talking of temperature when we stop talking of individual atomic velocities).

- Thermodynamic entropy is of practical, also economic interest to us, since the entropy change measures the losses of something valuable, the capacity to extract work from heat.

- There exist clear relationships between probability distributions of microvariables and the corresponding macrovariables. From this it is clear what information about the microlevel is hidden in respect to a given description of the macrolevel. E.g. if we speak only of temperature and density of a gas, the hidden information refers to the position and velocity of all gas molecules but not for instance to the kind of molecules present in the gas.

If we are now interested in estimating the negentropy that is produced by an economic activity or that is represented in an economic structure, the negentropy should refer to similar micro- and macro-levels: at the microlevel we should know the causal relationships between the variables and we should know which microvariables are of interestt and correspond to the variables of prime interest to us at the macro-level.

The term "information" as used in communication theory, also called entropy*, does not refer to analoguous levels of description. In communication theory the two levels of description that are related to each other by entropy are the level of signals (from a sender) and the level of messages, i.e. a certain number of successive signals (to the receiver). The reason why an analogy other than a mathematical analogy between entropy in thermodynamics and entropy in communication theory does not make sense is that the level of the messages is much too close to the level of the signals and much too far from a level that would be of direct interest to us. The measured information is completely neutral to any sense or content the message may have. It does not matter whether the message has

- any meaning
- is understandable or
- is useful.

A much better analogy in respect of the two levels of description exists between thermodynamic entropy and another term from communication theory, namely complexity. Complexity is the relationship between information that is necessary to describe a problem as generally and completely as is useful (so that a solution to the problem follows) to the relevant information that exists. The production of thermodynamic entropy can be seen as analogous to the increase of complexity, or the production of negentropy to the reduction of complexity. Complexity is however not easily quantifiable, the reduction of complexity can be considered as one of the central endeavours of society [9].

We understand now why Thoma's attempt to calculate the entropy of a piece of machinery simply by applying the entropy of communication theory does not relate well to the thermodynamic entropy dissipated in producing the machinery. On the other hand, a valuable machine will help us to reduce the complexity of a task but we have no way of calculating this reduction of complexity.

Bagno [10] simply equates negentropy production with money flow. He argues that a system that has incorporated much information is highly structured and reflects a high negentropy value. If there would be no economic structure at all, the probability one individual would encounter another with whom he could trade goods would be quite small.

----

* There is a considerable confusion in the literature whether one should consider information to be equal to entropy, or equal to negentropy. Information is equal to entropy, if, as advocated here, it measures the novelty passing from the sender to the receiver, i.e. before the receiver has taken charge of it. When the receiver has incorporated the information it becomes negentropy, because it has lost its novelty value.

The monetary economy represents an information network for consumer and producer to communicate with each other. The more the money flows the higher is, according to Bagno, who wrote his book in the early sixties, the negentropy value of the economy. The idea of looking at the economy as an information network is interesting, but taking the easy way-out by equating money flow and negentropy production is self-defeating. It does not help us further with our process chain.

There is still another way to consider the negentropy production in an economic system: we can again exclude from negentropy production the idea of complexity reduction but instead consider the system as an open system. Economic activities can then aim at furthering negentropy flow from the environment into the system. In the following section that approach is followed.

Negentropy gains

When Georgescu-Roegen views entropy as a generic term for decay - depletion of resources, pollution and the loss of fertile soil - what should the opposite be? Obviously the creation of resources, the clean-up of pollution and the increase of fertile top soil could by analogy be called gains of negentropy. These gains are of course possible. Technology contributes to the increase of entropy, but it can at the same time make the increase of negentropy possible.

Utterly useless rocks can become resources if a new technology can utilize one of their constituents. Bauxite became a resource about a hundred years ago, uranium ores 40 years ago, thorium-bearing ores may become resources in 20 years or so. Brine, that contains methane and lies several kilometers below the surface of the earth, is becoming a resource at present, due to its discovery and exploration.

Pollution is not inevitable. Recycling and the more efficient use of resources are possible. Biotechnology may help in closing production cycles. The exact requirement for zero pollution is not zero emission, but that the outputs of a system are not more polluted than the inputs.

Wise forest and soil management is another technology than can increase negentropy, i.e. these technologies can help to make more effective use of the solar negentropy flux, so that

$-d_e S$ (from photosynthesis) $> d_i S$ (from respiration and decomposition).

Top soil is only partially decomposed and its increase is equivalent to increasing the negentropy in the agricultural system.

Returning to the process chain, it is obvious that the net effect is an entropy increase, if the produced good is just thrown away. However, the purpose of the exercise was to make a useful object and the use of that object may lead to an increase in negentropy. The object may be used in agriculture or it may help to build an innovative society, in which technologies for making use of new resources and technologies for the more complete and sparing use of existing resources can be developed. Whether the overall entropy production of the process chain is positive or negative depends very much on the use of the object and cannot easily be judged, in any case not in a quantitative way.

The system's boundaries

The introduction of a general but also vague definition for the terms
entropy and negentropy should not at the same time lead to an unclear
definition of the boundaries of the discussed system.

The system's boundary for our process chain may be chosen to include
only the production steps. More particularly the boundaries may be
identical to the fences around the mine, the alumina plant, the reduction plant and the manufacturing plant. However, to be typical for
economic activities in general, the process chain has to include the
use of the produced object. This leads to a big enlargement of the
system's boundaries to include the entire economy with its resources.
To presuppose boundaries should be impermeable to negentropy influx
(for instance due to new resources) does no justice to the real situation. However, although the larger boundaries include a more representative system, they lead to making it utterly impossible to quantify
the entropy change. But they allow now for the entropy to decrease.

We may note here in passing that the entropy concept itself is an anthropocentric concept. The Carnot entropy is a measure related to what
is useful to man and the Boltzmann entropy is calculated by defining
the macrolevel such as to correspond to what is detectable by man's
senses. Should it then not be natural to choose anthropocentric boundaries for man's own economic system?

Clearer boundaries are given for the global system. It seems likely
that man today does indeed negatively influence the natural global
balance of thermodynamic entropy. Logging in tropical forests, and ex-
cessive use of fire wood in densely populated areas of the third world,
often causing the loss of top soil, lead to a reduction in the total
mass of the biosphere on the globe. It is not clear whether this reduction is fully compensated by irrigation and fertilization of crop
land and by forest management in temperate zones; indeed it is in principle not possible to calculate man's contribution separately because
the "natural" reference situation is unknown.

If it is assumed that the total mass of the biosphere stays approximately constant today, which may be the best guess available (more
data from the study of satellite pictures are expected to be available in the next years), the burning of fossil fuels is the activity
which will be responsible for a global entropy increase due to human
activities: the depletion of concentrated resources, the waste heat
production and the dispersal of CO_2 in the atmosphere. Burning fossil
fuels is however not a basic human need. A non-fossil area is already
in view, be it solar, nuclear or both.

The application of the second law of thermodynamics to the global entropy balance is of course inconclusive, because of the fact that the
system is open and there exists a large negentropy influx that can
sustain self-organizing mechanisms and compensate for irreversible
entropy increases within the system.

In the search for a closed system we might look at the planetary system and ask whether it could be treated as a closed system, to which
the second law of thermodynamics could be applied. Unfortunately in
that system we again run into the problem of a suitable definition of
entropy. Should we count the helium inside the sun as a store for negentropy and will our worry be the depletion of that resource? Or is
our problem the breakdown of the well-ordered motion of the planets?

For these two problems the micro- and macro-levels are entirely different: in the first case the levels are the micro- and macro-levels of thermodynamics, in the second case the macrolevel refers to the planetary system as a whole and the microlevel to single planets. It is interesting to study and to learn to understand these planetary problems, to worry about them in relation to human existence is certainly unnecessary.

We can draw the boundaries of the system even larger and include the entire universe. We mention here some results of calculations of the entropy of the universe [11] going back to Freeman Dyson not because they are in any direct way relevant to economic systems, but because they provide a nice and perhaps an appropriate image for the entropy situation we find ourselves in.

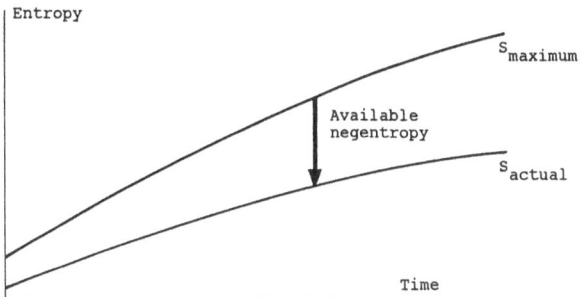

Fig. 2: Entropy of the universe: the actual entropy is increasing steadily although lagging behind the maximum possible entropy that increases even more quickly due to the expansion of the universe

In the universe, the largest forces are gravitational forces, and the main contribution to the entropy changes come from the formation and dissolution of black holes. It is speculated that, in an expanding universe, the possibility is given for larger and larger black holes to form. This allows for the entropy to increase forever. The emerging image is summarized in Fig. 2, in which the actual entropy increases all the time but lags behind the maximum possible entropy of the expanding universe. The difference between the two curves is the stock of negentropy, the necessary precondition for any life. Although being used up constantly, the stock of negentropy possibly increases rather than decreases over time.

The reason why this entropy picture of the universe may be an appropriate picture also for the economy is because the economy expands in two ways. For one thing there is no law of conservation of information, as incorporated information accumulates and organisation becomes more complicated all possible information and organisation also increases. Secondly, the economy expands in a real sense occupying the earth more and more completely and soon will also expanding beyond. Due to both facts, the maximum possible entropy is increasing, hopefully faster than the actual entropy.

Concluding remarks

The second law of thermodynamics does not allow for entropy to decrease in closed systems. However, entropy may decrease in open systems. It is the principle of self-organisation which allows a system to utilize a negentropy input to tip the entropy balance to the favourable side.

It does not make much sense to view the economic system as a closed
system. Although inside this system thermodynamic entropy increases,
between the economic system and its environment, other forms of entropy
are exchanged, mostly due to know-how development and the development
of more subtle forms of organisation.

All the same, it may be useful to think of entropy as a rather general
concept, and Georgescu-Roegen's concern about an increase of entropy
due to too much affluence and too much technological activities should
not be discarded lightly. Indeed a number of real and serious problems
can be subsumed under the idea of entropy increase, although the various problems cannot all be quantified in terms of entropy.

It has been suggested by this author before [12] that, at a certain
level of description, it may be useful to consider all required input
for a given task to consist of a mix of available energy, available
time and available information. Evidence is strong that to some degree, the mix can be changed by substituting the inputs for each other.
The idea is described in the energy-information-time triangle similar
to a 3-component phase diagram (Fig. 3). Note that the three inputs
are subject to different laws of conservation: the availability of
energy for doing work can only be used once (this is the second law of
thermodynamics), the availability of information, on the other hand,
may increase with the use of that information (if a particular know-
how is applied, this application publicises the know-how - this is the
inverse of what holds for thermodynamic entropy) and the conservation
laws of time are subtle and paradoxical: we cannot escape the steady
flow of time, the value of allocating more or less time to any particular activity does not depend only on the value of that activity, but
also on its alternatives.

Georgescu-Roegen's message can be translated into the image of this
triangle: We should be careful to perform tasks by dissipating as little
available energy (i.e. free energy or negentropy) as feasible, indicated by $-\Delta E$. This we can do by taking more time, $+\Delta t$, and using as

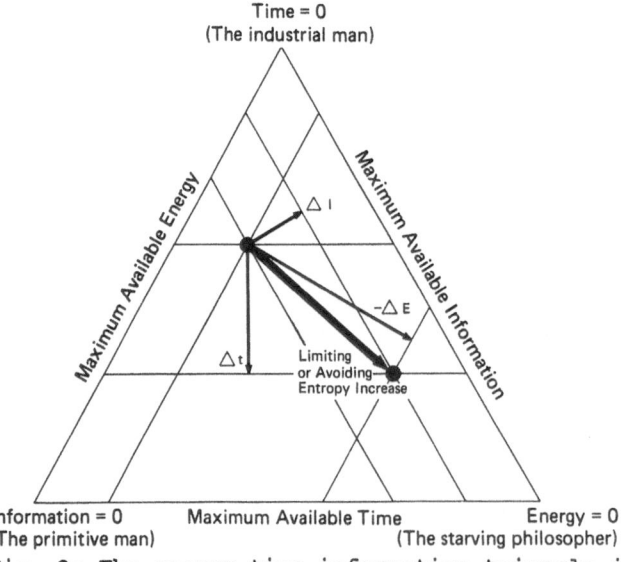

Fig. 3: The energy-time-information triangle illustrating the mutual
substitutability of the three basic inputs for any task

216

much available information, $+\Delta I$, as possible. Information is used here broadly to mean know-how, foresight, information, technololgy, as well as efficient organisation. We thus should perhaps perform tasks more in the way a starving philosopher would and less in the way of the industrial or the primitive man; this limits or even avoids an entropy increase.

Georgescu-Roegen's message is not as simplistic as that of some of his followers [3] and we may very well share his general concern. The purpose of this paper was to show that from a physics point of view our outlook does not necessarily have to be as pessimistic as his. In economic systems which are open and expanding systems, we may never run out of negentropy in spite of the second law of thermodynamics. Self-organisation is an important principle in this context in that it makes use of an influx of negentropy so that possibly we will never come close to Clausius' entropy maximum.

References

1. N. Georgescu-Roegen: The Entropy Law and the Economic Process
    (Harvard University Press, Cambridge, Mass. 1971)

2. D. Meadows et al.: The Limits to Growth
    (Universe Books, New York 1972)

3. J. Rifkin: Entropy - A New World View
    (The Viking Press, New York 1980)

4. A.M. Weinberg: "Avoiding the Entropy Trap",
    Bull. of the Atomic Scientist, Vol. 38, Nr. 8, Oct. 1982, p. 32

5. D.T. Spreng: "Energieaufwand und Recycling von Aluminiumverpackungen", Aluminium, 58. Jahrg., No. 10, p. 625 (1982)

6. H.W. Goeller and A.M. Weinberg: "The Age of Substitutability",
    Science, Vol. 191, p. 683-689 (1976)

7. J. Thoma: "Energy, Entropy and Information"
    Research Memorandum No. 77-32, International Institute for
    Applied Systems Analysis, Laxenburg, Austria (1977)

8. C.F. von Weizsäcker: "Evolution und Entropiewachstum" in
    Offene Systeme I, Beiträge zur Zeitstruktur von Information,
    Entropie und Evolution, E. v.Weizsäcker (Ed.),
    (Ernst Klett Verlag Stuttgart 1974, p. 200)

9. K.W. Deutsch und B. Fritsch: "Ueber Probleme und Methoden der
    Komplexitätsreduktion zur Informationsverarbeitung für Ueberlebensstrategien",
    Publication Series of the International Institute for Comparative
    Social Research, Wissenschaftszentrum Berlin, Pr 79-1, 1979

10. S. Bagno: The Angel and the Wheat - Communication Theory and
    Economics (Jonah Publishing Co., New York 1963)

11. Steven Frautschi: "Entropy in an expanding universe"
    Science, Vol. 217, No. 4560, p. 593 (1982)

12. D.T. Spreng: "On time, information and energy conservation"
    ORAU/IEA-78-22 (R), Institute for Energy Analysis, Oak Ridge
    TN (1978)

Part V

Complex Systems

Death – From Microscopic to Macroscopic Disorder

Friedrich Cramer

Max-Planck-Institut für Experimentelle Medizin, Abteilung Chemie
Hermann-Rein-Straße 3, D-3400 Göttingen, Fed. Rep. of Germany

1. Introduction: What is the Problem?

I would like to introduce the problem by looking at the famous drawing of Dürer, presenting his mother at the age of 63 years (Fig. 1). On the first glance, we are overwhelmed by the expression of this old, worn out senile woman, and stirred by the humane, even loving approach which the artist took and which is apparent through the seemingly naturalistic drawing. But let us set aside for this lecture any emotional attitude (can we?). From a scientist's point of view we can immediately notice several signs of ageing and approaching death. The woman has apparently lost her teeth which is easily noticed from the form of the mouth and chin. Destruction of teeth is a consequence of growth of symbiotic microorganisms which live in every mouth in symbiotic fashion. Somehow, at old age this symbiotic capability is lost. Secondly we notice the protruding and aberrant right eye. The diagnosis is obvious: Glaucoma. The internal pressure of the eye can no longer be regulated and leads to internal high pressure and destruction of the retina. This is a loss of regulatory capacity. Thirdly we notice the wrinkled skin. This is a change in protein composition. Indeed we know that in old age the keratine composition is different. Thus we can state three signs (or reasons?) of ageing and death: loss of symbiotic capability, loss of regulatory capability, change in proteins.

When asking what is death, we must first ask: What is life? I can of course not give a complete answer to this, I would rather prefer to give an operational defi-

Fig. 1:
Dürer's Mother; Drawing by A. Dürer, 1514

nition: on our earth, life is the self-reproducing system of proteins and nucleic
acids, which is capable of evolving or adapting. Thermodynamically this system is
extremely far away from equilibrium. Life forms a steady state within its semi-
permeable compartment. This system can exist only through a continuous input of
energy. A living organism, therefore, is a highly complex "limit cycle", highly
responsive to changes of conditions, adaptable to changes in environment, and con-
suming energy. The fact that we do not collapse immediately to a little heap of
ashes and a small cloud of CO_2 and water vapour is due to this enormous input.
The energy flux in the system life starts from radiation from the sun, goes through
chloroplasts, cellulose and starch, eating, oxidation of glucose and generation of
ATP. ATP is the general fuel of any living cell, for pumping, or muscle contraction,
or error correction. In higher organisms most of the ATP is generated by oxidation
of glucose in the mitochondria. When blocking mitochondria (respiration), any higher
cell is dead in a split second. It is, however, not this death which I want to discuss today, the statistical accident in this dangerous world. I shall speak about
death as an inherent limitation to the system life.

The "living" system of DNA → RNA → proteins is a feedback cycle in which the
information for the synthesis of proteins is transferred through RNA into the
protein-synthesizing system, which in itself is of protein nature. Thus the
correctness of the flow of information from DNA to protein determines the overall correctness and composition of the macromolecules. If the protein-synthesizing
proteins are made wrongly, this will lead to more incorrectly synthesized proteins
(Fig. 2). There is not much known on the accuracy of this system. In Table I a comparison of the genotype (DNA) and phenotype (proteins) is given. We are comparatively well informed about the events in DNA synthesis. Any single error in DNA
synthesis leads to a mutation which one can pick up very easily for instance in
bacterial cultures. Therefore replication errors in DNA can easily be measured.
A mutation can have an effect in two directions. Mostly it has a disadvantageous
influence on the organism if the mutation leads to a less functional protein. It
can, however, also lead to a better functioning protein which is then selected in
propagation of the organism. Thus mutation is the motor of evolution.

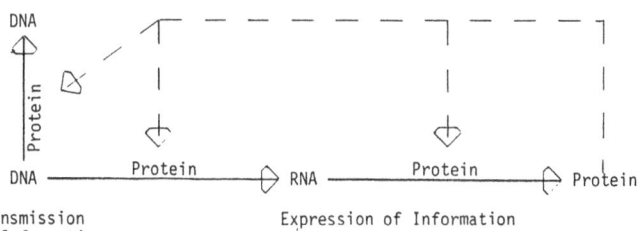

Transmission
of Information

Expression of Information

Fig. 2: Feedback of Errors during Transmission and Expression of Information

The role of errors in protein synthesis is much less obvious. The proteins form
the phenotype of an organism. Clearly, ageing and death is a property of the phenotype. The genotype and germ line cells are in principle immortal. Obviously, old
fathers generate young babies of the same appearance as young fathers do, the
corresponding mothers don't give birth to babies of adult or senile appearance. The
exact reason for ageing is not known as is the exact figure for the error in protein
biosynthesis (Table I). It has been proposed that ageing of a cell is due to an
accumulation of incorrectly synthesized proteins which finally lead to an error
catastrophe [1] (Fig. 3). This conjecture has so far not been verified. Only very
few data are known on the error rate in protein biosynthesis. We have therefore
set out to substantiate this problem by providing the necessary facts in protein
biosynthesis.

Table I. Comparison of Genotype and Phenotype

	Genotype	Phenotype
Material	DNA	Protein
Coded by	DNA	DNA
Biosynthesis	Enzymatic Replication	Ribosomal Synthesis
Information Flow	DNA → DNA	DNA → RNA → Protein
Function	Carrier of Linear Information	Three dimensional Interactions Catalysis, Catalytic Cycles
Effect of Error	Mutation	
Rate of Error	$1:10^3-10^9$	?
Function of Error	Random Generator	
"Sense"	Evolution	

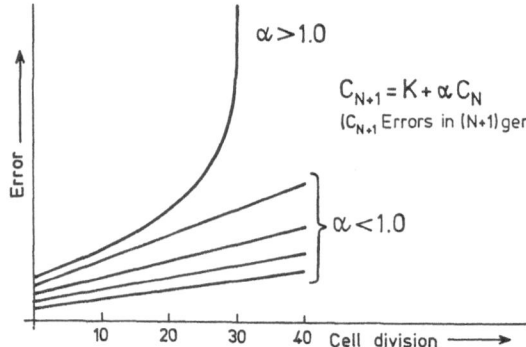

$\alpha > 1.0$

$$C_{N+1} = K + \alpha\, C_N$$

(C_{N+1} Errors in (N+1) generation)

$\alpha < 1.0$

Error

10   20   30   40   Cell division →

Fig. 3: Error catastrophe according to L. ORGEL [1]

2. Experimental Part: Substantiation of the Question

2.1. Qualitative Treatment

The problem of the correct production of the phenotype is to arrange the 20 amino acids in the sequence which is prescribed by the gene. The 20 amino acids are shown in Table II. They can be divided in subgroups, some of them are attached at the 2' position of the tRNA, some at the 3' and some do not show any specificity in this respect. Also, the 2' attaching amino acids are preferentially lipophylic, the 3' attaching preferentially hydrophilic. A conjecture to explain this phenomenon has been put forward recently [2]. In order to arrange the amino acids in the correct sequence, they are first attached to the tRNA (Fig. 4), the structure of which is known through X-ray structural analysis. This attachment is the key step in the information exchange between the "world of nucleic acids" and the "world of proteins", this process determines the correctness of the protein biosynthesis.

Table II. Specificity of Aminoacylation of "Deoxy" tRNAs

Aminoacylation at 2'	Arg, Gln, Ile, Leu, Met, Phe, Trp, Val
Aminoacylation at 3'	Ala, Gln, Gly, His, Lys, Pro, Ser, Thr
Aminoacylation at 2' and 3'	Asp, Asn, Cys, Tyr

The aminoacylation reaction is carried out by the enzyme aminoacyl-tRNA synthetase. For each amino acid there is one such synthetase [3]. The enzymatic reaction occurs in two steps (Fig. 5). The first is the so-called pyrophosphate exchange in which the amino acid is activated and the second the transfer of the amino acid to

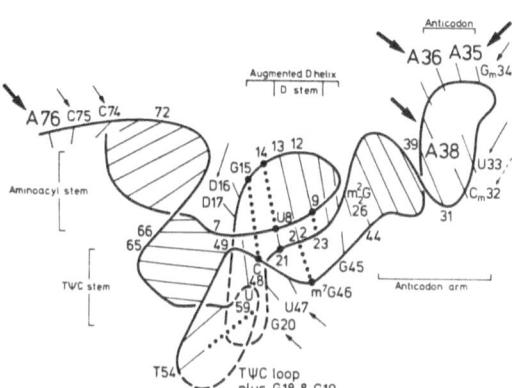

Fig. 4: Structure of tRNA

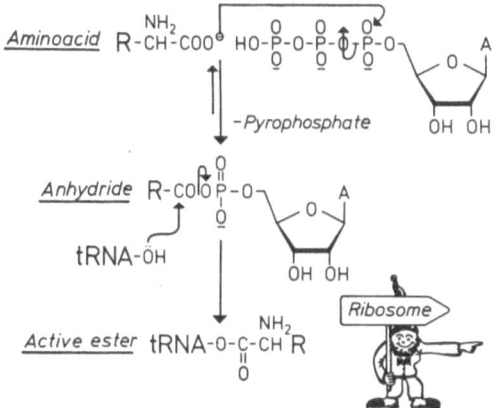

Fig. 5: Aminoacylation of tRNA

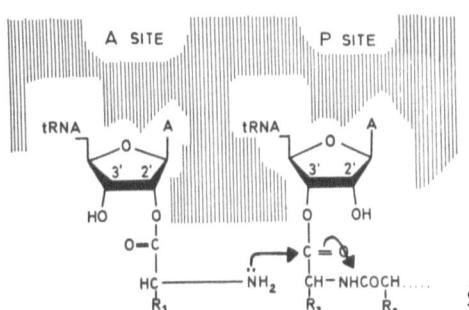

Fig. 6:
Scheme of Peptide Bond Formation at the Ribosome

the corresponding tRNA. Once the tRNA amino acids are acylated no further checking is possible. The aminoacyl-tRNA then goes to the ribosome (Fig. 6) where the incoming new amino acid is bound to the so-called A site from which a peptide bond is formed.

How large an accuracy is necessary? Normal enzymatic processes, which distinguish between similar substrates, are precise in the order of percent. For protein biosynthesis even a selection of the correct amino acid with a preference of 10^3:1 would by far not be sufficient. Assuming a chain length of the protein of 1000 amino acids, such an accuracy would mean an average of one error per protein. In order to have a 99% correct protein population the error in the individual incorporation must not be larger than one in 100 000. This accuracy requires very

special measures as we shall see. For these measures the cell has to pay a lot of
energy in form of ATP.

It might be justified, therefore, to assume that a high accuracy in the first
step, the aminoacylation, is followed by further steps of the same accuracy. Otherwise the investment at the beginning would be useless. There are no exact data
known about the fidelity of the ribosomal machinery. I would like to propose, however, that the fidelity at the ribosome is equal to the aminoacylation reaction.

One particular case has been discussed frequently in the past: the distinction
between isoleucine and valine (Fig. 7). L. Pauling has calculated for these very
similar amino acids an error of 5% [4,5]. Indeed, valine is activated by the isoleu-
cyl enzyme. It is, however, not transferred to tRNA. Instead, in the presence of
tRNA the activated valine is hydrolysed [6]. Some years ago we did a key experiment.
A tRNA[Ile] was chosen which was lacking the non-accepting 3'hydroxyl. This tRNA is
charged with the cognate amino acid isoleucine quite normally. With the non-cognate
amino acid valine a 100% mischarging occurs. No hydrolysis was observed. The conclusion is that the non-accepting hydroxyl group catalyses a proofreading process
[7]. This is shown in Fig. 8. Obviously the non-accepting hydroxyl is instrumental
in a proofreading hydrolytic process which rescues the tRNA from being mischarged.
The mechanism is depicted in Fig. 9. The non-accepting hydroxyl group is part of a
hydrolytic scheme, similar to the one in chymotrypsin or other esterases. It either
accepts the wrong amino acid first or hydrolyses it directly in a proofreading process.

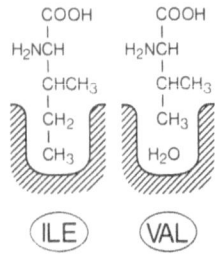

Fig. 7: Schematic Space Filling
by Isoleucine and Valine
at the Enzyme Pocket

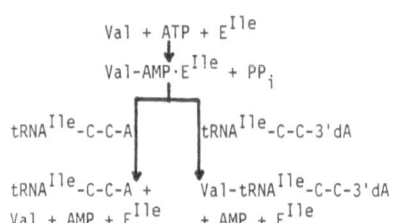

Fig. 8: Misaminoacylation with Valine
of tRNA[Ile] with and without
3'Hydroxyl Group

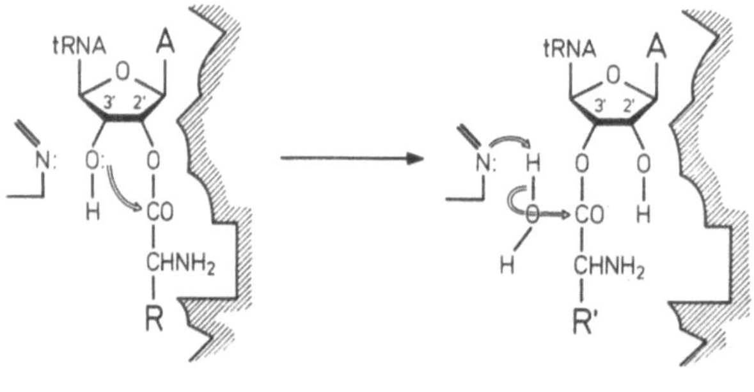

Fig. 9: General Mechanism of
Hydrolytic Action in Aminoacyl
Ligases

The important new feature in this mechanism is that the active site of the
enzyme is used twice; in this case the specificities multiply. The essential feature of this process is shown in Fig. 10. It is a double checking cycle.

In the first cycle the correctness of the substrate is checked in the usual
manner by an adsorption/desorption process in the pocket of the enzyme. The specificity there is determined by the binding energy and by kinetic parameters. These
processes are reversible. If the substrate is recognized as being wrong, it goes
back to the pool without loss of energy. In the second checking cycle, a chemical

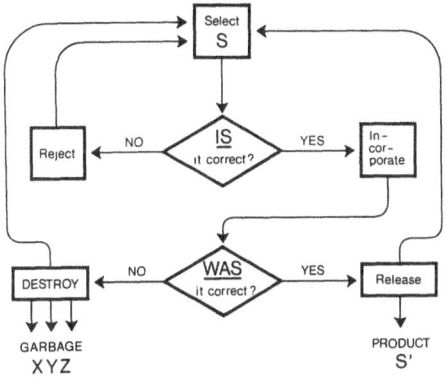

reaction had already occurred. If afterwards the incorrectness of the product is recognized, energy is consumed through hydrolysis since an irreversible chemical reaction had occurred before (Fig. 10).

For each misactivation and subsequent hydrolysis in proofreading at least one molecule of ATP, one unit of the general fuel of all cells, is used up. This is a clear demonstration that information must be paid for, in this case in molecular units.

Even with the sophisticated proofreading mechanism, protein biosynthesis is not error-free, there is a finite error of 1 in 10^5. One can demonstrate theoretically that with an higher input of energy the error can be reduced further. In principle, an absolute accuracy could be obtained with an infinite amount of energy [8]. Apparently, the observed figures are a compromise reached in evolution between costs and benefits.

2.2. Quantitative Treatment

One now can measure the accuracy of aminoacylation quantitatively. The accuracy of the primary reaction is measured by blocking the proofreading when introducing a 3' amino group into tRNA. By a competition experiment it turns out that the primary accuracy of the enzyme is 180:1 [3,9]. The additional accuracy of the proofreading is measured by following the ATP consumption. For each proofreading step one molecule of ATP is consumed which is the price for specificity. It turns out that 500 molecules of ATP are hydrolysed before one valine is incorrectly incorporated. Therefore the proofreading has a specificity of 800:1. Thus the overall specificity is about 100 000:1 [9,10].

Meanwhile we have measured a number of further accuracies. It turns out that the accuracy of incorporation of amino acids is of the order of 10^5, in some cases 10^6.

What can one do with such data? Is it possible to calculate the lifetime of a cell with these figures? In Table III it is shown which data in principle are required in order to calculate an error catastrophe. Besides the accuracy of protein biosynthesis one must know the lifetime of the individual protein and also the influence of an error on the functionality of the particular protein. All these figures can in principle be obtained. A calculation of such a system of hypercycles should in principle be carried out similarly as in the case of any other evolutionary process. Table IV summarizes the data.

Experimental research in this field can use a different, more biological approach, by introducing into microorganisms a mutation which leads to accumulation of errors. Such mutations are not yet known. We are working, however, on a problem of the following kind. In Neurospora crassa there exists a mutant which exibits a premature ageing. This mutant is temperature sensitive. At 30° it grows normally, at 40° it

Table III: Basic Figures for Calculation of Lifetime of "Network Life"

1) Measured Error Rates
   Yeast   Ile/Val $1:1.8\times10^5$
   E.coli Phe/Tyr $1:3.7\times10^5$
           Phe/Leu $1:9.2\times10^5$
           Phe/Met $1:7.6\times10^5$

2) Turnover of Key Proteins (measurable)

3) Influence of Mistake on Function (from Mutants)

Table IV. Comparison of Genotype and Phenotype

	Genotype	Phenotype
Material	DNA	Protein
Coded by	DNA	DNA
Biosynthesis	Enzymatic Replication	Ribosomal Synthesis
	DNA → DNA	DNA → RNA → Protein
Function	Linear Information	Three dimensional Interactions
	Carrier	Catalysis, Catalytic Cycles
Error	Mutation	Change in Function
Rate of Error	$1:10^3$–10^9	$1:5\cdot10^5$
Function of Error	Random Generator	Decrease in Efficiency
"Sense"	Evolution	Age? Death?

"ages". It is known that this mould has a mutation in the leucyl-tRNA synthesizing system [11]. We have recently found that the leucyl-tRNA synthetase of these organisms indeed is slightly different from that of the wild type. We believe that this mutant carries a "read-through mutation" in the leucyl-tRNA synthetase, which leads to its premature ageing [12]. We are now studying whether an error in leucine incorporation propagates into other proteins.

In more complex organisms, especially those which have a central nervous system, many even more complicated interconnections may lead to ageing and death (Fig. 11). Such a system is fundamentally complex [13].

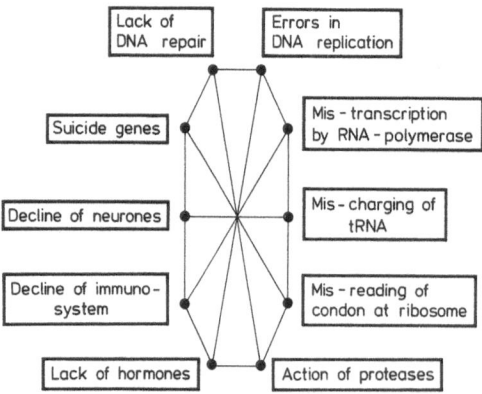

Fig. 11: Network of Molecular Events which Potentially can Contribute to Ageing and Death

3. Conjectures on Death

3.1. Death is a catastrophic event in a network system in which a limit cycle or the system of limit cycles can no longer be kept up. Such systems cannot be described in simple terms with normal transformations. They require transformational systems such as the BAKER transformation, which include unpredictable branching points [14]. There are, of course, probabilities but it can in principle not be predicted at which point in time or under which conditions the membrane potential of a nerve will break down nor at which time a heart will cease to beat.

3.2. Death and life are symmetrical. In this meeting we have heard how life could have been formed from simple molecules through an evolution of hypercycles yielding an evolutionary tree with branching points [15]. In this way, the formation of life both in Darwinian evolution and in embryonic evolution may be treated. For ageing and death, the same system of evolving hypercycles may be applicable, only with negative signs.

3.3. Death is a final consequence of the limitation to chemistry which in its more complicated structures is not error free. These limitations of the chemistry of life can be corrected by input of energy in principle. Death could be postponed with an large input of energy, immortality be reached through infinite input of energy. This, however, would not be in the interest of life as far as the genotype is concerned. What must be preserved is not the phenotype but the genotype. The genotype must be propagated.

3.4. Therefore, for the required continuation of the genotype the phenotype must be sacrificed sooner or later and it depends on the living conditions of the particular species what is the most advantageous life span for that purpose.

4. Concluding Remarks

I think there really is some progress in science. Thereby I do not mean the finding of new particles, of new chemical compounds, or of certain technical discoveries, I rather mean the discovery, the creating of fruitful new unifying concepts. And such a unifying concept seems to me to be the one which we are discussing in this meeting.

Whether we say:

> "Death is unpredictable"
>
> or
>
> "Amidst our life we are embraced by death" (Bible),

both sentences originate from the same basic knowledge, one expressed in the cool precise quantitative language of the scientist and the other in the more colourful language of the pathetic human being.

Whether we say:

> "Death is a consequence of the necessary propagation of life"
>
> or
>
> "Und solang Du das nicht hast,
> dieses Stirb und Werde,
> bist Du nur ein trüber Gast
> auf der dunklen Erde"   (Goethe)

both sayings express the same basic experience, which gives us after all the necessary motivation for executing our exact science.

I would like to conclude with a phrase of Søren KIERKEGAARD, characterising life as a dissipative structure 150 years ago: "Life can only be understood backwards but it must be lived forwards".

Citations

1 L.E. Orgel: Nature 243, 441-445 (1973); Proc. Natl. Acad. Sci. 49, 517-521 (1963)
2 F. Cramer and W. Freist: in press (1983/84)
3 F. Cramer, F. von der Haar and G.L. Igloi: Transfer RNA: Structure, Properties and Recognition, Eds. P.R. Schimmel, D. Söll, J.N. Abelson, pp. 267-279 (Cold Spring Harbor, USA, 1979)
4 L. Pauling: in Festschrift Arthur Stoll, pp. 597-602, Birkhäuser-Verlag, Basel
5 L.C.M. DeMaeyer: Ber. Bunsen Ges. 80, 1189 (1976)
6 A.N. Baldwin and P. Berg: J. Biol. Chem. 241, 839-845 (1966)
7 F. von der Haar: FEBS Letters 79, 225-228 (1977)
8 M.A. Savageau and R.R. Freter: Biochemistry 18, 3486-3493 (1979)
9 W. Freist and F. Cramer: unpublished
10 F. von der Haar, H.-J. Gabius and F. Cramer: Angew. Chem. 93, 250-256; Angew. Chem. Int. Ed. 20, 217-223 (1981)
11 P.M. Beauchamp, E.W. Horn and S.R. Gross: Proc. Natl. Acad. Sci. USA 74, 1172-1176 (1977)
12 F. Cramer, F. von der Haar and J. Schischkoff: unpublished
13 F. Cramer: Interdisciplinary Science Reviews 4, 132-139 (1979)
14 I. Prigogine: From Being to Becoming: Time and Complexity in the Physical Sciences (Freeman, Oxford, 1979/1980)
15 M. Eigen, W. Gardiner, P. Schuster and R. Winkler-Oswatitsch: Scientific American 244/4, 78-94 (1981); M. Eigen and P. Schuster: Naturwiss. 64, 541-565 (1977); Naturwiss. 65, 7-41, 341-369 (1978)

Local Synaptic Modification Can Lead to Organized Connectivity Patterns in Associative Memory

Günther Palm

Max-Planck-Institut für Biologische Kybernetik,
D-7400 Tübingen, Fed. Rep. of Germany

Abstract.

Without their cerebral cortex people seem to be unable to perform the more interesting types of behavior. On the other hand the cortex is anatomically and electrophysiologically surprisingly uniform. So all the different important capabilities that have been attributed to different cortical areas seem to be achieved by invoking almost the same machinery. How is this possible?

We propose that the cortex is merely a large memory. The organizing principle of this memory is simple and local: local correlations in cortical activity are stored by enhancing the local connectivity between the active elements. This principle (called Hebb's law) leads to the long-term storage of "preferred" global activity patterns in the cortex (called cell assemblies). Each of these patterns can be activated by any sufficiently large part of it.

Viewed as a retrieval procedure in a memory, this process is known as self-addressing or as autoassociation in the context of associative memories. As a data storage technique, Hebb's local rule or the corresponding global mechanism of autoassociation turn out to be indeed efficient, even for the purposes of today's large computer memories. As for the cerebral cortex, there is now experimental evidence for variable synaptic connectivities obeying Hebb's law.

1.An Introduction to the Brain for Physicists

To many people research on the human brain is of particular interest because the brain is believed to be responsible for the most complex typically human faculties. The individual motivation for this interest in the brain is often the hope to get some insight into the working of one's own brain, i.e., into one's own thinking. This individual motive is usually not mentioned in publications. The professional scientific attitude is to analyze the brain as a "very complex physical system". What is the nature of this physical complexity of the brain?

229

From the anatomical and from the electrophysiological point of
view the brain is surprisingly uniform. It contains of the order of
10^{10} neurons which - as far as it has been tested - have quite simple input-output relations (e.g., Katz [15]). A neuron has on the
average about 10^4 inputs and one output, that is distributed (via
synapses) to again on the average 10^4 neurons. It has been shown
already in 1943 by McCulloch and Pitts [22] that a sufficiently
large network of such neurons can in principle produce any kind of
input-output behavior however complicated, if the neurons are appropriately interconnected. Therefore, the complexity observed in
the working of the brain very simply arises from the sheer number of
neurons involved, not by big differences in the functioning of the
individual neurons.

This kind of complexity is at first sight not unknown to the physicist. In statistical mechanics, physicists have managed to deal
with 10^{20} or more particles. The basic idea behind the solution of
this kind of problem has been emphasized and formalized again by
H.Haken and others working in the theoretical field of synergetics
and in the mathematical field of bifurcation theory. From a large
number of differential equations one has to extract few "order parameters", which are basically sufficient to describe the dynamics of
the whole system of equations (for the exact meaning of this "slaving principle", see for example Haken [10] or Carr [6], Iooss and
Joseph [14]). Then one ends up with only two, three or four differential equations which can be analysed.

This scheme works quite well, when one starts with a given set of
equations and looks for a simplification of the mathematical analysis. In practice, however, this scheme can only be regarded as
one step in a cyclic procedure of theory building. One starts with
a certain set of approximative microscopical equations, containing
many ad hoc assumptions and also several parameters that have to be
decided upon. Then one does the mathematical analysis, compares the
results with some qualitative macroscopic knowledge about the system, and readjusts the microscopic equations accordingly.

In statistical mechanics, for example, the problem was to find
reasonable microscopic equations, and to define macroscopic variables like temperature and pressure in the model in such a way that
it becomes possible to deduce the known phenomenological laws of
thermodynamics from the microscopic equations. The starting point
of this whole program was the idea that temperature, pressure and
similar variables could be expressed as averages over appropriate

functions of the individual position and impulse variables of a large
number of particles, which interact basically as a billard gas.

The main point of this is that one usually needs some insight into
the macroscopic phenomenological level in order to obtain an intuition about the appropriate set of equations, the appropriate range
of parameters, and the appropriate type of expression for the macroscopically "interesting" variables in terms of the variables in
the "microscopic" equations. Then one can use mathematics, and in
some cases the slaving principle, to analyse whether this intuition
was right or wrong and where it should perhaps be corrected.

Now, when we consider the brain as a physical system, we can certainly set up a system of equations for the neuronal activity.
Clearly we will have at least 10^{10} microscopic equations (at least
one equation for every neuron). What is even worse, we will have
even more parameters to fix. Most of those parameters will be the
initial connectivities between the neurons, which can be up to 10^{20}
(one connectivity parameter for each pair of neurons). It would be
quite hopeless to try to measure all these parameters independently.
Therefore we need some additional heuristic ideas, probably from the
macroscopic, i.e., behavioral or psychological level, to get some
intuition about the kind of equations to use and the appropriate
choices of all the parameters.

In this situation a sober physicist will tend to analyse only
those aspects of human or higher animal's behavior for which the
phenomenology is comparatively simple and clear-cut. The latest
work of J. Cowan [7] is a good example for this kind of approach to
the brain. But, of course, if you are less sober, you will find
these aspects boring, and concentrate your interest on the "higher"
human faculties, like planning, decision-making, classifying, thinking. Here the phenomenology is not at all clear. What exactly are
the phenomena you are interested in? And what intuition can you
gain on the definition of corresponding macroscopic observables in
terms of microscopic variables, i.e., neuronal activities? Also the
aim of your analysis can no longer be a reduction of all these human
faculties to three or four equations for the corresponding "order
parameters". At least most people, and also most physicists, would
find it rather unsatisfactory to describe human behavior as that of
something like a random number generator. But still you can try to
use insight in the behavioral phenomena as an additional guideline
for setting up the right type of microscopic equations and finding
the right range of parameters.

I think that it is indeed possible to proceed along these lines, but that there is a dangerous trap that has to be avoided: namely to take your own introspections too literally. For example, my introspection tells me that I have an image or even picture of the world in my brain. But, of course, this cannot be literally so. Therefore I may speak of a "representation of the world" inside my head. But even this should be understood only in a highly abstract sense. It probably only means that a certain amount of knowledge about the world is represented or laid down in my brain somehow. This must be true, because without this knowledge I would not have been able to react appropriately in the many different situations I have faced. Another problem connected to this is the problem of the little man in the brain, who is the one who "looks" at this "picture" of the world that is produced in the brain. He also comes into the brain by way of introspection, but of course, there is no room for him in a physical theory of the brain.

An easy way of avoiding this trap, which can also lead you into the logical problems of self-reference (very nicely demonstrated and illustrated by Hoffstaedter [13]), is to obey the "rule of good manners": in a conversation with somebody you should avoid talking about yourself and about himself. So you talk only about things or persons to which both of you have roughly the same perspective (see also Praitenberg [31]). This simple rule helps a lot when you are dealing with emotions, drives and feelings like pain and pleasure, because they lose much of their metaphysical mystery when you do not observe them inside yourself, but calmly analyze the behavior of an animal (or even a person you are not emotionally related to) that makes you attribute these emotions to him. It turns out that this type of behavior need not be complex at all. In fact it is possible to build quite simple devices that would be judged by most observers as showing certain emotions or drives like anger or hunger (for example Gray Walter's "machina speculatrix" [35] or Braitenberg's "vehicles" [5]).

To summarize, even when you are interested in the more complex kinds of behavior, as a physicist you should simply analyse the observable behavior of other people or of animals and try to make inferences about the type of microscopic neuronal equations that are consistent with it.

This can be regarded as an approach of system analysis to behavior. It does not at all exclude teleological arguments like "he is trying to get the banana", on the macroscopic level, although there certainly is no room for them in the microscopic equations.

Teleological concepts can in fact be very helpful in descriptions of animal behavior. The systematic use of teleological arguments as a heuristic guideline for the analysis of complex systems is one of the basic ideas of cybernetics. It is clearly expressed in Wiener [36], and it makes it possible to transfer knowledge from the engineer to the problems of the natural scientist, who wants to "understand" biological organisms.

By the way, the biologist's attitude towards "understanding" has always been quite different to the physicist's: whereas for the physicist the final criterion for a real understanding of natural phenomena was always the ability to predict them, for the biologist "understanding" has a more teleological meaning: he wants to understand what a wing of a bird is for, and why exactly this particular wing is so well adapted to the purpose of enabling this particular bird to fly.

The idea is then to use the guideline of purpose and thus the biologist's approach for an analysis of behavior on the macroscopical level in order to get additional insight into the right type of neuronal equations on the microscopical level. This has been one of the basic ideas in the field of artificial intelligence. It is not sufficient simply to simulate neuronal networks on a computer and play around with the parameters, one also needs in advance a clear idea of task the network has to perform, in order to judge its performance.

2. The Cortex is an Associative Memory

If we try to follow this approach to an understanding of the brain, we can start again with the seemingly naive question: How can such a relatively uniform network of neurons as in the cerebral cortex be so well suited to the performance of all the different complex human capabilities that have been attributed to different cortical areas?

As a solution to this puzzle it has been proposed that the cortex serves mainly as a memory - of course an especially well-organized one. For example it has been said to be a "holographic" memory (Gabor [9]), because the memory seems to be distributed (not localised) in the sense that no single entry of the memory is lost, when one cell or even several cells die, only the quality of all the stored information slightly degrades.

In this talk I shall try to explain this proposal in more detail. I shall not use the word "holographic", however, because it can only be meant in a metaphorical way, and it provokes the danger of falling into the trap mentioned above, since it is too closely related to images and to the introspective conviction that we may have pictures of the world in our heads. Instead, I shall try to make explicit the abstract ideas behind this kind of organization of a memory, which has been postulated as a model for parts of the brain by many authors (e.g., Amari [1], Anderson [2], Kohonen [17], Malsburg [19], Marr [20], [21], Nass and Cooper [23], Uttley [34]).

But before we enter this discussion let us briefly see how the assumption that the cortex (or at least large parts of it) is merely a memory can account for the localization of so many typically human capabilities in it.

Indeed, for practically all the higher human capabilities memory is necessary, so the whole capability is disrupted if we destroy the corresponding memory. Moreover, many capabilities that have been investigated in the field of artificial intelligence can be performed by quite simple algorithms, when they are supplied with a well-organized memory. Such a good memory does already most of the necessary information processing. In [27] I have shown this in detail for the problem of chess-playing. The basic problem for a higher organism like a mammal is to adjust and optimize its behavior upon previous experience. For this it needs criteria which have to be optimized (probably provided by inborn drives and valuations) and a memory which stores previous situations together with the animal's reaction in these situations and the resulting evaluations some moments later - perhaps it should also contain the "next" situation one moment or some moments later. Since for higher animals it is quite improbable that exactly the same situation (in terms of sensory inputs and perhaps internal states of the animal, like hunger) will occur again, the memory should have the important property that it also associates similar responses to similar input situations.

Now we can define more exactly the task that a memory has to fulfill in order to be well suited for the optimization of behavior for a higher animal. This should be the task that the neuronal network of the cerebral cortex has to perform. There are two ways of formulating this task:

(i) Pattern mapping. Here the memory has to store a set S of pairs (x,y) of patterns x and y in such a way that in the recall the memory responds with the output y to the input x, for every pair (x,y) in S.

The patterns x and y may be sequences of symbols from an alphabet A, and then it would be convenient if the memory would not only respond with y to x, but also with something similar to y (in the Hamming distance) to something similar to x.

(ii) Pattern completion. Here the memory has to store a set S of
patterns x in such a way that in the recall it responds with the
output x to any input sequence x' that contains a sufficiently
large part of the sequence x, or that is sufficiently close to x in
the Hamming distance.

There are very simple solutions to these two tasks and these have
been discussed in the literature under such names as conditional
probability machine [34], Lernmatrix [33], associative memory [17],
holographic memory [9], simple memory [21]. The basic idea is always the same: Assume that the alphabet A consists of real numbers
- in many cases A is just $\{0,1\}$ or $\{-1,0,1\}$. Then you can store
in a matrix arrangement the correlations or products $s_{ij}=x_i \cdot y_j$ of
the entries of $x=(x_1,...,x_l)$ and $y=(y_1,..,y_k)$ for each pair (x,y) in
S. All these correlations are superimposed in one matrix for the
whole set S, i.e., the matrix $s_{ij}=\sum_{(x,y)\in S} x_i \cdot y_j$ is formed. To this
matrix $S=(s_{ij})$ we can give the input x and obtain $xS=\sum_i x_i s_{ij}$, which
turns out to be approximately equal to y for every pair $(x,y) \in S$.
In other words we simply store a superposition of all the local
correlation between the items x_i and y_j of the patterns x and y
that are to be stored $((x,y) \in S)$. And the storage of these local
correlations leads to a global association between the corresponding
patterns x and y for all (x,y) in S. Pattern completion is achieved
by exactly the same principle, simply by storing the pairs (x,x) in
the matrix for every $x \in S$. I have termed this variation of the idea
autoassociation.

It has been shown in many simulations that this principle of
storing information indeed works (e.g. Fukushima [8], Nass and Cooper
[23], Kohonen [17]). Moreover in some more theoretical papers the
conditions for its working have been examined (Amari [1], Kohonen
[17], Palm [24]). A very simple way of understanding how it works
is to consider A= $\{0,1\}$, and to visualize the matrix $S=(s_{ij})$ as a
switching matrix of switches s_{ij} that connect between the ith horizontal and the jth vertical wire of two sets of wires running
through the matrix (cf.fig.1) Then the storage of two patterns x and
y switches on the corresponding switches $s_{ij}=x_i \cdot y_j$ (see fig.2).
When we later put in x to the horizontal wires we get out y on the
vertical wires through exactly those switches which have been
switched on ($s_{ij}=1$). In general we do not exactly obtain y but
something very similar to it - the accuracy of the retrieval (of y)
can be made arbitrarily high, in principle (see Palm [24]).

In the framework of neurons this takes the following form:
assume that we have two populations M and N of neurons somewhere in

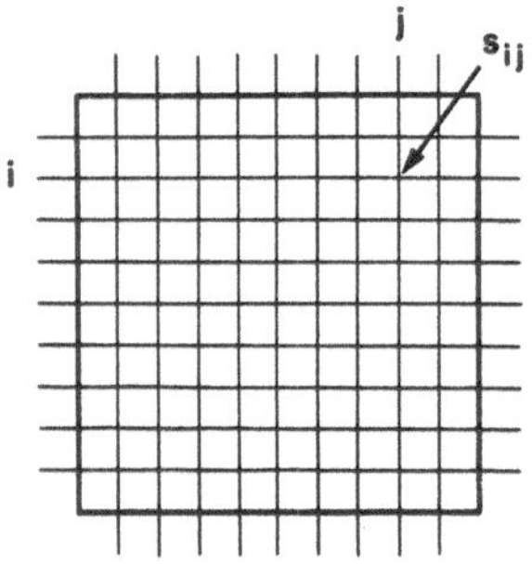

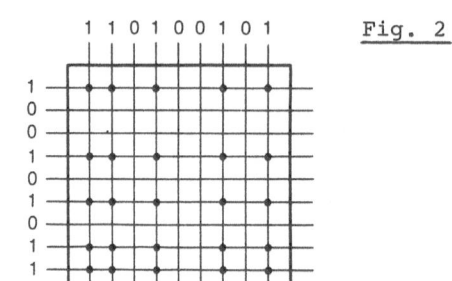

Fig. 1

Fig. 2

the brain, and that there are many connections (synapses) from the neurons in M to those in N. Then the activities x_i of the neurons in M could represent the pattern x and the activities y_j of those in N the pattern y. If we postulate that the synapse s_{ij} from neuron i in M to neuron j in N stores by its strength (synaptic weight) the local correlation between x_i and y_j, then the synaptic weights s_{ij} could form the matrix $\$=(s_{ij})$ of the memory and through those synapses the activities x_i of the neurons in M would later automatically lead to the activities $\sum_i x_i s_{ij} = (x\$)_j$ as required for the recall. (Here we assume a linear addition of activities in the dentritic tree of the neuron, which is not undebated (e.g.Koch, Poggio and Torre [16]) but still a very reasonable assumption). In the case of autoassociation one would simply need one population N of neurons with many feedback connections to itself, and this is probably the best description of the connectivity of the cerebral cortex.

The postulate that the weight or effectivity of a synapse should increase proportional to the correlation of the activities in the pre- and post-synaptic neurons is today well known as Hebb's law. There is indeed some experimental evidence for it in the cerebral cortex (Baranyi and Féher [3], Levy and Stewart [18], Rauschecker and Singer [32]). In the case of autoassociation, i.e., of one population N of neurons with many feedback connections, it leads to the storage of preferred activity patterns $x=(x_1,..,x_1)$ in the sense that each pattern x can be activated by any sufficiently large part of it. Sets of neurons whose activation has this property of self-completion have been termed cell assemblies by D.O.Hebb [11],[12].

In summary the hypothesis that the cortex is an associative memory reasonably solves the puzzle posed at the beginning of this section, it is grossly consistent with the experimental evidence, especially the anatomy and electrophysiology of the cerebral cortex and technically, associative networks are indeed working. Moreover it even has a certain introspective appeal as a description of our way of thinking and remembering (see Hebb [12]). But in the first section I have promised even more, namely that it might be possible to obtain heuristic restrictions on the possible equations for the neuronal microdynamics. I do believe that this becomes possible when we consider the information storage properties of associative memories in a more quantitative way. This is essentially my present research program and I have already obtained some first results in this direction. I have started by asking the following quantitative questions:

i) How effective is an associative memory as compared for example to computer memories? How many bits of information can be stored in an associative memory with n binary switches, that works like the switching matrix described above?

Of course, the answer cannot be more than n, but I could show that it lies well above 40% of n for concrete cases [24] and that asymptotically (i.e., for large n) it approaches $n \cdot \ln 2$.

ii) How many patterns can be stored in such a memory?

This depends critically on the accuracy that is required for the retrieval [24], therefore the answer cannot be given in one number. Asymptotically the number of patterns that can be stored is proportional to $n/\log n$. One simple fact should be pointed out in addition: if B is the amount of bits that can be stored and z the number of patterns that can be stored, then $B \approx n \cdot \ln 2$ and $B = (B/z) z$, i.e., the total number of bits stored in the memory is the number of patterns stored times the number of bits of information that are stored about an average pattern.

These results can only be applied to the brain with some caution. They have been derived for matrix memories where there is a possibility for a connection between any two wires. In the cortex, even if all synapses were Hebb synapses, we have only of the order of 10^{14} synapses compared to the possible 10^{20} pair of neurons that could in principle be connected. Therefore one has to consider networks describing the restrictions on the possible connections, which can be either random (as in Palm [25]) or designed (as in Palm [26]). These studies can be inspired by and inspiring for the analysis of the real cortical connectivity (as in Palm and Braitenberg [29], Palm and Schüz [30]), and there is still much work to be done.

Up to now I have obtained two intuitive ideas on the neuronal microdynamics from these considerations.

1) I have managed to classify the possible local rules (or equations) that govern the change in synaptic weight as a function of the pre- and postsynaptic activity, and there are arguments to believe that these rules should be Hebb-like (see Palm [28]).

2) In order to filter out one of the many superimposed patterns in an associative memory one can use threshold detection (e.g., in Palm [24]) on the output of the memory during recall. This leads to an argument for the possibility of varying the threshold of neurons in large areas of the brain in a uniform way, and to some simple algorithms for this threshold control (e.g.,Braitenberg [4] and Palm [27]).

3. Associative Memory can be Efficient as a Computer Memory

Finally let me show that the quite reasonable information storage capacity that can be achieved with associative memories is indeed comparable and in some cases superior to that of conventional computer memories. For an easy comparison I shall restrict myself here to the task of pattern mapping which is not uncommon for computer memories, and to the <u>switching</u> <u>matrix</u> type of associative memory, which as a hardware device has a certain similarity to available computer hardware.

We have a set S containing z pairs (x,y) of patterns $x \in A$ and $y \in B$ to store, i.e., x is a word of length l from the input alphabet A (with a elements) and y a word of length k from the output alphabet B (with b elements). For the storage the x's and y's have to be coded into 0,1 sequences and I will describe two codes for this. A set G containing $g=2^h$ elements can be coded without redundancy into $\{0,1\}^h$ (0,1 sequences of length h), and this can be done simply by binary counting. This code is called $c:G \longrightarrow \{0,1\}^h$. By counting G can also be coded with a lot of redundancy into $\{0,1\}^g$: the jth element of G is coded into a 0,1 sequence of length g that has only one 1 in the jth place. This code is called $d:G \longrightarrow \{0,1\}^g$.

Now the conventional storage method for this problem is described by Fig.3. For the storage it needs a $z(l \cdot \log_2 a + k \cdot \log_2 b)$ matrix of one-bit storage elements, where each pair (x,y) is coded by the code c into a 0,1 sequence of length $l \cdot \log_2 a + k \cdot \log_2 b$ and each sequence is stored in one row of the matrix. In the recall the rows of the matrix are addressed serially and every time the first part of the stored pattern is compared with the input x until the correct row is found and the corresponding output y is passed through the gates.

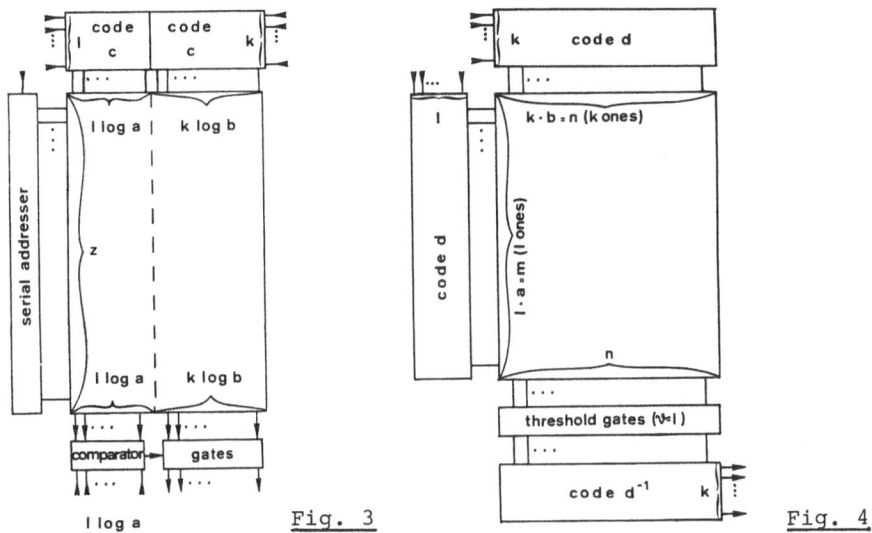

Fig. 3   Fig. 4

One possibility of using the switching matrix as an associative memory for this problem is described by figure 4. Here the patterns x and y of the pairs (x,y) in S are coded by the code d into 0,1 sequences of length l·a and k·b, respectively. These are then associated with each other in the switching matrix as described above. In the recall x is coded by d and given as an input to the matrix. The output is then passed through threshold gates with threshold l (this works because the coded pattern d(x) always has exactly l ones). The resulting 0,1 sequence is finally translated by the inverse of d into y. This method needs an mxn matrix with $m=l \cdot a$ and $n=k \cdot b$ for the storage. It does not work well for all values of l,a,k,b and z, and I have calculated some combinations of parameter values for which it does work well and for which the storage capacity is reasonably high (above, say, 30% of a bit per switch in the matrix). For example if a=2800, l=9, b=33, k=3, and z=62,500, the method works with an efficiency of about 40% of a bit per switch. Indeed, the information stored is $z \cdot k \cdot \log_2 b = 945,800$ and the size of the matrix is $l \cdot a \cdot k \cdot b = 2,494,800$. In this case the storage space needed for the conventional method is $z(l \cdot \log_2 a + k \cdot \log_2 b)$ 6,441,300+945,800 =7,387,100, which is much less effective. Generally one can expect that the associative storage method becomes preferable to the conventional one, when $l \cdot \log a > k \cdot \log b$.

There is one additional point that should be mentioned. The matrix considered in this example is fairly large (although it is still small compared to a possible $10^{10} \times 10^{10}$ matrix in the cortex). This is because about 10^6 bits have to be stored in the example.

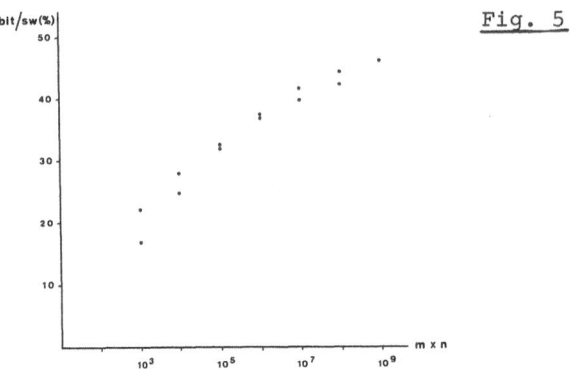

Fig. 5

With a smaller number of bits to be stored the example would have
been less favorable for the associative memory, because associative
memories work   more efficiently   the larger they are (see fig.5).
This may explain why associative memories have not been used for
computers up to now, because sufficiently large integrated computer
memories were not commercially available anyhow.

References
1. Amari S.-I. (1977): Neural theory of association and
      concept-formation. Biol.Cybern.26,175
2. Anderson J.A, (1972): A simple neural network generating an
      interactive memory. Math.Biosc.14,197
3. Baranyi A.and Feher O. (1981): Intracellular studies on corti-
      cal synaptic plasticity. Exp.Brain Res.41,124-134
4. Braitenberg V. (1978): Cell assemblies in the cerebral cortex.
      In: Theoretical Approaches to Complex Systems(Heim R.,Palm
      G.,eds.). Springer,Berlin Heidelberg New York
5. Braitenberg V. (1984): Vehicles. MIT Press (to appear)
6. Carr J. (1975): Applications of Centre Manifold Theory.
      Lecture Notes, Brown University,Providence
7. Cowan J. (1984): In Proceedings of the International Symposium
      on Synergetics. Springer, Berlin Heidelberg New York (to ap-
      pear)
8. Fukushima K. (1975): Cognitron: a self-organizing multilay-
      ered neural network . Biol.Cybern.20,121
9. Gabor D. (1969): Associative holographic memories. IBM
      J.Res.Dev.13,156
10. Haken H. (1977): Synergetics. An Introduction.
      Nonequilibrium Phase Transitions in Physics, Chemistry and Bi-
      ology. Springer, Berlin Heidelberg New York

11. Hebb D.O. (1949): The Organization of Behaviour. John Wiley, New York

12. Hebb D.O. (1958): Textbook of Psychology. Saunders, Philadelphia London Tonronto

13. Hoffstaedter D.R. (1979): Gödel, Escher, Bach. Basic books, Plenum Press, New York

14. Iooss G. and Joseph D.D. (1980): Elementary Stability and Bifurcation Theory. Springer, Berlin Heidelberg New York

15. Katz B. (1966): Nerve, Muscle and Synapse. Mc Graw-Hill, New York

16. Koch C., Poggio T.and Torre V. (1982): Retinal ganglion cells: a functional interpretation of dendritic morphology. Phil.Trans.R.Soc.London Ser.B 298, 227-264

17. Kohonen T. (1977): Associative Memory. Springer, Berlin Heidelberg New York

18. Levy W.B.and Steward O. (1979): Synapses as associative memory elements in the hippocampal formation. Brain Res.175,233-245

19. Malsburg von der C. (1973): Self-organization of orientation sensitive cells in the striate cortex. Kybernetik 14,85

20. Marr D. (1969): A theory of cerebellar cortex. J.Physiol.202,437

21. Marr D. (1971): Simple memory. Philos.Trans.R.Soc.London,Ser.B 262,23

22. Mc Culloch W.S. and Pitts W. (1943): A logical calculus of the ideas immanent in nervous activity. Bull.Math.Biophys.5,115

23. Nass M.M.and Cooper L.N. (1975): A theory for the development of feature detecting cells in visual cortex. Biol.Cybern.19,1

24. Palm G. (1980): On associative memory. Biol.Cybern.36,19

25. Palm G. (1981): Towards a theory of cell assemblies. Biol.Cybern.39,181

26. Palm G. (1981): On the storage capacity of an associative memory with randomly distributed storage elements. Biol.Cybern.39,125

27. Palm G. (1982): Rules for synaptic changes and their relevance for the storage of information in the brain. In: Cybernetics and Systems Research (Trappl R.,ed.). North-Holland Publishing Company

28. Palm G. (1982): Neural Assemblies.An Alternative Approach to Artificial Intelligence. Springer, Berlin Heidelberg New York

29. Palm G. and Braitenberg V. (1979): Tentative contributions
    of neuroanatomy to nerve net theories. In: Progress in cyber-
    netics and systems research, vol.III (Trappl R.,Klir G.J., Ric-
    ciardi L. eds.). Wiley, New York
30. Palm G. and Schüz A. (1984): (in preparation)
31. Praitenberg F. (1981): Berühmte Tiroler im Ausland. Arunda
    11,70
32. Rauschecker J.P. and Singer W. (1981): The effects of early
    visual experience on the cat's visual cortex and their possible
    explanation by Hebb synapses. J.Physiol.310,215
33. Steinbuch K. (1961): Die Lernmatrix. Kybernetik 1,36
34. Uttley A.M. (1956): Conditional probability machines and con-
    ditioned reflexes. In: Automata studies (Shannon L.E., Mc
    Carthy I.,eds.) Univ.Press,Princeton N.J.
35. Walter W.Gray (1953): The Living Brain. W.W.Norton and Co.,
    New York
36. Wiener N. (1948): Cybernetics. Wiley, New York

The Organization of Action in Time

Dietrich Dörner

Psychologie II, Universität Bamberg, D-8600 Bamberg, Fed. Rep. of Germany

1.   Introduction

Living beings - animal or human - normally are multimotivated, i.e., normally in every moment of time there exist a lot of active motives. An individual might be a bit hungry, a bit sleepy, a bit thirsty, there might be a want for social contact or a want for a certain information or for arousal.

Mostly such different motives exclude each other in respect to their execution. It is not possible to buy bread at the baker's shop, to read a book, and to sleep at the same moment.

The arrangement of different intentions in time is often done by conscious thinking. But more often it is the result of an unconscious functioning machinery. Just fifteen minutes before shop closing time one gets the idea: "Oh - I have to buy bread!" - Or one interrupts writing a letter to take a meal or to drink a cup of tea. Or one forgets to eat when reading an exciting book.

We have done a lot of empirical work on human planning and decision making behavior when solving problems in complex political or economical situations (see for instance DÖRNER et al. 1983). We learned from these investigations that there exist certain patterns of "intention regulation", which are specific for the person or specific for the situation. For instance you will find in situations of helplessness and time pressure rapid oscillations between different intentions. Subjects tackle one problem, work on it for a while, leave it, work on a second problem, leave it, work on a third one, on the first one again, without finishing any of them.

Another typical pattern is "encapsulation·behavior". A subject engages in solving a "problem" which is irrelevant and unimportant because he is able to solve it. Everybody knows such forms of encapsulation. One engages eagerly in cleaning the tobacco pipes or in rearranging one's library instead of writing a difficult paper.

Another point worth mentioning is the difference in time horizons which subjects use. Some individuals work with very short time horizons; they exhibit a kind of "muddling-through behavior", just working on the problem the deadline of which is nearest, not considering long-range problems. Other individuals however take into account long-range problems and coordinate the deadlines of the different problems with their problem-solving behavior.

With respect to different time horizons the regulation of the aspiration level is important. There exist many problems which could be solved perfectly - for which the costs in time are high - or which could be solved less perfectly saving time   that way. If other problems are urgent for their deadlines are near, it might be reasonable to lower the aspiration level of the actual problem to save time.

When studying complex decision making and planning behavior we found that the regulation of action in time - the arrangement of different intentions in time - is a very crucial matter. Different patterns of time management characterize cer-

tain individuals and certain types of situations as well. Therefore we began to
reflect upon the mechanisms of time management. We tried to develop an information processing machinery which could produce different patterns of time management and which could serve as a hypothesis on the form of information processing
underlying human time management. I shall now present this hypothesis.

As in human beings time management is influenced by the type of problem, by the
emotional state, by the strength of motivation, by certain aspects of the personality, by the forms of thinking processes, it should be clear that a theory of time
management must incorporate very different psychological entities. One has to take
into account motivational, emotional and cognitive processes and combine them with
the requirements of the situation. One has to take into account memory processes
and memory structures and traits of the specific personality. It follows that a
theory of time management must be an integrative one and must represent the interactions of different processes.

2.   Theory

2.1 The Intention Memory

At a certain moment an individual might have different intentions. Every intention
has a certain - more or less fixed - deadline, i.e., a point of time by which
the problem has to be solved. The individual has a certain idea about the goal state
and possesses a more or less vague estimation of the time it takes to solve the problem. Further on the individual has an idea about his efficiency in working on the problem. And last the subject usually has a feeling about the specific importance of the
problem.   In all one can think about an intention as represented by a memory structure of the following form:

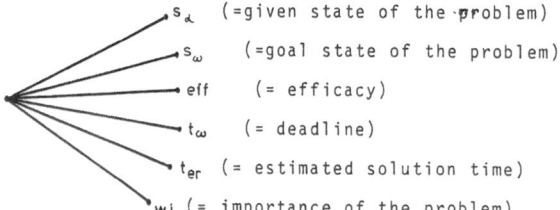

s_α (=given state of the problem)

s_ω (=goal state of the problem)

eff   (= efficacy)

t_ω (= deadline)

t_{er} (= estimated solution time)

wi (= importance of the problem)

Fig. 1: The main components of an
intention memory

Some of the ingredients of this structure are dependent on each other. So for
instance efficacy will influence the estimations about solution times. If the efficacy estimations diminish, the estimations of solution times will grow.

All the ingredients of the memory structure of an intention are variable, i.e.,
they will vary in the course of time. When working on a problem the "given state"
will continuously change. The efficacy estimations will be changed by success and
failure when working on the problem. The time estimations will change respectively.

2.2 Efficacy

Of great importance for our theory is the efficiency of an individual at problem-
solving. The efficacy estimations of an individual originate - this is our assumption - from two sources. Efficacy in problem solving means the combination of epistemic
and heuristic competence. Epistemic competence results from knowledge about the specific field of reality of the problem. It might for instance be the skill of re-
pairing cars or ability to cook.

Heuristic competence is not specific. It is - generally speaking - the competence
of an individual to gain knowledge, to find knowledge if this is not at hand. The

244

main method to gain new knowledge about a field of reality is thinking and reasoning. Therefore heuristic competence refers to the thinking abilities of an individual.

It is necessary to distinguish subjective and objective competence. Subjective competence is an important component of the self-concept of an individual whereas objective competence refers to the factual abilities. Objective and subjective competence must not always be identical or similar. You will find many individuals with a much better or worse subjective competence than factual competence. Such over- or underestimations of one's competence are mostly stable and not modifiable by experience. In short they are often the result of neurotic disorders of a person. The overestimation of competence might be a kind of self-defense of an individual.

The assumptions which an individual has about his efficacy in handling problems and challenges play a decisive role for the organization of action as we shall see. Some aspects of these interactions are depicted in the next figure. Very important is the relation of efficacy to the emotional state of an individual. A decrease of the efficacy of an individual in a certain problem situation will produce emotions like anger or fear - that depends. Anger or fear influence the level of cognitive processing. Generally they lead to an "externalization" of behavior. Thinking might for instance be replaced by overt aggressive behavior or there might be flight tendencies. A fearful individual or an individual in a state of anger normally is not capable of good reasoning or thinking. Therefore such emotional states again trigger the efficacy of an individual. The efficacy might again influence the emotional state and so on. This positive feed-back loop may lead to very fatal developments of the behavior of an individual.

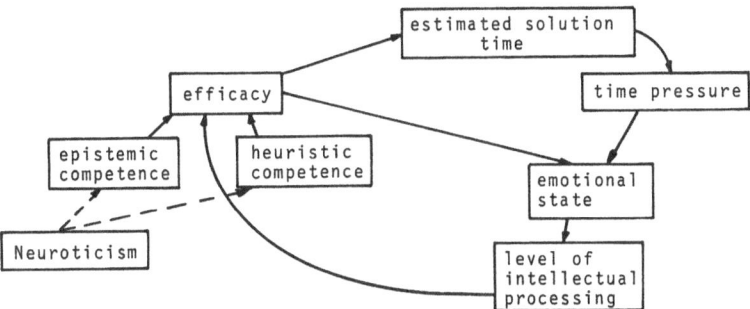

Fig. 2: The role of efficacy estimations for the processes of problem solving

2.3 The "Pressure" of an Intention

There exists at every moment a certain tendency of an intention to take control of the behavior of an individual. We call this tendency the "pressure" of an intention. The dynamics of the pressure of an intention are of central importance for our theory. The pressure of an intention results from its importance, its efficacy and the deadline of a problem and might be modelled mathematically in the following way:

(1) $\quad p_i = \text{MIN} (s_i, r_i) \quad \{= \text{"pressure" of intention } a_i\}$

(2) $\quad s_i = \dfrac{1}{1 + \exp (-a(1+\text{eff}_i) \cdot (t - (t\omega_i - \text{ter}_i \cdot at \cdot (2-\text{eff}_i)))) \cdot w_i}$

245

$$(3) \quad r_i = \frac{1}{1 + \exp\left(-v(2-\text{eff}_i) \cdot (t\omega_i-(t+\text{ter}_i \cdot (vt \cdot (2-\text{eff}_i))))\right)} \cdot W_i$$

eff_i = estimated efficacy of solution procedures,

t = time, $\qquad$ W_i = importance of a_i,

$t\omega_i$ = deadline,

ter_i = estimated solution time,

a, at, v, vt : constants.

s_i in (2) is the "activation value" of an intention. s rises towards the deadline of a problem. It rises late but with a high gradient when the efficacy is high, whereas it rises early and with a low gradient when efficacy is low. This means that a problem with high efficacy "jumps" into the mind all of a sudden but with high strength whereas an intention with low efficacy comes to mind rather early but reaches the maximum later than an intention coupled with high efficacy.

r_i in (3) is the "resignation value". r represents the rate of decay of the tendency of an intention to take control of the behavior. r will begin to diminish when time has transgressed the point $t\omega$-ter, that is the last point of time where one could begin the solution of a problem with the hope of success. r decays quickly and early when the efficacy is low, whereas it decays late and with a relatively low gradient when efficacy is high.

As indicated by formula (1) the "pressure" of an intention is the minimum of the r and s values. In fig 3 examples for the pressure values of two intentions are depicted.

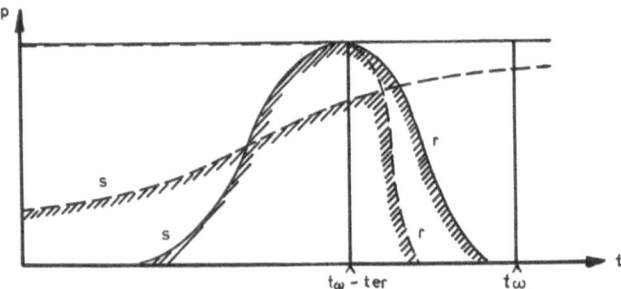

Fig. 3: The "pressure values of a problem with high efficacy (-) and with low efficacy (---) as minima of the r and s values

Now at a certain point of time an individual has several intentions. Which one takes control? The most simple answer to this question would be: the intention with the highest pressure at the moment.

It is easy to show that such a mechanism could not work. Such a simple mechanism would result mostly in very heavy oscillations of the control of behavior. If the behavior control of an individual were governed by such a simple "maximum principle" the control would oscillate between the different intentions. No reasonable behavior would result.

To establish a certain stability in the regulation of behavior it is necessary to introduce a kind of "lateral inhibition". The intention just active in controlling behavior must "defend" its position by inhibiting the other intentions.

Mathematically such an inhibition could most easily be effected by diminishing the pressure values p of the nonactive intentions in the way depicted in equation (4).

$$(4) \quad p_i' \;=\; p_i - (p_i \cdot p_{akt} \cdot inh) \quad \{\text{for } i \neq akt\}$$

$$akt = \text{"active" intention,}$$
$$inh = \text{inhibition factor.}$$

Usually the introduction of such an inhibition leads to a certain yet not absolute stability of the control of behavior.

A further extension seems necessary with respect to the different time horizons of individuals. When several intentions have the same deadline they will have high pressure values at nearly the same time according to the model described so far. This again will result in a very high,psychologically implausible,instability of behavior, but not always. Many individuals have a feeling for the requirements of their problems and begin with the solution rather early even if the deadlines are still far away. Other individuals with a low time horizon do not and begin their work only under high "pressure", i.e., when the deadlines come up very close.

Mathematically such differences can be modelled in the following way. There exists a certain "load of required time", represented by the letter b. b is simply the sum of the estimated solution times of the individual problems multiplied by a "time horizon factor" zsp. zsp is determined by a mechanism working according to equation (5).

$$(5) \quad zsp_i \;=\; \frac{1}{1 \;+\; \exp\left(-\frac{1}{zh}(t\omega_i - t - zh)\right)}$$

$$zh = \text{"time horizon".}$$

Psychologically b is the anticipated time load of an individual. b influences behavior in the following way: if b is larger than the time which is at disposal, the aspiration levels of some of the intentions diminish, i.e., the individual is no longer striving for a perfect solution but will be content with a subperfect one. Such a reduction of the aspiration level diminishes the time load.

Further on, the higher b is, the earlier the pressure values of the intentions rise. This results in an earlier and greater effort to solve the actual problems. The mathematics for that mechanism are simple and I shall not go into details here.

3.   Deductions

The most interesting question is whether the model described is sufficient to produce psychologically sound patterns of behavior. This can be investigated experimentally by measuring such parameters as self-efficacy, neuroticism and so on, to bring the individual into a certain situation with several demands or problems with certain deadlines, to observe the behavior of the individual and to compare the behavior with the predictions of the model. Until now we have not done such investigations. They have problems of their own for there are very difficult problems of measuring as always in psychology.

To get a first impression of the psychological plausibility of the system we investigated whether the model changes its behavior accordingly in a psychologically reasonable way, if one changes the parameters. I shall present now some examples of the behavior of the model under certain conditions. The examples of fig 4 were produced by computer simulations of the model. The computer model was given a certain problem situation with problems of different difficulties and deadlines. Fig. 4 shows the "pressure values" of the problems ($\propto \ldots \kappa$). A filled circle at the end of a line means the solution of a problem.

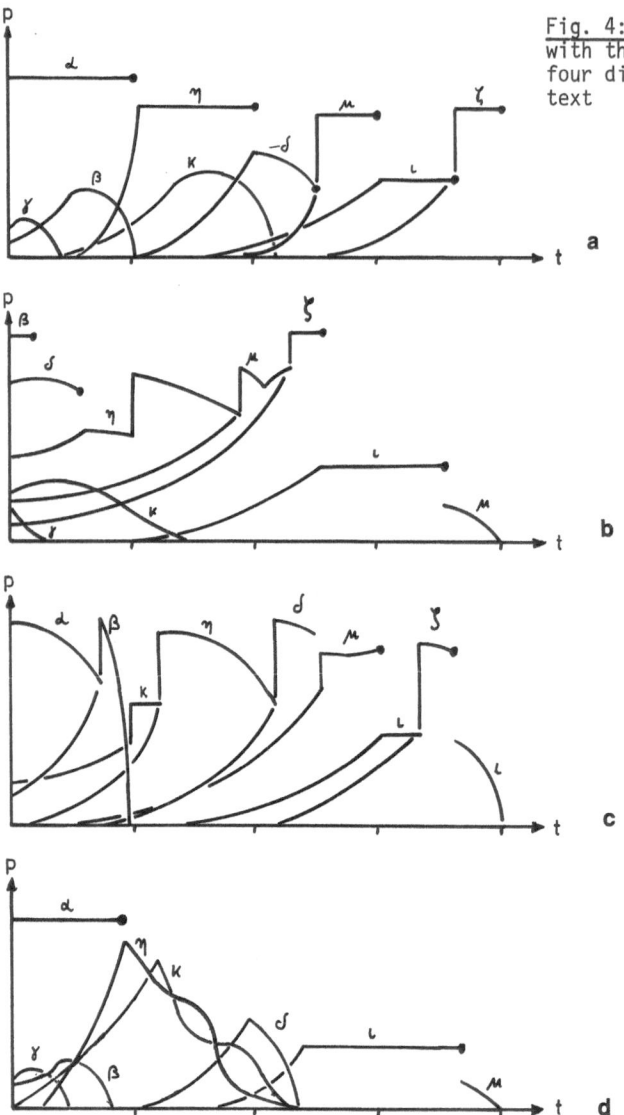

Fig. 4: Simulated time management with the same problem situations and four different "individuals". See text

The first example (4a) shows the behavior of a simulated subject with a rather high objective and subjective efficacy. A very clear and sound behavior is exhibited. First the system solves α, then η, then δ (with diminished aspiration level as indicated by the - sign). Then the problems μ, ι and ζ are solved. The problems γ, β and κ are not treated for reasons of lack of time.

The next example (4b) is produced by the same model just by changing one parameter, namely the heuristic competence. The heuristic competence is low in this example. This leads to a completely different pattern of behavior which is however psychological quite reasonable. Instead of problem α the system begins with β and δ. The reasons for this are that the system has a high confidence in its abilities to solve β and δ whereas its confidence in its abilities concerning α are low. Therefore the system solves problems which are not urgent at the moment but which

248

it is able to solve, whereas those problems which should be solved are skipped over. One can observe such forms of behavior rather often with humans of a low self esteem.

The example of 4c was produced by changing just one parameter of the individual of 4b. The 4c individual exhibits a strong overestimation of its heuristic abilities. Its factual heuristic abilities are low, as low as the abilities of the 4b individual. The overestimation leads to a pattern of behavior which is rather different from that of the 4b individual. The 4c person tackles difficult problems but always gives up very quickly. The simulated individual solves only 2 problems, whereas the 4b individual (identical with 4c with respect to the factual abilities and different only with respect to the image of its abilities) solves 4 problems. Again the simulated behavior seems to be in good agreement with the psychological reality of a subject overestimating its abilities.

Example 4d shows an individual identical to the 4a individual with regard to the factual abilities but different with regard to the assumptions about its abilities. The system in this case overestimates its epistemic abilities with respect to the problems γ, β and η. When tackling these problems it becomes aware of the difficulties of these problems and a period of quick oscillation begins. The period finds its end when the system "encapsulates" itself and takes up a problem (ι), which is neither urgent nor important. The system solves this problem. Then for a short time it tries to solve μ, but gives up rather quickly.

Summarizing these examples it seems as if the model produces patterns of behavior which with regard to the parameters are psychologically sound. The question whether the model permits predictions of "real" human behavior remains open and must be investigated.

References

Dörner, D., Kreuzig, H.W., Reither, F. & Stäudel, T. (Eds.):
            Lohhausen: Vom Umgang mit Unbestimmtheit und Komplexität,
            Huber, Bern 1983

New Problems for an Old Brain - Synergetics, Cognition and Evolutionary Epistemology

Gerhard Vollmer

Zentrum für Philosophie und Grundlagen der Wissenschaft
Justus-Liebig-Universität, Otto-Behaghel-Straße 10c
D-6300 Giessen, Fed. Rep. of Germany

Synergetics investigates the behavior of complex systems consisting of many interacting subsystems. It studies systemic properties such as cooperative behavior,
competition, self-organization, instabilities, phase transitions at critical points,
bifurcation, and symmetry breaking. Having grown from theoretical physics, notably
from laser theory, it is developing into a new comprehensive discipline which connects several previously independent fields of research.

This contribution looks at synergetics from the perspective of philosophy of
science. We shall investigate how synergetics and philosophy may be relevant to
each other (1). We shall then focus our attention on one particular aspect of this
relation, which might be characterized by the question "Why is synergetics so
difficult?" The answer refers to evolutionary epistemology and to the fact that
our cognitive abilities, being functions of a natural information-processing system,
our brain, are products of biological evolution (2). Therefore, our intuitive cognitive faculties are adapted to a world of medium dimensions, to the "mesocosm".
But theoretical knowledge can transcend these mesocosmic limits (3). We are also
adapted to simple systems only, to modest complexity, strong and linear causality.
Science in general, and physics in particular, had to start mesocosmically, to advance beyond this mesocosm gradually and to free themselves from the grip of mesocosmic intuition, natural language and everyday experience. Synergetics is a further
step in this direction (4).

1 Synergetics Relevant to Philosophy of Science?

There are several philosophical aspects of synergetics. With some of them, synergetics contributes to philosophical questions (a to d), with others, philosophy
of science might help synergetics to find its place in the wide spectrum of different sciences (e to g). We shall try to characterize those aspects by formulating
(and answering) explicit questions.

a) How does our brain work?

Synergetics might promote our understanding of the brain, of its structure, formation, evolution and functioning. There have been several steps towards such an understanding in terms of, say, mathematical biophysics. The first step is due to
McCulloch and Pitts (1943) who proposed to look at the brain as a network of binary
elements (the neurons) and to study the properties of such networks through the use
of Boolean algebra or propositional logic [1]. A second step was the attempt of Rapo-
port and Shimbel (1948) at investigating the brain as a hierarchy of random nets [2].
Similar methods have been applied to the spread of epidemics and to processes of
rumor spreading. More attempts are being made by various branches of neuroscience
and of artificial intelligence. Synergetics might well provide a further step by
viewing the brain as a self-organizing cooperative system where fluctuations and
phase transitions play a prominent role, both in its normal functioning and in
pathological cases such as hallucinations or epileptic seizures.

We suggest that the brain works essentially as a Turing machine, possibly
equipped with a lot of chance generators. We should stick to this idea as long as

250

possible because, if it is false, this might be the only chance to discover its falsity. Synergetics might well help to refute or to back up this hypothesis. And since thinking and knowing are functions of our brain, no cognitive science, no theory of knowledge, no epistemology can afford to ignore such findings.

b) *How do new ideas arise?*

One outstanding feature of human thinking is its creativity. Synergetics might help us to understand how new ideas arise, be it in individuals or in groups, be it by the interaction of different parts of the brain, by the cooperation or competition of several brains, or by the employment of cognitive tools such as computers. The processes of getting new ideas, fruitful concepts, seminal hypotheses, theories or models, powerful rules, algorithms or calculi, effective evaluation procedures, interesting problems, experiments or arguments, all these processes might be described as transitions from informational disorder to informational order. Synergetics seems well qualified to bring light into such processes.

Such knowledge might also suggest heuristic strategies for creating new ideas such as playing, dreaming, team-work, or brain-storming. It might be a good strategy, for instance, to look for a minimal algorithm reproducing all former observations and to use it for prediction-making. Since the minimality of an algorithm relates to its complexity (cf. section 4) and since complex systems are investigated by synergetics, the latter might provide valuable hints at more creativity.

Needless to say, such findings will also provide suggestions for educational problems. We should realize that the capacity of our brain is wasted most of the time. Several ideas on how to change this state of affairs are worth being discussed. As an example, we refer to the thoughtful book "Mindstorms" by Seymour PAPERT [3].

For a theory of knowledge which cares for facts and is meant to be in tune with science, none of these problems and hoped-for solutions can be irrelevant. On the contrary, our looking for and taking into account more factual knowledge might be the only way to solve some venerable problems of epistemology and philosophy of science.

c) *How are scientific theories accepted?*

Synergetics might help us to understand how opinions, prejudices, ideologies, religions, convictions and arguments are spread and how scientific theories are accepted. For a philosopher of science, the most challenging of these problems is the last one, usually running under the heading "evolution of science" or "theory dynamics".

HAKEN [4] has expressed the view that Kuhn's "revolutionary" conception of the history of science (or of the acceptance of theories) may most adequately be represented in synergetic terms such as modes, fluctuations, instabilities, order parameters, and slavery. Although this might be true, it would not render Kuhn's view any more acceptable. We might also succeed (and should try) to represent other views on the dynamics of theories by synergetic concepts and principles, for instance the more "evolutionary" views of POPPER [5] or TOULMIN [6]. It seems that synergetics would fit even better a conception of rational choice between theories which would be predictable in principle or could be, at least, rationally reconstructed. True, individual scientists might have irrational "revelations", go through sudden conversions, or experience gestalt switches. But such a description does not automatically apply to collective phenomena in scientific communities. Although there is intellectual "inertia" or "hysteresis" in theory change, what counts in the long run, are empirical facts and rational arguments.

d) *Does synergetics provide or favor unity of science?*

Unity of science might be discussed under several aspects: unity of method, unity of language ("weak" reduction by definitions), isomorphic descriptions, unity of laws (strong reduction by derivation). Synergetics has shown that complex systems, consisting of many interacting parts, exhibit remarkably simple behavior at critical points and may be described by a few physical quantities ("order parameters").

Typical examples are lasers, ferromagnets, heated fluids, atmospheres in motion, chemical reaction systems, competing biomolecules, neural nets, computer networks, animal and human societies. Moreover, though being completely different in composition, such systems show quite similar or even isomorphic behavior. It was just this unexpected fact which stimulated Hermann Haken to found a new discipline, synergetics, concentrating on cooperative phenomena and looking for "universal" behavior in totally different systems.

This is a clear case of structural identity or isomorphism. As such, it is also a case for unity of science, though not in its strongest sense. It achieves much more than unity of method or of language, but still less than unity of laws. It promotes unity of science, but doesn't achieve it. This peculiar character suggests a further question.

e) Is synergetics a branch of physics?

At first sight, the answer seems to be yes. Synergetics originated from laser theory. Its founder, Hermann Haken, is a theoretical physicist. And the standard examples of synergetics (lasers, crystals, fluids) are still taken from physics.

Even so, synergetics is not a physical discipline. It is an interdiscipline which connects several branches of factual science.

A comparison with cybernetics might be helpful. Following the subtitle of Norbert Wiener's epoch-making book "Cybernetics", this discipline studies "control and communication in the animal and the machine". Thus, looking for common structures in living and nonliving systems, cybernetics is neither a biological nor a physical discipline. In its strict sense, it is a structural science, to be compared with logic, mathematics, information science, automata theory, theory of formal languages etc., and with important applications in several, otherwise independent, disciplines. Typical cybernetic systems are information channels, feedback systems, control mechanisms, neural nets, or computers. Sometimes, cybernetics is mistaken for one of its fields of application, most regularly for computer science, where it actually came from. This is understandable, but inadequate.

Likewise, synergetics studies common structures in physical, chemical, biological, psychological, sociological systems. It is a structural science, again with important applications in many, if not all, empirical sciences. Typical synergetic systems are cooperative systems, where many subsystems interact. Again, we should not mistake synergetics for one of its fields of application, for laser theory, for hydrodynamics, or for physics in general.

It is obvious, however, why synergetics, though being an interdiscipline, took off from physics. Physics is the most mathematized empirical science, and physicists have the best chances to combine   factual knowledge on complex systems with the necessary mathematical tools and to have their theories empirically tested. But the origin and the history of synergetics and its personal and institutional bonds with physics should not mislead us to take it as a purely physical discipline.

f) Does synergetics reduce nonphysical theories to physics?

From what has been said in 1d and 1e, it should be clear that synergetics does not reduce other sciences to physics. The discovery of an isomorphism between two systems does not yet reduce either of them to the other. Isomorphisms refer only to structure, not to composition.

Let's study some examples. The fact that an electric circuit is isomorphic to a mechanical pendulum or to an oscillating mass suspended from a spring does not imply that it is a mechanical system or that electricity is reduced to mechanics. Diamond and silicon are, though isomorphic, still different materials. Water waves, sound, electromagnetic signals show striking similarities in their modes of propagation which are interesting, unsuspected, heuristically fertile and didactically helpful. Nevertheless, they remain completely different processes. That all basic interactions of nature are isomorphic in some sense (namely in being accounted for by non-Abelian gauge theories), does not, by itself, mean that they are identical. The fact that gravitational, electrostatic and magnetostatic forces all vary by an inverse square law with distance, doesn't make those forces identical, although it may correctly indicate some deep relation (which finds its expression in the Goldstone theorem).

And if synergetics has discovered that lasers, chemical systems, biomolecules, hydras, brains, social systems, and computer networks, show similar behavior patterns, this does very well signify isomorphisms between them, but does not necessarily signal their reducibility to physics.

Even so, structural similarities require an explanation. Behind those analogous surface phenomena, there might hide deep-rooted regularities. In fact, it is precisely those similarities which suggest further research. Synergetics in particular, though not reducing other sciences to physics, supplies good arguments for trying it.

g) Why is synergetics so difficult?

Theories which try to explain quarks or quasars are quite young. They are young because their objects came to our attention only lately. Synergetics is also a young discipline. But apart from lasers, its objects were known for centuries. Why, then, did it come so late?

The answer is rather simple. Synergetics came so late because it is so difficult. It is difficult because it investigates complex systems and because our intuitive knowledge and our cognitive strategies are not well adapted to such complex systems. The question, to which systems, structures and degrees of complexity our cognitive apparatus is adapted, is investigated by evolutionary epistemology.

2 Outline of Evolutionary Epistemology

Epistemology is a philosophical discipline which investigates human knowledge. But what is knowledge? For a thorough answer, we would need a lengthy chapter, book or library. Let us put up with a working definition: Factual knowledge is an adequate internal reconstruction and identification of outside objects in the cognizing subject.

Three levels of knowledge may be distinguished: perception, experience, and theoretical (or scientific) knowledge. In perception, the reconstruction and identification of objects is unconscious and uncritical, mostly even incorrigible, but at least visualizable. Experience, in contrast, makes use of natural language, simple logical inferences, observation and generalization, abstraction and conceptualization. In experience, the reconstruction of objects obtains consciously, though still uncritically. In science, finally, which includes formal logic, use of models, mathematical structures, artificial languages, external memories, artificial intelligence and instrumentally enriched experience, the reconstruction of external objects is conscious and critical. Quite frequently, however, the results cannot be visualized. A further characteristic trait of scientific knowledge is its youth. Whereas perception and experience have existed millions of years, science is a phenomenon of the last centuries or, at best, millenia. In our context, this difference will prove decisive.

Knowledge emerges from a constructive interaction of the cognizing subject with the object to be known. For such an interplay to be effective, the constitutive elements must fit each other. As an example, the human eye is sensitive to exactly that section of the electromagnetic spectrum where the radiation of the sun has maximum intensity and where the atmosphere is translucid. We can just see where there is something "to be seen".

In some, but not all cases this fit even amounts to congruence or isomorphism. Thus, the three-dimensional reconstruction of outside objects in our visual perception fits very well the fact, taught by physics, that the real world and all physical objects are, indeed, three-dimensional.

Our subjective cognitive structures can be innate or individually acquired. As a rule, both aspects, nature and nurture, are closely intertwined. We learn this from empirical science. It has been found that we are born with genetically conditioned information about the external world. These hypotheses are not always correct. But as a rule, they are quite reliable.

Why is that so? Why do our subjective cognitive structures fit the outside world? Historically, many answers have been given to this question. Evolutionary epistemology gives a new answer: *Our cognitive device is a product of biological evolution. Our subjective cognitive structures fit the world because they were adapted to this world in the course of evolution. And they are partially isomorphic to this world because otherwise we would not have survived* [7]. Or, to cite George Gaylord Simpson: "To put it crudely but graphically, the monkey who did not have a realistic perception of the tree branch he jumped for was soon a dead monkey - and therefore did not become one of our ancestors " [8].

Thus, our cognitive abilities are functions not of an immaterial soul or mind, but of a natural (physical, biological, neural) system, namely our central nervous system (sense organs included). Essentially, our brain has originated as an organ not for cognition, but for survival. This biological interpretation of our cognitive abilities has remarkable consequences for epistemology, anthropology and philosophy of science. We can study only a few of them.

3 Mesocosm

Every organism is adapted to a particular section of the real world, the section it copes with in perceiving and acting. Biologists call this segment the organism's "ecological niche". Quite analogously, we may refer to that section of the world which is known to an organism, that is, which is reconstructed and identified by it, as its "cognitive niche". And Man's cognitive niche we shall call "mesocosm". This mesocosm is a world of medium dimensions, of modest distances, time intervals, weights, and temperatures, a world of small velocities, accelerations and forces, but also a world of rather simple systems. Table I tries to characterize this mesocosm by its lower and upper limits and by some typical examples.

Our subjective cognitive structures are selected for and by this mesocosm, they are adapted to it and have proved good therein. This consideration applies to two of our three cognitive levels: perception and experience. In a Kantian terminology, we would have to state that our "forms of intuition" (space and time) and our "categories of experience" (causality, substance, etc.) are perfectly mesocosmic, adequate for the world of medium dimensions, but possibly defective in other areas. And modern science has, indeed, taught us that microscopic, megacosmic and highly complex structures are very often counter-intuitive. Table II collects some objects of modern science under this aspect.

It is of utmost importance, however, (and was not recognized by Kant) that in theoretical (or scientific) knowledge we are able to transcend this mesocosm. Using logic, language, mathematics, formal systems, we are able to formulate theories which dispense with or even contradict our mesocosmic prejudices. Moreover, we are able to test them by their mesocosmic consequences. We test Einstein's non-Euclidean theory of gravitation (general theory of relativity) with Euclidean means. We test noncausal quantum mechanics with perfectly causal setups. And if need be, we could test four-dimensional cosmologies by their observable three-dimensional consequences. Thus, we can have empirical knowledge of things we cannot visualize intuitively or experience mesocosmically.

Table I: The mesocosm, our cognitive niche

Quantity	reaching from	examples	up to	examples
times t	seconds s	heartbeat	decades	lifetime
distances s	millimeters mm	dust 0.05 m hair 0.1 mm	kilometers km	horizon 20 km, a day's walk 30 km, earshot (thunder)
velocities $v = \frac{\Delta s}{\Delta t}$	standstill v = 0		v = 10 m/s = 36 km/h	sprinter, missiles, animals
accelerations $a = \frac{\Delta v}{\Delta t}$	uniform motion a = 0		a ≈ 10 m/s^2 ≈ g	sprinter, free fall
masses, weights	grams	sand, dust	tons	rocks, trees, beasts, cars
temperatures	-10^{0}C	freezing-point	100^{0}C	boiling point of water
electric and magnetic fields	———		———	
complexity	zero com-plexity	static, iso-lated, point-like systems	linear systems	linear growth linear causality

Table II: Not mesocosmic, hence counter-intuitive systems

A R E A	S Y S T E M S	T H E O R I E S
very small distances (and short times)	quarks elementary particles atoms, molecules	quantum chromodynamics particle physics quantum mechanics
very large distances	stars, galaxies universe as a whole	astrophysics cosmology
very large velocities	electrons in atoms particles in accelerators and in cosmic radiation	special theory of relativity
very large accelerations (strong gravitational fields)	white dwarfs neutron stars black holes	general theory of relativity
very complex structures	biomolecules, organisms central nervous systems computers feedback and control chaotic systems, e.g., atmosphere (weather) magnetic field of earth	biology neurology information science cybernetics synergetics meteorology geophysics
random systems	roulette chance generators	probability theory and statistics

This reasoning also applies to the study of highly complex systems. But they deserve a separate section.

4 Understanding Complex Systems

In characterizing a system, we must specify its parts and their mutual connections. Therefore, we explicate the concept of "system" as an ordered pair:

```
system  : =  〈components  | interactions〉
        =  〈substance   | structure  〉
        =  〈elements    | relations  〉
```

As an example, let's look at three different systems, namely special pieces of diamond, of graphite, and of silicon. In our terminology, they are represented as follows:

```
diamond : =  〈C atoms    | hexagonal lattice in〉
                         | homeopolar binding

graphite: =  〈C atoms    | honeycomb layers in 〉
                         | metal binding

silicon : =  〈Si atoms   | hexagonal lattice in〉
                         | homeopolar binding
```

Whereas diamond and graphite have identical parts (or substances), diamond and silicon have identical structure. This shows that a system is not sufficiently characterized by a description of its components. Although the brain consists of neurons,
it is not just a wild assembly of nerve cells, but rather a highly connected and
organized system. Knowledge of the parts is necessary, but clearly not sufficient
for our understanding of a system.

The concept of complexity refers to both   the number of components and to the
number, kinds and strengths of their interactions. Now, we are quite successful in
using the concept of complexity as a comparative concept, that is, in identifying a
system as more complex than another one. There is, however, as yet no universally
agreed upon quantitative measure of complexity. At least, there is complexity theory
[9], a branch of mathematics, especially of computer science, with a rather useful
explication. There, the complexity of a system is understood as the minimal number
of bits necessary to describe the system [10].

Descriptions of systems may be rather lengthy. But they may be shortened by making
use of symmetries, regularities, repetitions, periodicities, invariances, in short,
by the elimination of redundancy. If a system is not redundant at all, if no detail
of its structure can be inferred from any other, then it is its own simplest description. In such a case, no description can be shorter than the system described. Such
a system would be completely chaotic or random.

Fortunately, our world is different. There are regularities and similarities, there
is law and order. Our world is highly redundant. And it is the central task of science
to make use of the world's redundancy for describing it simply [11]. Quite often, scientific progress may be interpreted as the discovery that some natural system, state
or process allows of an even simpler description. This is especially true for steps
towards the unity of science. Great scientists such as Newton, Maxwell, Darwin, or
Einstein have also been great unifiers [12].

Synergetics is perfectly in tune with this cognitive aim of science. It helps to
eliminate redundancy in two ways. First, it formulates laws for the behavior of complex physical systems. And laws are obviously less redundant than the processes they
describe (or "rule"). Second, synergetics shows how these laws can be applied to
other (nonphysical) systems and disciplines. This transfer leads to further reduc-
tions of redundancy.

It is evident that the description of a system cannot be compressed indefinitely.
There is always a lower limit to the elimination of redundancy. This lower limit is
given by the intrinsic complexity of the system, that is, by its minimal description.
Any attempt to lower this limit even further must lead to a distortion.

On the other hand, we have seen in section 3 that our intuitive understanding of
complex systems has an upper limit. This limit, being given by our genetically conditioned visualizing abilities, is modestly low. In fact, it is restricted to linear
systems, strong linear causality, linear growth. Therefore, we have trouble to grasp
more complex relations, nonlinear behavior, long causal chains, causal bifurcations
and other ramifications, side effects, causal networks, feedback structures, cyber-

netic causality, self-regulation, exponential or hyperbolic growth, weak causality, chance events, time series, etc.

It would be tempting, though impossible for lack of space, to illustrate those failures of our intuitive cognitive faculties by some case studies concerning, for instance, simultaneous birthdays, compound interest, world population, the concept of "status quo", (ill-)treatment of ecosystems, roulette, decisions under uncertainty and risk, etc. [13].

It is equally enlightening to study, under this aspect, the history of physics, of science, of human knowledge in general. In the perspective of evolutionary epistemology, we expect to find that any science had to start from the mesocosm and had to find its way beyond this mesocosm slowly. This conjecture is nicely borne out by the history of physics. Aristotle's mechanics is not just false in some queer way, it rather mirrors mesocosmic experiences quite economically. The same is true of impetus theory, a medieval theory of motion [14]. Thus, evolutionary epistemology can provide a better understanding of the beginnings of science.

And yet, as we all know, scientists didn't stop there. They discovered that our intuitive view of the world is not necessarily correct. Newton's mechanics contains already counter-intuitive elements, such as the law of inertia or gravitation as action at a distance. And this applies all the more to modern physics. Under such an evolutionary aspect - which includes the evolution of our cognitive abilities and the evolution of scientific knowledge- it is to be expected that all sciences advance from mesocosmic dimensions to areas outside.

Synergetics has left our mesocosm neither for the extremely small, as did elementary particle physics, nor for the very large, as did cosmology, nor for the very fast or heavy, as did relativity theory, but rather for the very complex. And whereas it seems plausible that owing to the limiting nature of their object systems, elementary particle physics and cosmology could come to an end and meet some kind of completion, scientists may find consolation with the fact that there is no end in sight for the investigation of complex systems [15].

Theories of such complex systems will be difficult to frame and difficult to grasp. We should not reproach science in general, or synergetics in particular, for framing and using counter-intuitive theories. If their constructs are not visualizable, it is not their fault, but ours. More precisely, it is the "fault" of evolution which did not equip us with an intuitive understanding of the whole world. It was just not necessary for survival. And yet, such knowledge is not completely inaccessible to us. However, in trying to attain objective knowledge, we have to sacrifice intuition for theoretical knowledge, and simplicity for truth.

5 Conclusion

Synergetics is a young discipline. It came so late because it is so difficult. And it is so difficult because it investigates complex systems. Systems may be complex by consisting of many components with many interactions, showing nonlinear effects and weak instead of strong causality.

Many properties of such systems are counter-intuitive. They defy intuition because our intuition is adapted to our cognitive niche, the mesocosm, to a world of medium dimensions in general, of low complexity in particular. Understanding complex systems is therefore difficult, though not impossible.

When I was a student I learned that the physics of the future would be nonlinear physics. Looking at synergetics, it seems to me that the future has finally begun.

References

1   W. McCulloch, W. Pitts: "A Logical Calculus of the Ideas Immanent in Nervous Activity", Bull. Math. Biophysics 5, 115-133 (1943)
2   N. Rashevsky: Mathematical Biophysics, vol.II (Dover, New York ^{3}1960), ch.XX
3   S. Papert: Mindstorms. Children, Computers, and Powerful Ideas (Harvester, Brighton 1980)   ch.16
4   H. Haken: Erfolgsgeheimnisse der Natur (Deutsche Verlagsanstalt, Stuttgart 1981),
5   K.R. Popper: Objective Knowledge. An Evolutionary Approach (Clarendon, Oxford 1972), ch.II
6   S. Toulmin: "The Evolutionary Development of Natural Science", Am. Scientist 55, 456-471 (1967); Foresight and Understanding. An Enquiry into the Aims of Science (Hutchinson, London 1961), ch.6; Human Understanding (Princeton University Press 1972)
7   For an introduction to evolutionary epistemology see G. Vollmer: Evolutionäre Erkenntnistheorie (Hirzel, Stuttgart ^{3}1981); "Mesocosm and Objective Knowledge. On Problems Solved by Evolutionary Epistemology", in F.M. Wuketits (ed.): Concepts and Approaches in Evolutionary Epistemology (Reidel, Dordrecht 1983), pp. 69-121.
8   G.G. Simpson: "Biology and the Nature of Science", Science 139, 81-88 (1963),p.84
9   An introduction is given by N. Pippenger: "Complexity Theory", Sci.Am. 238, 90-100 (June 1978).
10 G.J. Chaitin: "Randomness and Mathematical Proof", Sci. Am. 232, 47-52 (May 1975)
11 This point is made by H.A. Simon: The Sciences of the Artificial (MIT Press, Cambridge/London 1969, ^{2}1981) pp. 219-222
12 Cf. G. Vollmer: "The Unity of Science in an Evolutionary Perspective", Twelfth Int. Conf. on the Unity of the Sciences, Chicago 1983 (to be published)
13 Such case studies are made in G. Vollmer: "Das alte Gehirn und die neuen Probleme - Aspekte und Folgerungen einer Evolutionären Erkenntnistheorie", in M. Horvat (Hrsg.): Das Phänomen Evolution (Literas-Verlag, Wien 1984)
14 That the impetus theory is a mesocosmic or intuitive theory of motion is shown in M. McCloskey: "Intuitive Physics", Sci.Am. 248,114-122(April 1983)
15 For the question whether physics, or science in general, can be brought to an end, see G. Feinberg: "Physics and the Thales Problem", J. Philosophy 63, 5-17 (1966); G.S. Stent: Paradoxes of Progress (Freeman, San Francisco 1978),esp. pp. 47-58; S.W. Hawking: Is the End in Sight for Theoretical Physics?

Index of Contributors

M. Eigen, P. Schuster

The Hypercycle
A Principle of Natural Self-Organization

1979. 64 figures, 17 tables. VI, 92 pages
ISBN 3-540-09293-5
(This book is a reprint of papers which were
published in "Die Naturwissenschaften"
issues 11/1977, 1/1978, and 7/1978)

Contents: Emergence of the Hypercycle: The
Paradigm of Unity and Diversity in Evolution.
What is a Hypercycle? Darwinian System.
Error Threshold and Evolution. - The
Abstract Hypercycle: The Concrete Problem.
General Classification of Dynamic Systems.
Fixed-Point Analysis of Self-Organizing Reac-
tion Networks. Dynamics of the Elementary
Hypercycle. Hypercycles with Translation.
Hypercyclic Networks. - The Realistic Hyper-
cycle: How to Start Translation. The Logic of
Primordial Coding. Physics of Primordial
Coding. The GC-Frame Code. Hypercyclic
Organization of the Early Translation Appara-
tus. Ten Questions. Realistic Boundary Con-
ditions. Continuity of Evolution.

G. Eilenberger

Solitons
Mathematical Methods for Physicists

2nd corrected printing. 1983. 31 figures.
VIII, 192 pages
(Springer Series in Solid-State Sciences,
Volume 19)
ISBN 3-540-10223-X

Contents: Introduction. - The Korteweg-de
Vries Equation (KdV-Equation). - The
Inverse Scattering Transformation (IST) as
Illustrated with the KdV. - Inverse Scattering
Theory for Other Evolution Equations. - The
Classical Sine-Gordon Equation (SGE). - Sta-
tistical Mechanics of the Sine-Gordon System.
- Difference Equations: The Toda Lattice. -
Appendix: Mathematical Details. - Refer-
ences. - Subject Index.

M. Toda

Theory of Nonlinear Lattices

1981. 38 figures. X, 205 pages
(Springer Series in Solid-State Sciences,
Volume 20)
ISBN 3-540-10224-8

Contents: Introduction. - The Lattice with
Exponential Interaction. - The Spectrum and
Construction of Solutions. - Periodic Systems.
- Application of the Hamilton-Jacobi Theory.
- Appendices A–J. - Simplified Answers to
Main Problems. - References. - Bibliography.
- Subject Index. - List of Authors Cited in
Text.

Solitons

Editors: **R. K. Bullough, P. J. Caudrey**
1980. 20 figures. XVIII, 389 pages
(Topics in Current Physics, Volume 17)
ISBN 3-540-09962-X

Contents: *R. K. Bullough, P. J. Caudrey:* The
Soliton and Its History. - *G. L. Lamb, Jr.,
D. W. McLaughlin:* Aspects of Soliton Physics.
- *R. K. Bullough, P. J. Caudrey, H. M. Gibbs:*
The Double Sine-Gordon Equations: A Physi-
cally Applicable System of Equations. -
M. Toda: On a Nonlinear Lattice (The Toda
Lattice). - *R. Hirota:* Direct Methods in Soli-
ton Theory. - *A. C. Newell:* The Inverse
Scattering Transform. - *V. E. Zakharov:* The
Inverse Scattering Method. - *M. Wadati:*
Generalized Matrix Form of the Inverse Scat-
tering Method. - *F. Calogero, A. Degasperis:*
Nonlinear Evolution Equations Solvable by
the Inverse Spectral Transform Associated
with the Matrix Schrödinger Equation. -
S. P. Novikov: A Method of Solving the Period-
ic Problem for the KdV Equation and Its
Generalizations. - *L. D. Faddeev:* A Hamilton-
ian Interpretation of the Inverse Scattering
Method. - *A. H. Luther:* Quantum Solitons in
Statistical Physics. - Further Remarks on
John Scott Russel and on the Early History of
His Solitary Wave. - Note Added in Proof. -
Additional References with Titles. - Subject
Index.

Springer-Verlag Berlin Heidelberg New York Tokyo

Critical Phenomena

Proceedings of the Summer School Held at the
University of Stellenbosch, South Africa
January 18–29, 1982
Editor: F.J.W.Hahne
1983. VII, 353 pages
(Lecture Notes in Physics, Volume 186)
ISBN 3-540-12675-9

Contents: *M.E.Fisher:* Scaling, Universality and
Renormalization Group Theory. – *H.Thomas:*
Phase Transitions and Instabilities. – *A.Aharony:*
Multicritical Points. – *M.J.Stephen:* Lectures on
Disordered Systems. – *A.L.Fetter:* Lectures on Cor-
relation Functions.

H.N.Shirer, R.Wells

Mathematical Structure of the Singularities at the Transitions Between Steady States in Hydrodynamic Systems

1983. XI, 276 pages
(Lecture Notes in Physics, Volume 185)
ISBN 3-540-12333-4

Contents: Introduction: Transitions in Hydrodyna-
mics. Modeling Observed Transitions. – Introduc-
tion to Contact Catastrophe Theory: The Stationary
Phase Portrait. The Definitions of Mather's Theory.
Mather's Theorems. Altering Versal Unfoldings.
The Lyapunov-Schmidt Splitting Procedure. Vector
Spaces and Contact Computations. Classification of
Singularities. Summary. – Rayleigh-Bénard Convec-
tion: Classification of the Singularity. Physical
Interpretation of the Unfolding. – Quasi-Geostro-
phic Flow in a Channel: Heating at the Middle
Wavenumber Only. Singularities in the Vickroy and
Dutton Model. Butterfly Points in the Rossby
Regime. – Rotating Axisymmetric Flow: The But-
terfly Points. Unfolding about the Butterfly Point:
The Hadley Problem. Unfolding about the Butterfly
Point: The Rotating Rayleigh-Bénard Problem.
Dynamic Similarity. – Stability and Unfoldings:
Invariant Sets of Matrices. Smooth Submanifolds of
R^n. Transversality and Tangent Space. Versal
Unfoldings and Contact Transformations of the First
Order. Stability and First-Order Versal Unfoldings
and Contact Transformations. First-Order Mather
Theory. Conclusion. – Appendix: Summary of Spec-
tral Models: The Lorenz Model. The Vickroy and
Dutton Model. The Charney and DeVore Model.
The Veronis Model. – References.

Stochastic Processes Formalism and Applications

Proceedings of the Winter School Held at the
University of Hyderabad, India
December 15–24, 1982
Editors: G.S.Agarwal, S.Dattagupta
With contributions by numerous experts
1983. VI, 324 pages
(Lecture Notes in Physics, Volume 184)
ISBN 3-540-12326-1

The aim of this book is to introduce research workers at the pre- and postdoctoral levels to the basic
concepts, techniques and applications of stochastic
theories; some elementary topics are also treated.
An attempt is made to give the important results,
sometimes without derivations. The book is, however, self-contained. The most important aspect of
the book is the application of stochastic processes to
a wide range of systems and it is here that the book
contains many new results. In some cases it is
shown how some ot he known results can be derived elegantly by using some recently developed techniques.

B.-O.Küppers

Molecular Theory of Evolution

**Outline of a Physico-Chemical Theory of the Origin
of Life**
Translated from the German by P.Woolley
1983. 76 figures. IX, 321 pages
ISBN 3-540-12080-7

Contents: Introduction. – The Molecular Basis of
Biological Information: Definition of Living
Systems. Structure and Function of Biological Macromolecules. The Information Problem. – Principles of Molecular Selection and Evolution: A Model
System for Molecular Self-Organization. Deterministic Theory of Selection. Stochastic Theory of Selection. – The Transition from the Non-Living to the
Living: The Information Threshold. Self-Organization in Macromolecular Networks. Information-Inte-
grating Mechanisms. The Origin of the Genetic
Code. The Evolution Hypercycles. – Model and
Reality: Systems Under Idealized Boundary Condi-
tions.Evolution in the Test-Tube. Conclusions: The
Logic of the Origin of Life. – Mathematical Appendices. – Bibliography. – Index.

Springer-Verlag Berlin Heidelberg NewYork Tokyo